AF834617

Ancient Carved Ambers in the J. Paul Getty Museum

Ancient Carved Ambers in the J. Paul Getty Museum

Faya Causey

With technical analysis by Jeff Maish, Herant Khanjian, and Michael R. Schilling

THE J. PAUL GETTY MUSEUM, LOS ANGELES

This catalogue was first published in 2012 at http://museumcatalogues.getty.edu/amber. The present online version was migrated in 2019 to https://www.getty.edu/publications/ambers; it features zoomable high-resolution photography; free PDF, EPUB, and MOBI downloads; and JPG downloads of the catalogue images.

First edition 2012
Paperback and ebook editions 2019
https://www.github.com/gettypubs/ambers

Published by the J. Paul Getty Museum, Los Angeles
Getty Publications
1200 Getty Center Drive, Suite 500
Los Angeles, California 90040-1682
www.getty.edu/publications

First edition:
Marina Belozerskaya and Ruth Evans Lane, *Project Editors*
Brenda Podemski and Roger Howard, *Software Architects*
Elizabeth Zozom and Elizabeth Kahn, *Production*
Kurt Hauser, *Cover Design*

2019 editions:
Zoe Goldman, *Project Editor*
Greg Albers, *Digital Manager*
Maribel Hidalgo Urbaneja, *Digital Assistant*
Suzanne Watson, *Production*

Distributed in the United States and Canada by the University of Chicago Press

Distributed outside the United States and Canada by Yale University Press, London

Library of Congress Cataloging-in-Publication Data
Names: Causey, Faya, author. | Maish, Jeffrey, contributor. | Khanjian, Herant, contributor. | Schilling, Michael (Michael Roy), contributor. | J. Paul Getty Museum, issuing body.
Title: Ancient carved ambers in the J. Paul Getty Museum / Faya Causey ; with technical analysis by Jeff Maish, Herant Khanjian, and Michael Schilling.
Description: Los Angeles : The J. Paul Getty Museum, [2019] | Includes bibliographical references. | Summary: "This catalogue provides a general introduction to amber in the ancient world followed by detailed catalogue entries for fifty-six Etruscan, Greek, and Italic carved ambers from the J. Paul Getty Museum. The volume concludes with technical notes about scientific investigations of these objects and Baltic amber"—Provided by publisher.
Identifiers: LCCN 2019016671 (print) | LCCN 2019981057 (ebook) | ISBN 9781606066348 (paperback) | ISBN 9781606066355 (epub) | ISBN 9781606060513 (ebook other)
Subjects: LCSH: J. Paul Getty Museum—Catalogs. | Amber art objects—Catalogs. | Art objects, Ancient—Catalogs. | Art objects, Etruscan—Catalogs. | Art objects—California—Los Angeles—Catalogs. | LCGFT: Collection catalogs.
Classification: LCC NK6000 .J3 2019 (print) | LCC NK6000 (ebook) | DDC 709.0109794/94—dc23
LC record available at https://lccn.loc.gov/2019016671
LC ebook record available at https://lccn.loc.gov/2019981057

Front cover: *Pendant: Divinity Holding Hares* (detail, 77.AO.82, cat. no. 4).

Every effort has been made to contact the owners and photographers of objects reproduced here whose names do not appear in the captions. Anyone having further information concerning copyright holders is asked to contact Getty Publications so this information can be included in future printings.

URLs cited throughout this catalogue were accessed prior to first publication in 2012; during preparation of the present editions in 2019, some electronic content was found to be no longer available. Where URLs are no longer valid, the author's original citations have been retained, but hyperlinks have been disabled in the online and ebook editions.

Contents

Introduction

Amber and the Ancient World

The J. Paul Getty Museum collection of amber antiquities was formed between 1971 and 1984. Apart from the Roman *Head of Medusa* (figure 1), which Mr. Getty acquired as part of a larger purchase of antiquities in 1971, all the other ancient amber objects were acquired as gifts. The collection is made up primarily of pre-Roman material, but also includes a small number of Roman-period carvings, of which the *Head of Medusa* is the most important. The pre-Roman material includes a variety of jewelry elements that date from the seventh to the fourth centuries B.C.: fifty-six figured works and approximately twelve hundred nonfigured beads, fibulae, and pendants. This volume examines the fifty-six objects of pre-Roman date representing humans, animals, and fantastic creatures, plus a modern imitation. The Getty's nonfigured pre-Roman objects and the Roman works are not included in this catalogue.

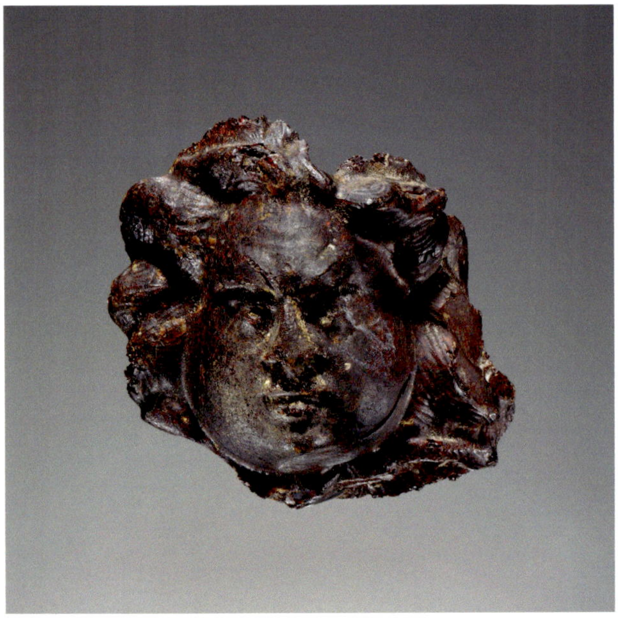

Figure 1 *Head of Medusa*, Roman, 1st–2nd century A.D. Amber, H: 5.8 cm (2 3/10 in.), W: 5.8 cm (2 3/10 in.). Los Angeles, J. Paul Getty Museum, 71.AO.355.

The ambers were acquired by their donors on the international art market. The loss of any artifact's context is immeasurable, and any attempt to discuss ambers without their original context is, to borrow an analogy from Thorkild Jacobsen, "not unlike entering the world of poetry." Poetry plays a part in locating the cultural ambients in which the ambers of this catalogue once performed. In addition to ancient literary sources, the work here is examined via a large interdisciplinary toolkit, including art history, archaeology, philology, pharmacology, anthropology, ethnology, and the history of medicine, religion, and magic.

At a critical moment in writing this introduction, I read two of Roger Moorey's final contributions, his 2001 Schweich Lectures, published as *Idols of the People: Miniature Images of Clay in the Ancient Near East* (2003), and his *Catalogue of the Ancient Near Eastern Terracottas in the Ashmolean* (2004). Both were important to the final shaping of my text. (It is from the latter publication that I borrowed Jacobsen's quotation.) Certain of Moorey's observations played critical roles; among them is his cautionary note in the *Catalogue*: "Even if it may be possible to identify who or what is represented, whether it be natural or supernatural, that does not in itself resolve the question of what activity the terracotta was involved in."[1]

Indeed, in what "activity" were these carved ambers involved? This catalogue attempts to address this question. Keeping in mind the challenges presented when working with decontextualized artifacts, I make comparisons to scientifically excavated parallels, to documented works in museums, and, with extra care, to unprovenanced material in other collections, public and private. The evidence suggests that amber was dedicated primarily to female divinities, and that most pre-Roman amber objects were buried with women and children. Individually and as a whole, the Getty Museum's amber objects are important witnesses to the larger social picture of the people who valued the material.[2]

My interest was first sparked by the peculiar nature of the carved amber on display in the British Museum and by Donald Strong's masterful 1966 catalogue of the material.[3] Strong duly noted the magical aspects of the subjects of Italian Iron Age ambers, and I took as a challenge one comment: "Many of the more enigmatic subjects among these carvings probably have a meaning that is no longer clear to us."[4]

NOTES

1. Moorey 2004, p. 9.

2. White 1992, p. 560: "We have seen in the ethnographic record that material forms of representation are frequently about political authority and social distinctions. Personal ornaments, constructed of the rare, the sacred, the exotic, or the labor/skill intensive, are universally employed, indeed essential to distinguish people and peoples from each other." White's work on Paleolithic technology, the origins of material representation in Europe, and the aesthetics of Paleolithic adornment have informed this study more than any specific reference might indicate. Throughout his work, White underlines the variety, richness, and interpretive complexity of the known corpus of prehistoric representations. It is through his work that I began to understand the nonverbal aspects of adornment and to consider systems of personal ornamentation. See R. White, "Systems of Personal Ornamentation in the Early Upper Paleolithic: Methodological Challenges and New Observations," in *Rethinking the Human Revolution: New Behavioural and Biological Perspectives on the Origin and Dispersal of Modern Humans*, ed. P. Mellars et al. (Cambridge, 2007), pp. 287–302; and R. White, *Prehistoric Art: The Symbolic Journey of Humankind* (New York, 2003), p. 58, where he cites the innovative G. H. Luquet, *L'art et la religion des hommes fossiles* (Paris, 1926). In the 2007 article, White publishes the earliest known amber pendant (the amber is almost certainly from Pyrenean foreland sources), from the Archaic Aurignacian level 4c6 at Isturitz, France.

3. The watershed British Museum catalogue of carved amber by Strong was published in 1966 (Strong 1966). Since that time, there has been considerable research on amber in the ancient world and related subjects, and a significant number of amber-specific studies have been published during the last several years. These range in type from exhibition and collection catalogues, excavation reports, and in-depth studies of individual works to broader sociocultural assessments. Still, many finds and investigations (including excavation reports) await publication, and the study of amber objects is behind that of other contemporary visual arts media. There are many reasons for this lag, including the nature of the material itself. Only a small number of carved amber objects are on display in public collections; relatively few are published or even illustrated; and too few come from controlled contexts. Many important works are in private collections and remain unstudied. Moreover, under some burial conditions, and because of its chemical and physical structure, amber often suffers over time. Poorly conserved pieces are friable, difficult to conserve and sometimes even to study; they can be handled only with great care and therefore are notoriously difficult to photograph, illustrate, or display. Much more remains to be learned about amber objects from a uniform application of scientific techniques, such as neutron activation analysis, infrared spectrometry, isotope C^{12}/C^{13} determination, and pyrolysis mass spectrometry (PYMS), as recent research has demonstrated. For the various methods of analysis, see the addendum to this catalogue by Jeff Maish, Herant Khanjian, and Michael Schilling; also Barfod 2005; Langenheim 2003; Serpico 2000; Ross 1998; and Barfod 1996. C. W. Beck's lifetime of work on amber is indicated in the bibliographies of these publications.

   To date, only a very small percentage of pre-Roman ancient objects have been analyzed. Several key projects specifically related to the study of amber in pre-Roman Italy were completed in recent years, including the cataloguing of amber in the Bibliothèque nationale, Paris (D'Ercole 2008), and that in the National Museum, Belgrade, and in Serbia and Montenegro (Palavestra and Krstić 2006). In addition, two recent exhibitions of amber from the Italian peninsula, the 2007 *Ambre: Trasparenze dall'antico*, in Naples, and the 2005 *Magie d'ambra: Amuleti et gioielli della Basilicata antica*, in Potenza, have added much to the picture of amber consumption, especially for pre-Roman Italy. In 2002, Michael Schilling and Jeffrey Maish of the Getty Conservation Institute identified thirty-five ambers in the Getty collection as Baltic amber (see the addendum to this catalogue).

4. Strong 1966, p. 11. Strong also comments: "Etruscan necklaces include a wide range of amulets of local and foreign derivation and the whole series of 'Italic' carvings consist largely of pendants worn in life as charms and in death with some apotropaic purpose. The big necklaces combined several well-known symbols of fertility, among them the ram's head, the frog, and the cowrie shell. The bulla which is common in amber was one of the best-known forms of amulet in ancient Italy." (For the bulla, see n. 152.)

some cases, deceased girls may have been adorned as brides—a common aspect of funerary ritual.

How these objects might have functioned in reference to clanship or other social identities, during either life or the rituals surrounding death, should also be considered. Among certain populations, there might have been a generally accepted role for amber, in the range of subjects into which it was formed and/or the objects it embellished. Some subjects might have been pertinent to clans or larger communities, in the way that shield emblazons might be. Some imagery might have been special to family groups, who may have traced their origins, names, or even good fortune to a particular deity, animal, totem, or myth. If an elite person whose family's founder was a divinity or Homeric hero was buried with a ring with an engraved gem representing, say, Herakles (figure 7), Odysseus, or Athena, might the same have been done with figured ambers?

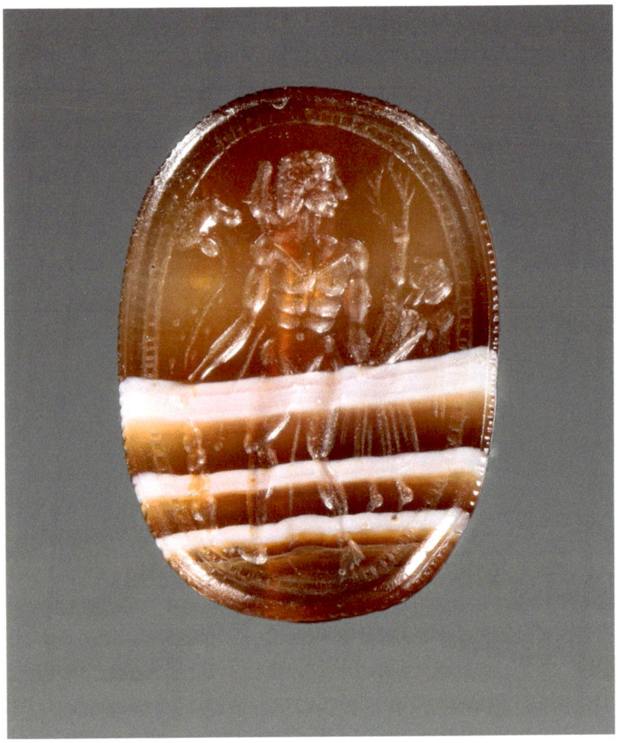

Figure 7 *Engraved Scarab with Nike Crowning Herakles*, Etruscan, 400–380 B.C. Banded agate, H: 1.8 cm (3/4 in.), W: 1.4 cm (9/16 in.), D: 0.9 cm (3/8 in.). Los Angeles, J. Paul Getty Museum, 85.AN.123.

The extent to which some of these ornament-amulets had a role in established cult or folk religion is difficult to ascertain, but it should not be either exaggerated or denied. The diversity of subjects that appear in figured amber over time suggests that the material was used within many different symbol systems, but always for its protective or regenerative aspects. Some pieces do incorporate elements relating, for instance, to Dionysos or Artemis, but as such, they occupy a hazy territory between identifiable religious practices and what Einar Thomassen calls "the appropriation of ritual power for personal ends."[18] The use of these amulets may have been dictated to some extent by skilled practitioners, but it is likely that the original, specific use of a protective amulet often would have eroded into a more generalized *portafortuna*, or good-luck, role over time.[19] The generally feared evil eye might have been warded off with any amber amulet.[20]

Worked amber and amber jewelry were well in evidence in northern Europe from the fourth millennium B.C. onward. The earliest evidence for worked amber in Italy is from the Bronze Age. We do not know where the amber found in graves dating to circa 1500 B.C. in Basilicata (near Melfi and Matera) was carved. In the later Bronze Age, Adriatic Frattesina, a typical emporium of the protohistoric era, was a place of manufacture. Already by this time, variety in style, subject, technique, and function was evident. Some of these early ambers are the work of highly skilled artisans; others are rudimentary in manufacture and indicate work by other kinds of amber-workers/amulet-makers, perhaps even priestesses, physicians, or "wise women." It is tempting to think of multiple ritual specialists involved in amber-working and amuletmaking, though perhaps in not so pronounced a fashion as in contemporary Egypt—although there is evidence for widespread amuletic usage in Italy even into modern times. We might well envision a scenario that includes simple gem cutters, sculptors, multiple ritual specialists—from healers to hacks—those with fixed locations in urban settings, and itinerants. Such a variety of practitioners offering objects and ritual expertise is likely, especially for amulets in a material as inherently magical as amber.[21]

NOTES

8. Reference from E. Thomassen, "Is Magic a Subclass of Ritual?" in Jordan et al. 1999, pp. 55–66.

9. Strong 1966, pp. 10–11, considers the amuletic and the magical aspects of amber separately from its medical uses. He distinguishes between early Greek and later (presumably Classical) Greek attitudes: "In early Greece the amuletic values of amber seem to have been recognized.... But in the Greek world generally the principal attraction of amber was its decorative qualities." Strong also differentiates Italic Iron Age usage from the Greek: in that period, the "amber carvings ... underline the magical aspects of the use of amber."

   Waarsenburg 1995, p. 456, successfully undertakes a religious interpretation in his study of the seventh-century B.C. Tomb VI at Satricum, countering the "viewpoint that Oriental or Orientalising figurative amulets had only a very generic apotropaic function in Italy ... and [that] they would not have been understood by the native population. Related to this viewpoint is an explicit reluctance against any interpretation which takes nonmaterial, *sc.* religious, aspects into account. Even the symbol of the nude female is frequently denied a specific meaning." D'Ercole 1995, p. 268, n. 19, suggests that beliefs surrounding amber, other than fashion or taste, might explain the long-continuing repetition of subjects among certain groups of figured ambers. Mastrocinque 1991, p. 78, n. 247, notes the supranormal aspects of figured amber, drawing attention to the relationship of the subject and the animating, electrical properties of amber. The amuletic, magical, or apotropaic properties of pre-Roman amber objects are noted by S. Bianco, A. Mastrocinque, A. Russo, and M. Tagliente in *Magie d'ambra* 2005, passim; Haynes 2000, pp. 45, 100 ; A. Russo in *Treasures* 1998, p. 22 ; Bottini 1993, p. 65; Negroni Catacchio 1989, p. 659 (and elsewhere); Fuscagni 1982, p. 110; Hölbl 1979, vol. 1, pp. 229ff., who (as quoted by Waarsenburg 1995) sees "all amulets [as having] had a similar, not exactly defined magic power; possibly they served against natural dangers such as animal bites, or against supranatural dangers such as the evil eye"; La Genière 1961; Richter 1940, pp. 86, 88; and RE, vol. 3, part 1, esp. cols. 301–3, s.v. "Bernstein" (by Blümner). For the Mycenaean period, see Bouzek 1993, p, 141, "who rightly insists first on the quasimagical properties of amber (not just the prestige)," as A. Sherratt notes in "Electric Gold: Reopening the Amber Route," *Archaeology* 69 (1995): 200–203, his review of Beck and Bouzek 1993. Compare, however, the more cautious opinion of Hughes-Brock 1985, p. 259: "Most amber is in ordinary bead form; since it is consistently found alongside standard beads of other materials, we cannot prove that the Mycenaeans thought of it as having any special amuletic value."

10. Eichholz 1962 is the edition used throughout this text.

11. J. Evans, *Magical Jewels of the Middle Ages and the Renaissance, Particularly in England* (Oxford, 1922), p. 13.

12. The subjects and forms of many pre-Roman figured ambers have precedents thousands of years older. The earliest surviving animal and human subjects in amber from northern Europe are dated to the eighth millennium; see, for example, M. Iršenas, "Stone Age Figurines from the Baltic Area," in *Proceedings of the International Interdisciplinary Conference: Baltic Amber in the Natural Sciences, Archaeology and Applied Art,* ed. A. Butrimas (Vilnius, 2001), pp. 77–86; M. Ots, "Stone Age Amber Finds in Estonia," in Beck et al. 2003, pp. 96–107; M. Irinas, "Elk Figurines in the Stone Age Art of the Baltic Area," in *Prehistoric Art in the Baltic Region,* ed. A. Butrimas (Vilnius, 2000), pp. 93–105; and I. Loze, "Prehistoric Amber Ornaments in the Baltic Region," in *Baltica* 2000, pp. 18–19. An amber duck found in a Danish Paleolithic context of 6800–4000 B.C. is the earliest example of a pendant type popular in Greece and Italy in the seventh century B.C. and first known in the eighth. (See n. 194 for further discussion of ducks in amber.) Such objects support the hypothesis that amber was traded with the south in both finished and unfinished forms. H. Hughes-Brock, "Mycenaean Beads: Gender and Social Contexts," *Oxford Journal of Archaeology* 18, no. 3 (August 1999): 293, suggests, "Some imports probably arrived with the specialist processes already completed nearer the source, e.g., preliminary removal of the crust of Baltic amber." Why not finished objects?

13. S. Eitrem, *Opferritus und Voropfer der Griechen und Röme* (1915; repr., Hildesheim and New York, 1977), p. 194, discusses the amuletic virtues of amber in Rome.

14. K. Benzel, in *Beyond Babylon* 2008, p. 25, with reference to pp. 350–52 in the same catalogue. Benzel cites J. Spacy, "Emblems in Rituals in the Old Babylonian Period," in *Ritual and Sacrifice in the Ancient Near East: Proceedings of the International Conference Organized by the Katholieke Universiteit Leuven, 17–20 April 1991,* Orientalia Lovaniensia Analecta 55, ed. J. Quaegebeur (Leuven, 1993), pp. 411–20; Z. Bahrani, "The Babylonian Visual Image," in *The Babylonian World,* ed. G. Leick (New York and London, 2003), pp. 155–70; and Z. Bahrani, *The Graven Image: Representation in Babylonia and Assyria* (Philadelphia, 2003), p. 127. See also H. Wildberger, *Isaiah 1–12: A Commentary,* trans. T. H. Trapp (1991; repr., Minneapolis, 2002).

15. Amber itself, and most of the subjects of figured amber, have fertility aspects. Modern Westerners tend to discuss the fertility and fecundity beliefs and rites of earlier peoples in the context of an increase of humans, hunt animals, edible botanics, agricultural products, and domesticated crops, which limits our understanding of fertility imagery, both its making and its use. That fertility magic was used to *control* reproduction (via, e.g., birth spacing) as well as spur procreation was first brought to my attention by R. White (public lecture 1999). See White 2003 (in n. 2, above), p. 58, where he cites G. H. Luquet, *L'art néo-calédonien: Documents recueillis par Marius Archambault* (Paris, 1926), and P. Ucko and A. Rosenfeld, *Paleolithic Cave Art* (London, 1967). Luquet was among the first to raise doubts about the idea that Paleolithic peoples were motivated to increase human fecundity through magical acts. Ucko and Rosenfeld were among the first to write that hunters and gatherers are generally more interested in limiting population growth than in increasing it. Compare the discussion by J. Assante, "From Whores to Hierodules," in *Ancient Art and Its Historiography,* ed. A. A. Donohue and M. D. Fullerton (Cambridge, 2003), p. 26, where she contrasts "Yahweh's command to be fruitful and multiply, and the Bible's emphasis on progeny in general," with the Mesopotamian "gods of prebiblical flood myths who did not destroy mankind because they sinned but because they overpopulated and made too much noise." Assante cites A. Kilmer, "The Mesopotamian Concept of Overpopulation and Its Solution as Reflected in the Mythology," *Orientalia,* n.s., 41 (1972): 160–77.

16. D. Frankfurter, *Bryn Mawr Classical Review* 1995.04.12 (review of Kotansky 1994).

17. The literature on gifts and gift giving in the ancient world is extensive. Although previous ownership of excavated objects is ordinarily difficult to establish, two Etruscan finds and one Etrusco-Campanian find might be seen as exempla of presentation, parting, and exchange articulated around banquets. Were these items exchanged among guests/friends? Were they components of a dowry or bride wealth, ransom or prizes, or funerary tributes? Haynes 2000, p. 69, cites the silver vessels deposited circa 660 B.C. with an aristocratic lady in the Regolini-Galassi Tomb at Cerveteri, inscribed with a male name in the genitive, and suggests that these luxury objects were the property of her husband. The seventh-century gold *fibula,* with its inscription in granulation, from Casteluccio-La Foce (Siena), in the Louvre (Bj 816), is a gift-ornament that recalls the fibulae of the peplos offered to Penelope (*Odyssey* 18.292–95). For the Louvre pin, see Cristofani, in Cristofani and Martelli 1983, no. 103; and Haynes 2000, p. 6809, fig. 47. The inscription on an Etrusco-Campanian bronze *lebes* found in Tomb 106 at Braida di Vaglio, which belonged to a woman of about sixty (the tomb also included two amber figured pendants, a satyr's head and a Cypriote-type Herakles), is another important example; for the inscription, see M. Torelli with L. Agostiniani, in Bottini and Setari 2003, p. 63, and appendix I, pp. 113–17. These inscriptions are further evidence of networked elites taking advantage of their literacy.

18. Thomassen 1999 (n. 8, above), p. 65.

19. Compare Faraone 1992, p. 37: "There is a tendency for all protective images, regardless of their 'original' purpose or the specific crisis that led to their manufacture, to assume a wider and wider role in the protection of a place, until they achieve a status as some vague 'all-purpose' phylactery against any and all forms of evil."

20. See n. 152.

21. The scenario of multiple ritual specialists recorded by the tenth-century A.D. compiler Ibn al-Nadim, who pronounced Egypt "the Babylon of the magicians," might provide a later model for pragmatic ritual expertise at all levels and the range of activities of itinerant artisans and healers in pre-Roman Italy. He records, "A person who has seen this state of affairs has told me that there still remain men and women magicians and that all of the exorcists and magicians assert that they have seals, charms of paper ... and other things used for their arts": Ibn al-Nadim, *Kitāb al-Fihrist,* trans. Bayard Dodge, *The Fihrist of al-Nadim: A Tenth-Century Survey of Muslim Culture* (New York, 1970), p. 726 (quoted in D. Frankfurter, "Ritual Expertise in Roman Egypt and the Problem of the Category 'Magician,'" in *Envisioning Magic: A Princeton Seminar and Symposium,* ed. P. Schäfer and H. G. Kippenberg [Princeton, 1997], p. 30).

What Is Amber?

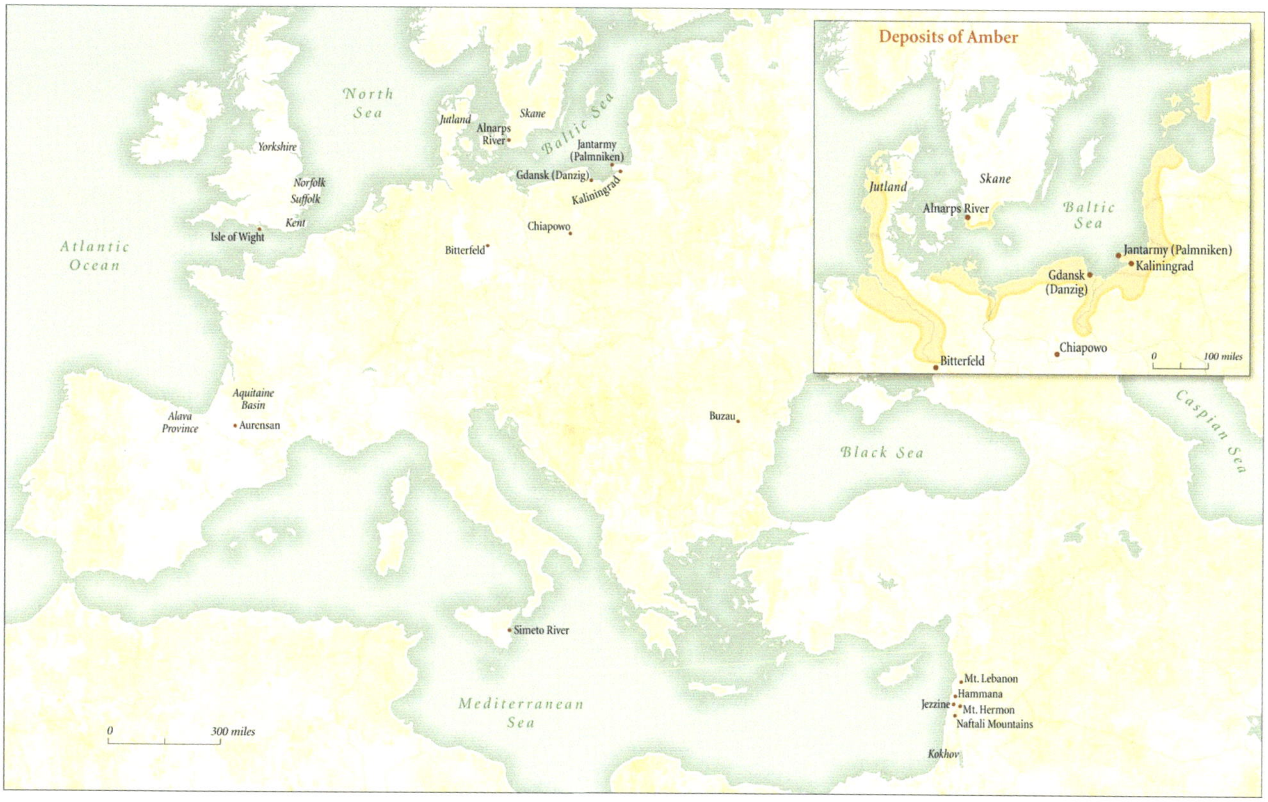

Figure 8 Sources of amber in the ancient world. Map by David Fuller.

It is important to say that amber is much studied but still not fully understood. The problems begin with the names by which the material is known: amber, Baltic amber, fossil resin, succinite, and resinite. Although all these terms have been used to describe the material discussed in this catalogue, they have confused as much as they have clarified. It is generally accepted that amber is derived from resin-bearing trees that once clustered in dense, now extinct forests.[22] Despite decades of study, there is no definite conclusion about the botanical source of the vast deposits of Baltic amber, as Jean H. Langenheim recently summarized in her compendium on plant resins:

> *It is clear that the amber is not derived from the modern species of Pinus, but there are mixed signals from suggestions of either an araucarian* Agathis*-like or a pinaceous* Pseudolarix*-like resin producing tree.... Although the evidence appears to lean more toward a pinaceous source, an extinct ancestral tree is probably the only solution.*[23]

Geologically, amber has been documented throughout the world (figure 8), with most deposits found in Tertiary-period sediments dating to the Eocene, a few to the Oligocene and Miocene, and fewer still to later in the Tertiary. Amber is formed from resin exuded from tree bark (figure 9), although it is also produced in the heartwood. Resin protects trees by blocking gaps in the bark. Once resin covers a gash or break caused by chewing insects, it hardens and forms a seal. Resin's antiseptic properties protect the tree from disease, and its stickiness can gum up the jaws of gnawing and burrowing insects.[24] In the primordial "amber forest," resin oozed down trunks and branches and formed into blobs, sheets, and stalactites, sometimes dripping onto the forest floor. On some trees, exuded resin flowed over previous flows, creating layers. The sticky substance collected detritus and soil and sometimes entrapped flying and crawling creatures (figure 10). Eventually, after the trees fell, the resin-coated logs were carried by rivers and tides to deltas in coastal regions, where they were buried over time in sedimentary deposits. Most amber did not originate in the place where it was found; often, it was deposited and found at a distance from where the resin-producing trees grew. Most known accumulations of amber are redepositions, the result of geological activity.[25]

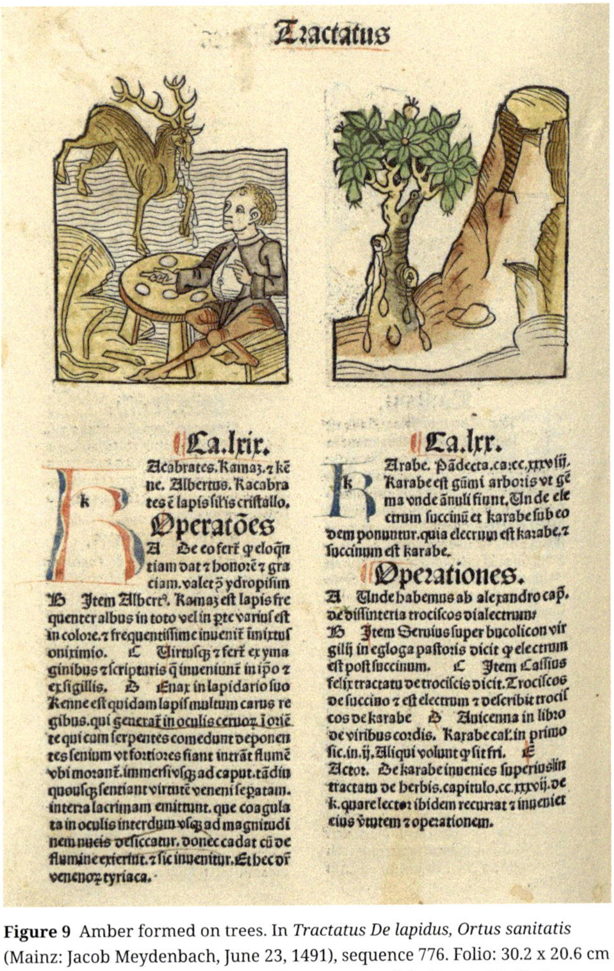

Tractatus

Ca.lxix.

Acabrates. Kamaz. ⁊ kene. Albertus. Racabrates ē lapis sil'is cristallo.

Operatões

A De co fert̄ q̄ eloq̄n tiam dat ⁊ honorē ⁊ graciam. valet ꝯ ydropisim

B Item Albert⁹. Kamaz est lapis frequenter albus in toto vel in pte varius est in colore. ⁊ frequentissime inuenit̄ imixtus onixinio. C Virtusq; ⁊ fert̄ ex ymaginibus ⁊ scripturis q̄ inueniunt̄ in ipo ⁊ ex sigillis. D Enax in lapidario suo Kenne est quidam lapis multum carus regibus. qui generat̄ in oculis ceruoꝝ i oriē te qui cum serpentes comedunt deponentes senium vt fortiores fiant intrāt flumē vbi morant̄. immersiq; ad caput. tādiu quousq; sentiant virtutē veneni sepatam. interna lacrimam emittunt. que coagulata in oculis interdum vsq; ad magnitudinem nucis desiccatur. donec cadat cū de flumine exierint. ⁊ sic inuenitur. Et hec cō veneno ꝝ tyriaca.

Ca.lxx.

Arabe. Pādecta. ca. cc. xxxviij. Karabe est gūmi arboris vt gēma vnde ānuli fiunt. Vnde electrum succinū et karabe sub eodem ponuntur. quia electrum est karabe. ⁊ succinum est karabe.

Operationes.

A Vnde habemus ab alexandro cap. de dissinteria trociscos dialectrum.

B Item Seruius super bucolicon virgilij in egloga pastoris dicit q̄ electrum est post succinum. C Item Cassius felix tractatu de trociscis dicit. Trociscos de succino ⁊ est electrum ⁊ describit trociscos de karabe D Auicenna in libro de viribus cordis. Karabe cal. in primo sic. in. ij. Aliqui volunt q̄ sit fri. E Actor. De karabe inuenies superius in tractatu de herbis. capitulo. cc. xxxvij. de k. quare lector ibidem recurrat ⁊ inueniet eius v̄tutem ⁊ operationem.

Figure 9 Amber formed on trees. In *Tractatus De lapidus, Ortus sanitatis* (Mainz: Jacob Meydenbach, June 23, 1491), sequence 776. Folio: 30.2 x 20.6 cm (11⅞ x 8⅛ in.). Handcolored woodcut. Courtesy of the Boston Medical Library in the Francis A. Countway Library of Medicine.

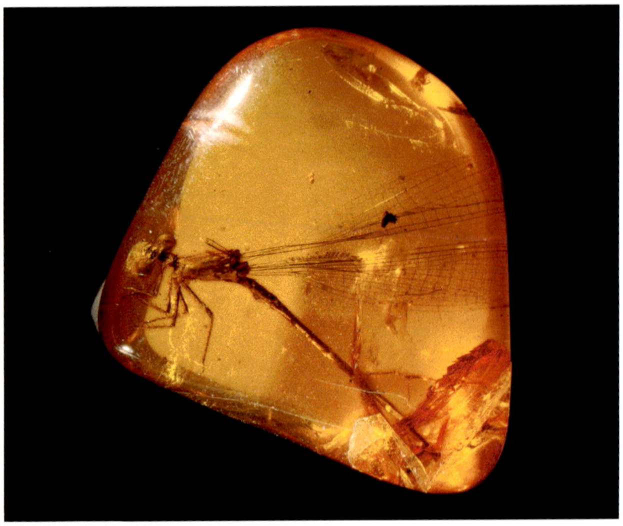

Figure 10 Damselfly in Dominican amber, L: 4.6 cm (1⅘ in.). Private collection. Photo: D. Grimaldi / American Museum of Natural History.

Chemically, the resin that became amber originally contained liquids (volatiles) such as oils, acids, and alcohols, including aromatic compounds (terpenes) that produce amber's distinctive resinous smell.[26] Over time, the liquids dissipated and evaporated from the resin, which began to harden as the organic molecules joined to form much larger ones called polymers. Under the right conditions, the hardened resin continued to polymerize and lose volatiles, eventually forming amber, an inert solid that, when completely polymerized, has no volatiles.[27] Most important, the resins that became amber were buried in virtually oxygen-free sediments.

How long does it take for buried resin to become amber? The amberization process is a continuum extending from freshly hardened resins to rocklike ones, and, as David Grimaldi points out, "No single feature identifies at what age along that continuum the substance becomes amber."[28] Langenheim explains: "With increasing age, the maturity of any given resin will increase, but the rate at which it occurs depends on the prevailing geologic conditions as well as the composition of the resin.... Changes appear to be a response primarily to geothermal stress since chemical change in the resin accelerates at higher temperatures."[29]

While some experts maintain that only material that is several million years old or older is sufficiently cross-linked and polymerized to be classified as amber, others opt for a date as recent as forty thousand years before the present.[30] Much depends on the soil conditions of the resin's burial. In its final form, amber is much more stable than the original substance. Amber is organic, like petrified wood or dinosaur bones, but, unlike those substances, it retains its chemical composition over time, and that is why some experts resist calling it a fossil resin (a nevertheless useful term).[31] Amber can also preserve plant matter (figure 11), bacteria, fungi, worms, snails, insects, spiders, and (more rarely) small vertebrates. Some pieces of amber contain water droplets and bubbles, products of the chemical breakdown of organic matter. It is not entirely understood how resins preserve organic matter, but presumably the chemical features of amber that preserve it over millennia also preserve flora and fauna inside it.[32] It must be that amber's "amazing life-like fidelity of preservation ... occurs through rapid and thorough fixation and inert dehydration as well as other natural embalming properties of the resin that are still not understood."[33] The highly complex process that results in amber formation gave rise to a wealth of speculation about its nature and origins. Whence came a substance that carried within it the flora and fauna of another place and time, one with traces of the earth and sea, one that seemed even to hold the light of the sun?

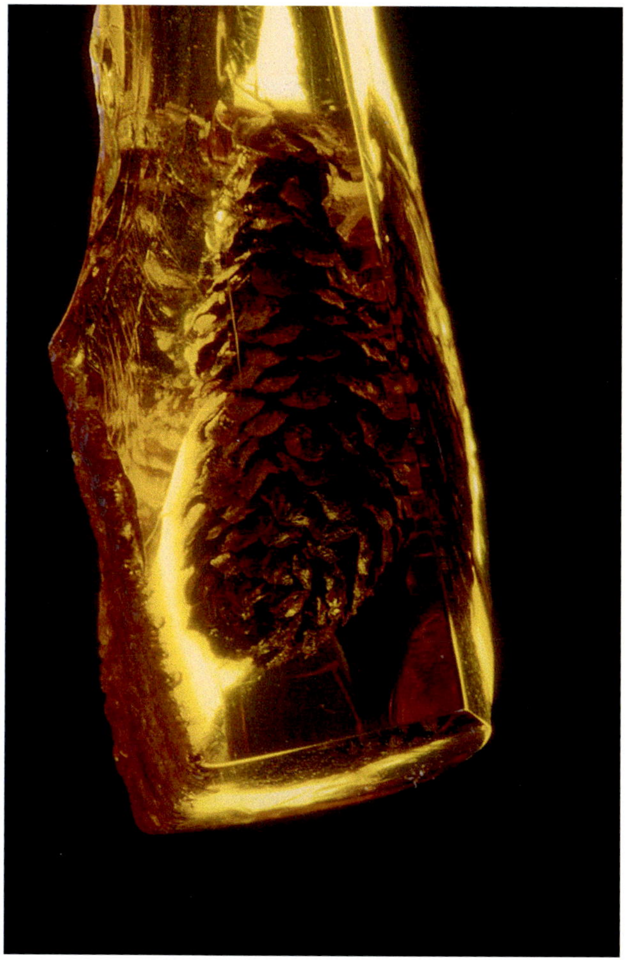

Figure 11 Cone in Baltic amber, L: 15.2 cm (6 in.). Private collection. Photo: D. Grimaldi / American Museum of Natural History.

NOTES

22. Recent sources consulted include E. Trevisani, "Che cosa è l'ambra," in *Magie d'ambra* 2005, pp. 14–17; E. Ragazzi, *L'ambra, farmaco solare: Gli usi nella medicina del passato* (Padua, 2005); Langenheim 2003; Weitshaft and Wichard 2002; Pontin and Celi 2000; Poinar and Poinar 1999; Ross 1998; *Bernstein* 1996; Grimaldi 1996; Å. Dahlström and L. Brost, *The Amber Book* (Tucson, AZ, 1996); Anderson and Crelling 1995; B. Kosmowska-Ceranowicz and T. Konart, *Tajemnice bursztynu (Secrets of Amber)* (Warsaw, 1989); Beck and Bouzek 1993; and J. Barfod, F. Jacobs, and S. Ritzkowski, *Bernstein: Schätze in Niedersachsen* (Seelze, 1989). The late C. W. Beck's lifetime of work on amber analysis is critical to any study of the material.

23. Langenheim 2003, p. 169.

24. Ross 1998, p. 2.

25. Nicholson and Shaw 2000, p. 451, with reference to Beck and Shennan 1991, pp. 16–17.

26. Ross 1998, p. 3: "The polymers are cyclic hydrocarbons called terpenes.... Amber generally consists of around 79% carbon, 10% hydrogen, and 11% oxygen, with a trace of sulphur."

27. Ross 1998, p. 3.

28. Grimaldi 1996, p. 16.

29. Langenheim 2003, pp. 144–45.

30. Langenheim 2003, p. 146, following Anderson and Crelling 1995.

31. Ross 1998, p. 3, in describing the amberization process, points to the critical element of the kinds of sediments in which the resin was deposited: "but what is not so clear is the effect of water and sediment chemistry on the resin." In the ancient world, amber does not seem to have been considered a fossil like other records of preserved life—petrified wood, skeletal material, and creatures in limestone. See A. Mayor, *The First Fossil Hunters: Paleontology in Greek and Roman Times* (Princeton, 2000).

32. Ross 1998, p. 12.

33. Langenheim 2003, p. 150.

Where Is Amber Found?

Deposits of amber occur throughout both the Old and the New Worlds, and many varieties are recognized. Of the many kinds of amber found in the Old World, the most plentiful today, as in antiquity, is Baltic amber (figure 12), or succinite (so called because it has a high concentration of succinic acid). This early Tertiary (Upper Eocene–Lower Oligocene) amber comes mainly from around the shores of the Baltic Sea, from today's Lithuania, Latvia, Russia (Kaliningrad), Poland, southern Sweden, northern Germany, and Denmark. The richest deposits are on and around the Samland peninsula, a large, fan-shaped area that corresponds to the delta region of a river that once drained an ancient landmass that geologists call Fennoscandia. This ancient continent now lies beneath the Baltic Sea and the surrounding land. Although this area has the largest concentration of amber in the world, it is a secondary deposition. Amazingly, the fossil resin "was apparently eroded from marine sediments near sea level, carried ashore during storms, and subsequently carried by water and glaciers to secondary deposits across much of northern and eastern Europe" over a period of approximately twenty million years.[34] In antiquity, most amber from the Baltic shore was harvested from shallow waters and beaches where it had washed up (once again, millennia later), especially during autumn storms that agitated the seabeds. It was only in the early modern period that amber began to be mined. With the introduction of industrial techniques, huge amounts have been extracted since the nineteenth century. It is estimated that up to a million pounds of amber a year was dug from the blue earth layer of the Samland peninsula in the first decades of the twentieth century.[35]

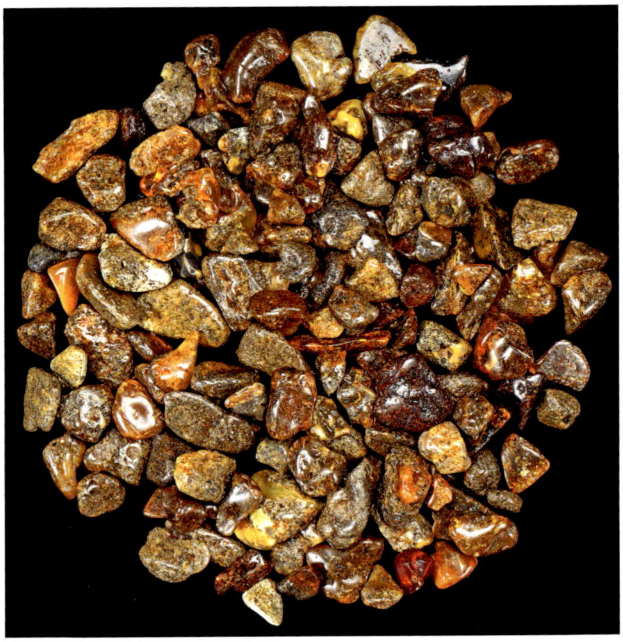

Figure 12 Baltic amber, L: 2.2 cm (7/8 in.). Private collection. Photograph © Lee B. Ewing.

Other kinds of amber used by ancient Mediterranean peoples have been identified with sources in today's Sicily,[36] Lebanon, Israel, and Jordan.[37] In addition to northern European sources, ancient accounts mention amber from Liguria, Scythia,[38] Syria, India, Ethiopia, and Numidia. However, of the varieties used in antiquity and known today, only succinite, or Baltic amber, is found in the large, relatively sturdy, jewelry-grade pieces such as were used for the sizable objects of antiquity, like the pre-Roman pendants of this catalogue, or for the complex carvings, vessels, and containers of Roman date. Small pieces of amber and the wastage of larger compositions could have been used for tiny carvings and other purposes. Non-jewelry-grade amber would also have been employed in inlay, incense and perfume, pharmaceuticals, and varnish, as is still the case in the modern period. Burmite (found in Burma, now Myanmar) and some amber from China, types also found in large, high-grade pieces, have long histories of artistic and other uses in Asia.[39]

NOTES

34. Langenheim 2003, p. 164.

35. For the modern mining of Baltic amber, see the overview in Rice 2006, chap. 3.

36. On Sicilian amber, see Trevisani in *Magie d'ambra* 2005, p. 16; Schwarzenberg 2002; Grimaldi 1996, p. 42; C. W. Beck and H. Harnett, "Sicilian Amber," in Beck and Bouzek 1993, pp. 36–47; Strong 1966, pp. 1–2, 4; and Buffum 1900. Pliny and the sources he consulted, including Theophrastus, discuss amber from Liguria. Ligurian deposits may indeed have been known in antiquity. Larger deposits may have been exhausted in antiquity. The ancient boundaries of Liguria include areas where non-jewelry-grade amber is known, as Trevisani maps. If it was dug up rather than originating in an oceanic or riverine source, it may not have had the same value. Moreover, the proximity of the material to its consumption point might have undermined its value. See n. 110 for more on amber's value.

37. In addition to the sources listed in n. 36, above, see J. M. Todd, "The Continuity of Amber Artifacts in Ancient Palestine: From the Bronze Age to the Byzantine," in Beck and Bouzek 1993, pp. 236–46, and J. M. Todd, "Baltic Amber in the Ancient Near East: A Preliminary Investigation," *Journal of Baltic Studies* 16, no. 3 (1985): 292–302. On Lebanese amber, see G. O. Poinar, Jr., and R. Milki, *Lebanese Amber: The Oldest Insect Ecosystem in Fossilized Resin* (Corvallis, OR, 2001), p. 15, who describe a few fist-sized pieces of "quite durable" Lebanese amber found in modern times, although generally Lebanese amber is collected in small, highly fractured pieces less than a centimeter in diameter. See also Grimaldi 1996, pp. 35–36.

38. On Scythian amber, see E. H. G. Minns, *Greeks and Scythians: A Survey of Ancient History and Archaeology on the North Coast of the Euxine from the Danube to the Caucasus* (1913; repr., New York, 1971), pp. 7, 440, with reference to Pliny, *Natural History* 33.161, 37.33, 37.40, 37.64, 37.65, and 37.119.

39. The geological source of Ming- and Ching-dynasty amber carvings is not assured. The amber might have come from Myanmar (Burma) or possibly from European, "Syrian," or Chinese sources. "China does have some large natural deposits of amber in Fushun, but these appear not to have been exploited" (Grimaldi 1996, p. 194). See also B. Laufer, "Historical Jottings on Amber in Asia," *Memoirs of the American Anthropological Association* 1 (1907): 3. On amber from Myanmar, see Langenheim 2003, p. 279: "Amber was collected from shallow mines in the Nagtoimow Hills in northern Burma and the major portion was sent to trade centers such as Mandalay and Mogaung ... and then brought by traders to Yunnan province in China where it was used by Chinese craftsmen from as early as the first Han dynasty (206 B.C. to A.D. 8)." Langenheim draws from H. L. Chibber, *The Mineral Resources of Burma* (London, 1934). See also D. A. Grimaldi, M. S. Engel, and P. C. Nascimbene, "Fossiliferous Cretaceous Amber from Myanmar (Burma): Its Rediscovery, Biotic Diversity, and Paleontological Significance," *Novitates* 3361 (March 26, 2002): 1–7; V. V. Zherikhin and A. J. Ross, "A Review of the History, Geology, and Age of Burmese Amber (Burmite)," *Geology Bulletin* 56, no. 1 (2000): 1–3; V. V. Zherikhin and A. J. Ross, "The History, Geology, Age and Fauna (Mainly Insects) of Burmese Amber, Myanmar," in *Bulletin of the Natural History Museum,* ed. A. J. Ross (London, 2000); Ross 1998, p. 15; *Bernstein* 1996; Grimaldi 1996, pp. 40–42, 194–208; and S. S. Savkevich and T. N. Sokolova, "Amber-like Fossil Resins of Asia and the Problems of Their Identification in Archaeological Contexts," in Beck and Bouzek 1993, pp. 48–50. In the annals of the Han and later dynasties, amber is mentioned repeatedly as one of the notable products of Roman Syria; see F. Hirth, *China and the Roman Orient: Researches into Their Ancient and Mediaeval Relations as Represented in Old Chinese Records* (Shanghai and Hong Kong, 1885), pp. 35–96.

    Pliny (*Natural History* 37.11) cites authors who attest to amber from Syria and India as well as to other sources east and south of Italy. Poinar and Milki, 2001 (n. 37, above), p. 77, suggest that many "nineteenth and twentieth century reports of amber finds in western Syria probably referred to localities within the confines of present-day Lebanon, since the latter had been a republic within the borders of Syria for a number of years." For amber from the ancient Near East, see M. Heltzer, "On the Origin of the Near Eastern Archaeological Amber," in *Languages and Cultures in Contact,* Orientalia Lovaniensia Analecta 96, ed. K. van Lerberghe and G. Voet (Leuven, 1999), pp. 169–76; S. M. Chiodi, "L'ambra nei testi mesopotamici," *Protostoria e storia del 'Venetorum Angulus': Atti del XX Convegno di studi etruschi ed italici, Portogruaro, Quarto d'Altino, Este, Adria, 16–19 ottobre 1996* (Pisa and Rome, 1999); and J. Oppert, "L'Ambre jaune chez les Assyriens," *Recueil de travaux relatifs à la philologie et à l'archéologie à égyptiennes et assyriennes* 21 (1880): 331ff.

The Properties of Amber

Amber is a light material, with a specific gravity ranging from 1.04 to 1.10, only slightly heavier than that of water (1.00). Amber may be transparent or cloudy, depending on the presence and number of air bubbles (figure 13). It frequently contains large numbers of microscopic air bubbles, allowing it to float and to be easily carried by rivers or the sea. White opaque Baltic amber may contain as many as 900,000 minuscule air bubbles per square millimeter and floats in fresh water. Clear Baltic amber sinks in fresh water but is buoyant in saltwater. Baltic amber has some distinguishing characteristics rarely found in other types of amber: it commonly contains tiny hairs that probably came from the male flowers of oak trees, and tiny pyrite crystals often fill cracks and inclusions. Another feature found in Baltic amber is the white coating partly surrounding some insect inclusions, formed from liquids that escaped from the decaying insects.[40]

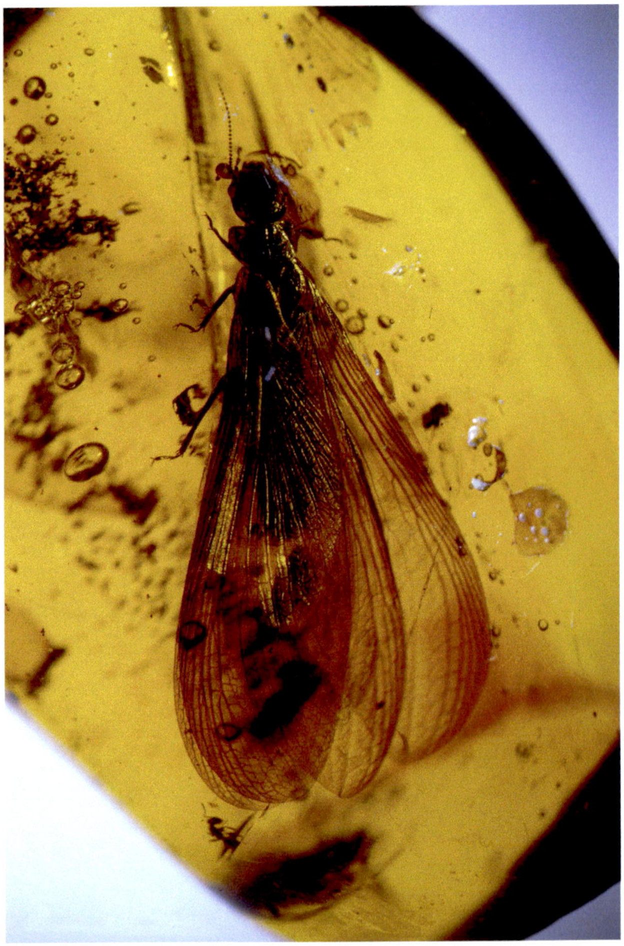

Figure 13 Extinct termite, *Mastotermes electrodominicus*, in Dominican amber, L: 4.6 cm (1⅘ in.). Photo: D. Grimaldi / American Museum of Natural History.

Amber's hardness varies from 2 to 3 on the Mohs scale (talc is 1 and diamond 10). This relative softness means that amber is easily worked. It has a melting-point range of 200 to 380°C, but it tends to burn rather than melt. Amber is amorphous in structure and, if broken, can produce a conchoidal, or shell-like, fracture. It is a poor conductor and thus feels warm to the touch in the cold, and cool in the heat. When friction is applied, amber becomes negatively charged and attracts lightweight particles such as pieces of straw, fluff, or dried leaves. Its ability to produce static electricity has fascinated observers from the earliest times. Amber's magnetic property gave rise to the word *electricity*: amber (Greek, *elektron*) was used in the earliest experiments on electricity.[41] Amber's natural properties inspired myth and legend and dictated its usage.

In antiquity, before the development of colorless clear glass that relies on a complex technique perfected in the Hellenistic period, the known clear materials were natural ones: water and some other liquids; ice; boiled honey and some oils; rock crystal; some precious stones; and amber.[42] Transparent amber is a natural magnifier, and, when formed into a regularly curved surface and given a high polish, it can act as a lens.[43] A clear piece of amber with a convex surface can concentrate the sun's rays. One ancient source suggests that such polished ambers were used as burning lenses.

Once amber is cleaned of its outer layers and exposed to air, its appearance—its color, degree of transparency, and surface texture—eventually will change. As a result of the action of oxygen upon the organic material, amber will darken: a clear piece will become yellow; a honey-colored piece will become red, orange-red, or red-brown, and the surface progressively will become more opaque (figure 14).[44] Oxidation commences quite quickly and starts at the surface, which is why some amber may appear opaque or dark on its surface and translucent at breaks or when subjected to transmitted light. However, the progress of oxidation is variable and depends on the time of exposure and other factors, such as the amount and duration of exposure to light. In archaeologically recovered amber, the state of the material is dependent upon burial conditions, and the degree of oxidation can vary widely, as the Getty collection reveals. The breakdown of the cortex causes cracking, fissuring, flaking, chipping, and, eventually, fractures. Only a very few ancient pieces retain something of their original appearance, in each case because of the oxygen-free environment in which it was buried. For instance, two fifth-century B.C. female head pendants that were excavated at waterlogged Spina are remarkable for their clear, pale yellow color (figure 15).[45] A large group of seventh-century B.C. amber-embellished objects from the cemeteries of Podere Lippi and Moroni-Semprini in Verucchio (Romagna) were preserved along with other perishable objects by the stable anaerobic conditions of the Verucchio tombs, which had been sealed with a mixture of water and clay (figure 16).[46] Various colors and degrees of transparency are in evidence, from pale, clear yellow to clear orange or red to opaque yellows, oranges, reds, and tans. Inclusions are common.

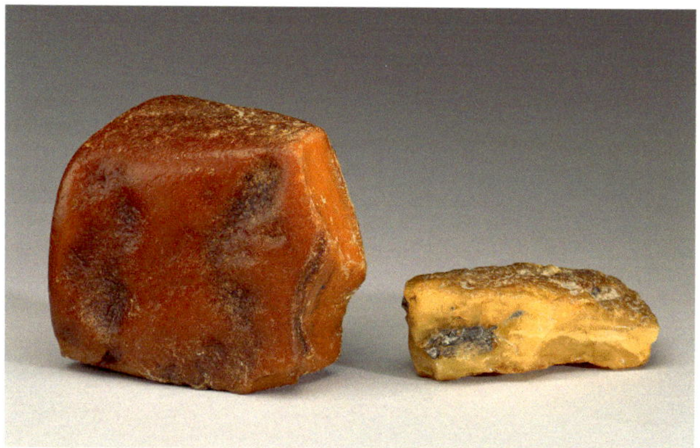

Figure 14 Two typical pieces of Baltic amber. Pale yellow amber was preferred by the ancient Greeks and Etruscans. Opaque orange amber was especially fashionable in Imperial Rome. L (orange amber): 9 cm (3½ in.). L (yellow amber): 5 cm (2 in.). Private collection. Photograph © Lee B. Ewing.

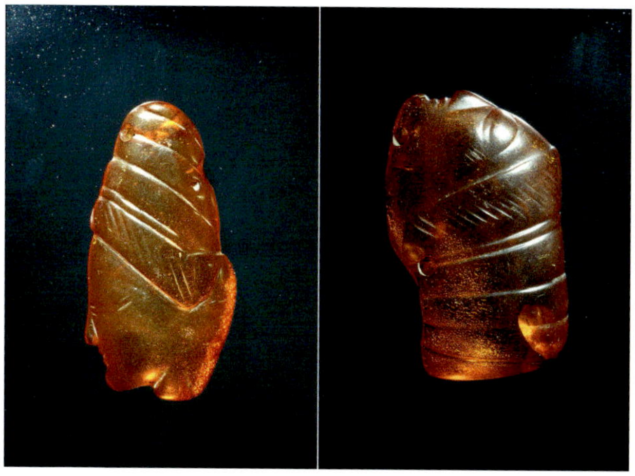

Figure 15 Female Head Pendants, from Tomb 740 B, Valle Pega, Spina, a tomb dating to the end of the 5th century B.C. Amber, H: 4.8 cm (1⅞ in.), W: 2.8 cm (1⅛ in.), D: 1.2 cm (½ in.) and H: 4.5 cm (1¾ in.), W: 2.8 cm (1⅛ in.), D: 1.4 cm (½ in.). Ferrara, Museo Archeologico Nazionale, 44877 and 44878. Ferrara, Museo Archeologico Nazionale / IKONA.

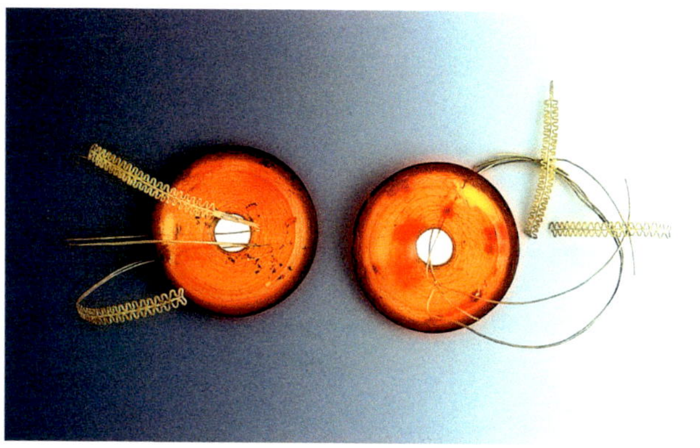

Figure 16 Earrings, from Tomb 23, Podere Lippi, Verucchio. First half of the 7th century B.C. Amber and gold, Diam. (amber, max): 6 cm (2⅜ in.). Verucchio, Museo Civico Archeologico, 8410-850. Verucchio, Museo Civico Archeologico / IKONA.

Many pre-Roman figured ambers exploit the material's transparency, offering the possibility of reading through the composition: the back is visible from the front and vice versa, albeit blurrily. This is a remarkable artistic conception, iconographically powerful and magical. Two extraordinary examples are the Getty *Lion* (see figure 54) and the British Museum *Satyr and Maenad* (figure 17).[47] From its top, the underside of the lion can be discerned. In the multigroup composition of the London amber, the large snake on the reverse appears to join in reveling with the figures on the front.

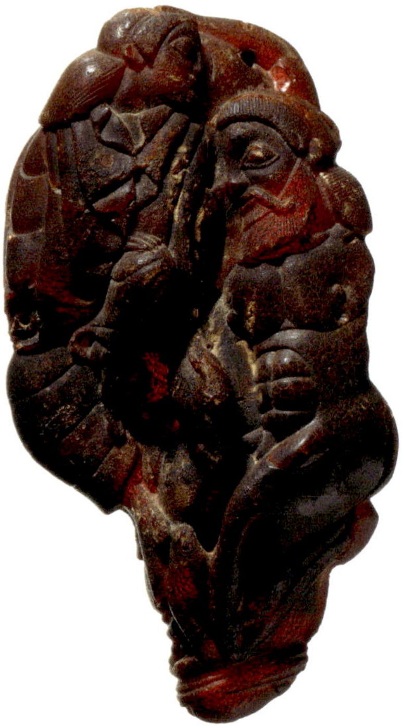

Figure 17 *Satyr and Maenad* pendant, Etruscan or Etrusco-Campanian, late 6th century B.C. Amber, H: 17.3 cm (6 4/5 in.), W: 9.5 cm (3 3/4 in.), D: 4.5 cm (1 3/4 in.). London, British Museum, 1865,0103.46. © The Trustees of the British Museum.

A number of seventh-century Greek, Etruscan, and Campanian objects include amber set into precious metal mounts or backed with silver or gold foil.[48] Some are internally lit by foil (or possibly tin) tubes. Amber's glow, its brilliance and shine, would be immeasurably enhanced in this way.[49] Simply shaped amber pieces set into gold and silver are mirrorlike, emanating radiance and banishing darkness.[50] Amber faces once mounted on polished metal, the Getty *Heads of a Female Divinity or Sphinx* (figures 18 and 45) might even seem to issue light, like the principal astral bodies, or to capture the shimmer of light on water.[51]

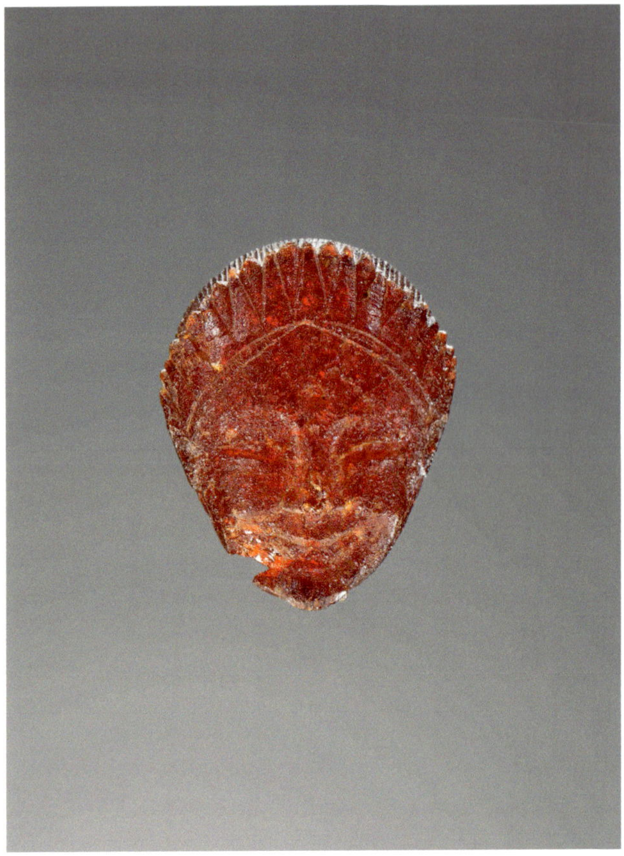

Figure 18 *Head of a Female Divinity or Sphinx* pendant, Etruscan, 550–525 B.C. Amber, H: 3.2 cm (1 1/4 in.), W: 2.6 cm (1 in.), D (face): 1.2 cm (2/5 in.), D (back): 0.5 cm (1/5 in.), D (joined): 1.7 cm (7/10 in.). Los Angeles, J. Paul Getty Museum, 76.AO.85.1 and 76.AO.86. Gift of Gordon McLendon. See cat. no. 10.

NOTES

40. Ross 1998, p. 11.

41. For the basic properties of amber, see Ross 1998, p. 4. The word *electricity* was coined by W. Gilbert, a physician at the court of Queen Elizabeth I, to describe this property in his 1600 book *On the Magnet, Magnetic Bodies and That Great Magnet the Earth.*

    The early Greek philosopher Thales of Miletos is credited by Diogenes Laertius as the first to recognize amber's magnetism: "Arguing from the magnet and from amber, he attributed a soul or life even to inanimate objects" (*Diogenes Laertius* 1.24, vol. 1, ed. and trans. R. D. Hicks, Loeb Classical Library 184 [London, 1993]). E. R. Caley and J. C. Richards, *Theophrastus on Stones* (Columbus, 1956), p. 117, argue that this claim rests on shaky ground; that Thales was the first to mention the property can be inferred only indirectly from Diogenes Laertius's statement: "Aristotle and Hippias say that, judging by the behaviour of the lodestone and amber, he also attributed souls to lifeless things." Caley and Richards consider the possibility "that it was Hippias who said that Thales understood the attractive property of amber, but there is no way of confirming such an inference because the works of Hippias are not extant." Plato (*Timaeus* 80c) alludes to amber's magnetism but denies that it is a real power of attraction. Aristotle does not mention amber in the relevant section of *On the Soul* (*De Anima* 1.2.405A). Thus, following Caley and Richards, Theophrastus is the earliest extant account. If Thales *did* describe amber's static electricity, he may have done so based on his observation of wool production, which used amber implements: distaff, spindle, and whorls. I owe this observation to Schwarzenberg 2002, who calls attention not only to the famous wool of Miletos, but also to the number of extant seventh-century spinning tools. Pliny notes that Syrian women used amber whorls in weaving and that amber picks up the "fringes of garments," and also comments on amber's electrostatic property. But, unlike Plato, he thinks its magnetic property *is* like that of iron. Plutarch (*Platonic Questions* 7.7) explains that "the hot exhalation released by rubbing amber acts in the same ways as the emanations from the magnet. That is, it displaces air, forming a vacuum in front of the attracted object and driving air to the rear of it": *De Lapidibus,* ed. and trans. D. E. Eichholz (Oxford, 1965), p. 200, n.b.

42. Clear colorless glass (with antimony used as the decolorizing agent) is documented in the eighth century B.C. in western Asia and again in the fifth and fourth centuries B.C. in Greece. In Egypt, the use of manganese as a decolorizing agent became common in the first century B.C.; see E. M. Stern and B. Schlick-Nolte, *Early Glass of the Ancient World, 1600 B.C.–A.D. 50: Ernesto Wolf Collection* (Ostfildern, Germany, 1994), p. 20.

43. For an excellent overview of lenses and their ancient employment, see Plantzos 1999, pp. 39–41, 110; and Plantzos 1997, pp. 451–64. According to Plantzos 1999, p. 41, "The discovery of crystals that could have served as magnifying lenses has been reported from Bronze Age sites, and although no similar objects can be dated to the Hellenistic period, some exist from Roman contexts." He also points out that "developments in optics already in the Classical period suggest the possibility of magnifying lenses." Various ancient authors describe the magnification of objects: Aristotle (*Posterior Analytics* 1.31) and Theophrastus (*On Fire* 73) observe "that convex pieces of glass can concentrate the sunrays, and light fire ... and an earlier reference in Aristophanes (*Clouds* 766–75) indicates how well observed [this] was." "For a lens to be able to contract light, a piece of glass with [a] regularly curved surface and a minimum diameter of around four centimeters was needed. Such a lens will have a short focus (between six and nine millimeters) and will therefore be quite useless as a general eye aid, but quite appropriate for a magnifying glass" (ibid.). Although no ancient literary source mentions amber's natural magnifying property, it is difficult to imagine that it went unnoticed. Many bulla-shaped amber pendants (of as early as seventh-century date) have regularly curved surfaces and are the right size to use as magnifiers, especially if the resin were clear. (On amber bullae, see n. 152.) The various techniques necessary to make a clear magnifying or burning lens from amber apparently were available by the first century A.D. The carving and polishing tools and technology were age-old, and as for the clarification process, Pliny relates a technique for "dressing" amber by boiling it in the fat of a suckling pig, a necessary step in making imitation transparent gemstones from amber, which Pliny also describes. A section of an entry (*Hualê*) in the Byzantine *Suda* may not refer to a glass lens, but rather to an amber one: "[A glass] is a round-shaped device of amber glass, contrived for the following purpose: when they have soaked it in oil and heated it in the sun they introduce a wick and kindle [fire]. So the old man is saying, in conversation with Socrates: if I were to start a fire with the amber and introduce fire to the tablet of the letter, I could make the letters of the lawsuit disappear." See "*Ὑάλη*," trans. David Whitehead, March 19, 2006, *Suda On Line*, www.stoa.org/sol-entries/upsilon/6 (accessed November 27, 2009).

Processed (boiled, molded, and then ground) amber lenses are described by the end of the seventeenth century. In 1691, C. Porshin of Königsberg invented an amber burning glass, which was said to be better than the glass kind; he also used amber to make spectacles. See O. Faber, L. B. Frandsen, and M. Ploug, *Amber* (Copenhagen, 2000), p. 101. For illustrations of amber lenses of the early modern period, see *Bernstein* 1996.

44. See Ross 1998, pp. 18–19; and Strong 1966, p. 14 (with reference to M. Bauer, *Precious Stones* [London, 1904], p. 537).

45. Ferrara, Museo Archeologico Nazionale 44877–78, from Tomb 740 B at Spina: C. C. Cassai, "Ornamenti femminile nelle tombe di Spina," in *Due donne* 1993, pp. 42–47; *Spina: Storia di una città tra greci e etruschi,* exh. cat. (Ferrara, 1993); and Negroni Catacchio 1989, fig. 470.

46. For splendid photographs of the Verucchio material, see *Verucchio* 1994.

47. Strong 1966, pp. 61–62, pl. XV.

48. See Plantzos 1999, p. 41, on the importance of color to ancient gemologists; he remarks that the "contrast of the translucent stone against the golden background of the ring was thought to be a merit of the jewel." "A gold tube lining the perforation of a transparent or translucent material such as amber or rock crystal has a marked effect on the brightness and thus appearance of the bead and is, in effect, a form of foiling": J. Ogden, "The Jewelry of Dark Age Greece: Construction and Cultural Connections," in *The Art of the Greek Goldsmith,* ed. D. Williams (London, 1998), pp. 16–17, also nn. 19–21 (in reference to objects from Lefkandi, the Tomb 2 jewelry from Tekke, the Elgin group, and an eighth-century tomb from Salamis).

49. *Agalma* is a Greek word used to describe the quality of brilliance; it is perhaps related etymologically to *aglaos* (shining). See Stewart 1997, p. 65. On *agalma* and *agalmata,* see n. 6.

50. The three gold pendants inlaid with amber from the Regolini-Galassi Tomb are superb examples of this mirrorlike quality (Vatican, Museo Gregoriano Etrusco 691, from the Sorbo Necropolis, Cerveteri): Cristofani and Martelli 1983, p. 262, no. 31; and L. Pareti, *La tomba Regolini-Galassi del Museo gregoriano etrusco e la civiltà dell'Italia centrale nel secolo VII a.c.* (Vatican City, 1947). The ivory handle of an Orientalizing ceremonial axe was inlaid with amber rectangles, circles, and triangles mounted on tinfoil, making them appear like tiny mirrors (Florence, Museo Archeologico Nazionale 70787): Bartoloni et al. 2000, p. 238, no. 268, where M. C. Bettini calls attention to the technique and notes parallels from Casale Marittimo and Verucchio.

    How an amber "mirror," however tiny, worked for the living or for the dead is worth reflection. That all documented mirrors from Etruria, and most from the rest of the circum-Mediterranean, come from graves (many with evidence of use wear) is critical to their interpretation. J. Lerner, "Horizontal-Handled Mirrors: East and West," *Metropolitan Museum Journal* 31 (1996): n. 3, compares the ancient disk mirror-fibula to the large amber-decorated fibulae found in Etruscan tombs (with reference to the "Morgan Amber" in New York [see figure 24]; she acknowledges J. Mertens for the observation). On reflection and mirror symbolism, see G. Robins, "Dress, Undress, and the Representation of Fertility and Potency in New Kingdom Egyptian Art," in Kampen 1996, pp. 32–33; A. Stewart, "Reflections," in Kampen 1996, pp. 136–54; J. Neils, "Reflections of Immortality: The Myth of Jason on Etruscan Mirrors," in De Puma and Small 1994, pp. 190–95; Pinch 1994; G. Pinch, *Votive Offerings to Hathor* (Oxford, 1994), pp. 235–38; A. Kozloff, "Mirror, Mirror," *Bulletin of the Cleveland Museum of Art* 71, no. 8 (1984): 271–76. For mirrors in the history of art, see *Source* 4, nos. 2–3 (1985); L. O. K. Congdon, *Caryatid Mirrors of Ancient Greece* (Mainz, 1981); G. F. Hartlaub, *Zauber des Spiegels: Geschichte und Bedeutung des Spiegels in der Kunst* (Munich, 1951); and H. Schwarz, "The Mirror in Art," *Art Quarterly* 13 (1952): 96–118. G. Robins's comments have relevance beyond Egypt (Robins in Kampen 1996, p. 32): "Mirrors, therefore, were not simply items in which one could see one's reflection, but were overlaid with symbolism relating to fertility and also health, and were surely believed to protect the user in this life. However, like the fertility figurine, they usually had a funerary function, too. Many mirrors have been found in burials, and it can be deduced that their positive symbolism would also have been regarded as helping the deceased to achieve rebirth into the afterlife." The mirror was a type of object closely equated with the disk of the sun as well as with that of the moon, but its distinctive Egyptian form is most like that of the visible sun. See n. 161 on the connection of mirrors and the sun and the possibility of drawing down the power of the sun.

51. Winter 1994, p. 123. Here and in later studies, I. J. Winter describes "the quality of intense light, or radiance, emanating from a particular work" as "one of the most positive attributes in descriptions of what we would call Mesopotamian 'art.'" She underlines that it is "the combination of light-plus-sheen yielding a kind of lustrousness" that was particularly positive, auspicious, and sacral, not only in Mesopotamia, but also in other cultures. This is borne out by many of the forms and subjects of amber and amber-enhanced objects of ancient Greece and Italy.

Ancient Names for Amber

The words used for amber in antiquity often suggest not only the qualities for which it was valued, but also theories of its origin and the uses to which it was put. Today, although amber is still widely sought out for jewelry, magic, and medicine, its floral and faunal inclusions may be its greatest attraction (as reflected in the title of the 1996 exhibition and book *Amber: Window to the Past* [Grimaldi 1996]). There is scarce textual evidence before Roman times to indicate an ancient fascination with the creature and plant remains interred within amber; however, its use in burials may be evidence enough.

The standard Greek word for amber was *elektron.*[52] The derivation of this word is uncertain, although scholars have suggested that it might have connections with *helko*, meaning "to draw or attract," or with *aleko*, meaning "to ward off evil."[53] The word is certainly associated with *elektor*, used in the *Iliad* to mean "the beaming sun,"[54] and is most likely derived from an Indo-European verb with the root meanings "brilliant" or "to shine." This quality of beaming, or reflecting the sun, is also suggested by the Germanic word for amber, *glaes* or *glese*, recorded in some ancient Latin sources as *glaesum*, the same word used for glass in this period.[55] The Indo-Germanic root for this word, *ghel*, means "lustrous, shimmering, or bright" and gives us words such as *glisten*, *glitter*, *glow*, and *yellow* in English. The current German word for amber, going back to thirteenth-century Middle Low German, is similarly evocative: *Bernstein* means "burning stone."[56]

When Pliny the Elder or one of his contemporaries admired a valuable piece of amber, the first thing to strike their eyes would have been the suggestion of fire (*imagine igneam*) or the material's gentle glow (*mollis fulgor*). The amber's color was certainly evocative—of wine, honey, wax, embers, or fire—but was of secondary importance to its shine. This glow had been the defining characteristic of amber for centuries.

Brilliance in amber, ice, rock crystal, or any stone was possible only because of its transparency. The ancients believed that transparency was possible because light was *let* through a material: thus transparent materials had performative powers.[57] The brilliance of amber, enhanced by the rich connotations of its names, ensured it a place in ancient literature alongside other rare, prized, and luminous materials—sight-arresting materials such as gold, silver, and ivory, whose magnificence often was associated with something beyond the merely human, with the heroic or divine.[58] This association is evident from the first extant occurrences of *elektron*, in Homer's *Odyssey.*[59] When Telemachus visits Menelaus's palace in Book 4, he is awestruck: "Mark the flashing of bronze throughout the echoing halls, and the flashing of gold, of amber, of silver, and of ivory. Of such sort, methinks, is the court of Olympian Zeus within, such untold wealth is here; amazement holds me as I look."[60]

It is the flashing of the jewels, more so than the jewels themselves, that puts Telemachus in mind of Zeus; the word he uses is *steroph*—the flash of a lightning bolt. Telemachus's association of the brightness, the shine, the *brilliance* of Menelaus's palace with divinity seems almost instinctive.

Elektron occurs two other times in the *Odyssey*: once in Book 15, when the swineherd Eumaeus, telling the story of his kidnapping to Odysseus, remembers the cunning Phoenician mariner who turned up at his ancestral home with an eye-catching golden necklace strung with amber pieces.[61] In addition, in Book 18, when the suitors vie with one another in the extravagance of their gifts to Penelope, Eurymachus's contribution is "a richly crafted necklace of gold adorned with sun-bright amber" (figure 19).[62] Another early occurrence of *elektron* is in the Pseudo-Hesiodic *Shield of Herakles*. In this passage, as in Homer's description of Menelaus's palace, amber takes its place in a list of rare and precious materials, to dazzling effect: "He took his glittering shield in his hands, nor had anyone ever broken it or damaged it with a blow; it was a marvel to see. The whole orb glowed with enamel, white ivory, and amber, and it shone with gleaming gold."[63]

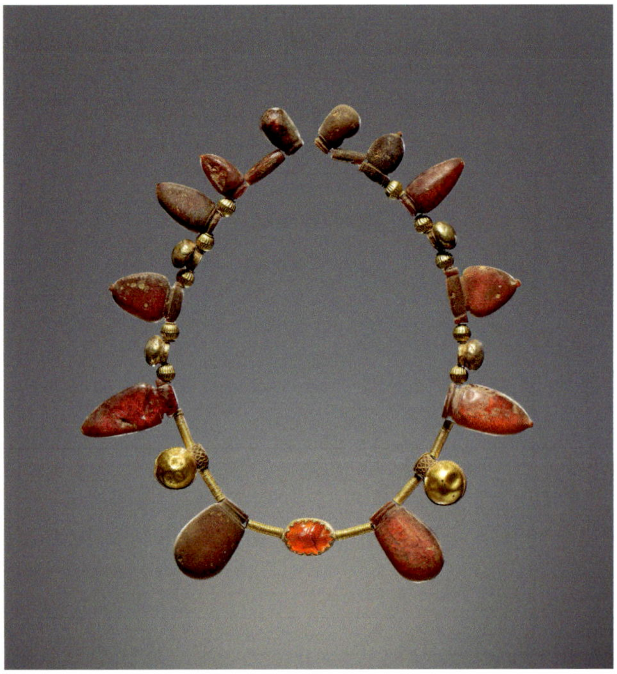

Figure 19 Necklace with a pendant scarab, Italic or Etruscan and Greek, 550–400 B.C. Amber, gold, and carnelian. L: 39.5 cm ($15\frac{9}{16}$ in.). Los Angeles, J. Paul Getty Museum, 77.AO.77.1. Gift of Gordon McLendon.

In each of these passages referring to the use of amber—the ornamentation of a seemingly Olympian palace, necklaces intended for elite women, and the shield of a hero—amber is inextricably bound up with the light of the sun, and it is associated with gods, heroes, and a social elite. The reflection of sunlight, in the halls of a king or on the armor of a hero, was a powerful reminder of the heavens and the heavenly; brilliance and luster were primary qualities to be looked for in a precious material such as gold, ivory, silver, or amber. The brilliance of the amber and other materials in Herakles' shield, combined with the perfect craftsmanship that it represented, called attention to its *poikilia*, the adornment and embellishment all fine works should display, and made it a *thauma idesthai*, a "marvel to behold"—what Raymond Prier has defined as "an intermediation between the polarities of men and gods, visually linguistic symbols of power."[64]

Although the most common, *elektron* was not the only Greek name for amber. It is likely that the substance referred to as *lyngourion* (there are other variants of the spelling—*liggourion*, for example) was a form of amber. Its derivation and its relationship to amber (*elektron*) were much discussed in antiquity and continue to be debated today. The earliest evidence for *lyngourion* is in Theophrastus's late-fourth-century B.C. lapidary, where he notes similarities between *lyngourion* and *elektron* but does not consider them the same material.[65] He seems to have had direct knowledge of some amber, which was dug up in Liguria and which he apparently considered a nonorganic substance. Theophrastus's *lyngourion* is as hard as amber, which he includes among stones possessing a power of attraction, and possesses the same powers of magnetism, but, according to him, it has a different origin: it is the hardened urine of wild lynxes, which "is discovered only when experienced searchers dig it up" (figure 20).[66] This origin story is doubtless the result of a fanciful attempt to explain the etymology of the word (*lyngourion* = lynx urine), a story that would have been additionally convincing because of the substance's color.

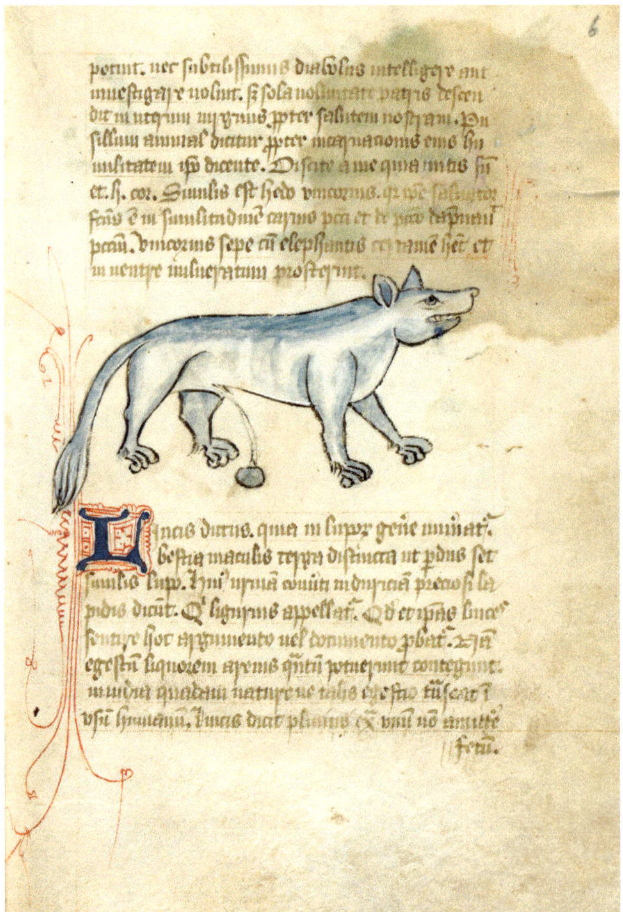

Figure 20 Lynx urine hardens into a stone. In *Bestiarius* GKS 1633 4°, 6r, English, 15th century. Parchment, H: 21 cm ($8\frac{1}{4}$ in.), W: 13.5 cm ($5\frac{3}{10}$ in.). Courtesy of The Royal Library of Denmark.

It was probably another attempt at etymology that persuaded Strabo that excessive quantities of amber could be found in Liguria.[67] Strabo makes no distinction between *lyngourion* and *elektron*, using the terms interchangeably. Pliny the Elder is as unimpressed with Strabo's talk of Liguria as he is with the lynx-urine story. Pliny lists a variety of sources containing variations on one or both of these themes, but his final word on *lyngourion* is that "the whole story is false, and no gemstone bearing this name has been known in our time." Although Pliny may have been justified in his skepticism (Liguria was no more a producer of amber than the lynx was of gemstones), *lyngourion* appears to be a term applied to highly transparent varieties of amber, while *elektron* was used more generally. Gemstones of *lyngourion* are first attested in third-century inventories of the Asklepieion on the south slope of the Acropolis and in the shrines of Artemis and Eileithyia (goddesses associated with childbirth, light, and the moon) at Delos.[68]

Several other terms for amber occur in Pliny the Elder's treatise: he cites Philemon as referring to a white, waxen form of amber from Scythia as *electrum*, and a tawny variety (from another part of Scythia) as *sualiternicum*. Pliny also attributes to his contemporary Xenocrates of Aphrodisias the claim that *sucinum* and *thium* are the Italian words for amber, and *sacrium* the Scythian word. Nicias, Pliny tells us, says that the Egyptians called amber *sacal* (perhaps meaning simply "rock"), and that the Syrian word was *harpax* (because of its magnetic qualities; the Greek *harpax* means "a thief" or "one who snatches").[69] Pliny also singles out Callistratus as the first to distinguish *chryselectrum*, or "gold amber."[70] Dioscorides, in his A.D. first-century *Materia Medica*, describes two types of amber: *elektron chrysophoron* (golden amber) and *elektron pteruyophoron* ("because it draws feathers to it"); and he uses the word *aigeiros*, which means "poplar," as a synonym for amber.[71] The poplar is associated not only with Herakles (the hero brought back poplar branches from the underworld), but also with the tale of Phaethon—the most prevalent myth about the origin of amber (see "Ancient Literary Sources on the Origins of Amber," below). Some authors, such as Pliny, use more than one term for the material, depending on the context.

NOTES

52. The word *elektron* was also used in antiquity to describe the alloy of silver and gold (modern *electrum*). Both the fossil resin and the alloy are found in the Shaft Graves at Mycenae, but the earliest surviving source to discuss both materials is Herodotus. Independently, Hughes-Brock 1993, p. 224, postulated that *elektron* was originally used for the resin and then transferred to the metal because the two materials shared certain optical properties. Much has been written on the relationship of resin and metal; these references are noted and reviewed in Hughes-Brock 1993 and Fuscagni 1982. See also C. L. Connor, *The Color of Ivory: Polychromy on Byzantine Ivories* (Princeton, 1998), p. 106, nn. 9–10; and H. G. Liddell, R. Scott, and H. Stuart-Jones, *Greek-English Lexicon*, 9th ed. (Oxford, 1968), s.v. "elektron" (in Greek), p. 768.

    Gold and silver alloys have been known as long as the individual metals. Naturally occurring alloys likely were used alongside human-made ones. The electrum alloy is much harder than either gold or silver. Pliny (*Natural History* 33.23.80) says, "All gold contains silver in various proportions.... Whenever the proportion is one-fifth, the ore is called electrum." J. Ogden, "Metals," in Nicholson and Shaw 2000, pp. 162–64, discusses the makeup of gold alloys in Egypt over time and the range of color in surviving objects made from gold-silver alloys. Traditionally, an alloy with more than 75 percent gold present is described as gold. If it is a gold-silver alloy with less than 75 percent gold, it is electrum, and, according to Stos-Gale and Gale's more recent nomenclature (Z. Stos-Gale and N. H. Gale, "Sources of Galena, Lead and Silver in Predynastic Egypt," *Revue d'Archéométrie* 3, suppl. [1981]: 285–96), "gold-silver alloys with 5–50 percent gold should be termed aurian silver (those with less than 5 percent gold are simply termed silver with low gold)." They go on to state: "The traditional division between electrum and gold at 75 percent gold level falls most inconveniently at just about the median composition for much Egyptian gold-work. Also the variable copper presence will have a major effect on colour" (ibid.). Compare Evely 2000, p. 401: "Electrum is a light-coloured alloy, though the precise percentage of silver required to constitute it varies according to authorities: as low as 8–10% or over 20% or even over 40%.... The commonest natural impurity of any degree is silver: anything up to 50% being called gold, thereafter the alloy is seen as basically a silver. It is largely a matter of semantics how such mixtures are termed, there being no hard and fast definition.... Pure gold probably never occurs naturally.... It is rare to find 98–99% purity." See also J. F. Healy, *Mining and Metallurgy in the Greek and Roman World* (London, 1978), pp. 201ff.

    Neb hedj, or "white gold," was long known in Egypt; its dual nature "meant that it was used sometimes with the significance of gold and at other times as if it were identical with silver," which early on was associated with the moon (Wilkinson 1994, p. 84). For discussion of early electrum usage in Mesopotamia, see P. R. S. Moorey, "The Archaeological Evidence for Metallurgy and Related Technologies in Mesopotamia, ca. 5500–2100," *Iraq* 44, pt. 1 (Spring 1982): 13–38; and P. R. S. Moorey, *Materials and Manufacture in Ancient Mesopotamia: The Evidence of Archaeology and Art*, BAR International Series 237 (Oxford, 1985).

    For other sources on amber's ancient names, see Schwarzenberg 2002; J. Puhvel, "On Terms for Amber," in *Studia Celtica et Indogermanica: Festschrift für Wolfgang Meid zum 70. Geburtstag*, eds. P. Anreiter and E. Jerem (Budapest, 1999), pp. 347–50; G. M. Catarsi, "Ambra: Mito e realtà," *Padusa*

Color and Other Optical Characteristics: Ancient Perception and Reception

We may imagine that when Zeus revealed his true form to Cadmus's daughter Semele at her rash request, his blinding brilliance was enough to reduce her to ashes even if he had left his thunderbolts behind. The name *Zeus* has associations of luminosity (it is derived from a word that means "to shine"), as do many of the common epithets for Greek deities: *Phoebus* Apollo means "radiant Apollo," and the goddess Athena is often described in Homer as *glaukopis*, which can be translated as "with gleaming eyes." Apollo appears to his worshippers at Delphi in a blaze of flame and brilliant light in the Homeric *Hymn to Apollo*. Similarly, the great heroes of ancient Greece often are depicted with a bright glow about them—like Achilles in the *Iliad*, "shining in all his armor like the sun."[72]

In Quintus Smyrnaeus's fourth-century A.D. *Fall of Troy*, the mourners at Ajax's funeral lay "gleaming gold" and "lucent amber-drops" around his body.[73] This connection between the radiance of precious jewels and the brilliance of heroes and gods was established in Greece as early as Homer. Given the strong associations among the dazzling, the divine, and the heroic, the choice of amber for a piece of jewelry or a work of art indicated a divine or heroic subject. For example, Pausanias mentions in his *Description of Greece* the amber statue of the emperor Augustus.[74] The image must have been a "marvel to behold."

When amber was considered in terms of its hue (instead of its brilliance), the images it evoked were no less striking. The most sought-after pieces ranged from yellow to red—colors that were associated with fire and the precious metal gold (figure 21).[75] The fiery and glowing colors were important to life, marriage, and death and were linked with divine forces. Yellow and red were redolent of fire (and consequently the sun) and of light itself, and were symbolic of life and regeneration.[76] In the Roman writings of Martial and Juvenal, gold was often referred to as being red.[77] And since Homer, amber and gold had been paired, and both were symbols of the sun.

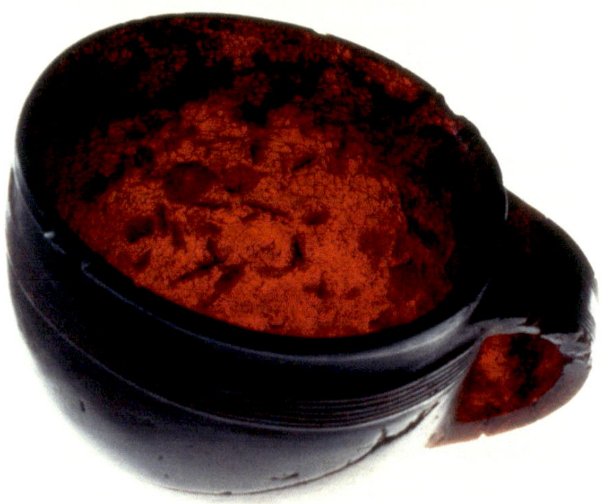

Figure 21 Hove tumulus cup, Wessex culture, Bronze Age. Amber, D: 8.9 cm (3½ in.). Brighton & Hove, Royal Pavilion & Museums.

The various images that a gemstone's color conjured up could sometimes, as in the case of *elektron*, determine its name. As we noted, etymologically the word is probably connected with *elektor*, "the beaming sun," the root meaning being "brilliant." Pliny the Elder, for instance, talks about a variety of jasper that was called *boria* (meaning "northern") "because it is like the sky on an autumn morning."[78] And when Pliny discusses the different colors of amber, his terminology is almost invariably metaphorical. "The pale kind," he writes, "has the finest scent, but, like the *waxy* kind, it has no value. The tawny is more valuable and still more so if it is transparent, but the color must not be too *fiery*; not a fiery glare, but a mere suggestion of it, that is what we admire in amber. The most highly approved specimens are the *'Falernian,'* so called because they recall the color of the wine; they are transparent and glow gently, so as to have, moreover, the agreeably mellow tint of honey that has been reduced by boiling."[79]

The metaphorical resonance of the colors associated with amber, like the divine and heroic associations of its brilliance, would doubtless have played an instrumental role in the kinds of subjects carved in amber and in its use. In ancient gemstones, a correspondence between color and subject was desired. According to an ancient epigram, the Nereid Galene was cut into an Indian beryl because the stone's blue color was appropriate for this personification of the calm sea.[80]

Amber's fragrance—it is the only "stone" that is both shining and fragrant—is enhanced through rubbing.[81] Amber is thus a perfect material for a divine image, especially when we recall that "statues were regularly polished with perfumed oils, perhaps matching the emanation of fragrance that forms so regular a part of divine ephiphanies."[82] Not only the fragrance, but also the great age of the material, its mysterious origins, its transmuted nature, and its electromagnetic, optical, and other properties, as well as its divine and heroic epithets, would have evoked a variety of ideas in its beholders—radiant Apollo, the fiery sun, Olympian honey, Falernian wine.

NOTES

72. *Iliad* 19.398 (R. Lattimore, trans., *The Iliad of Homer* [Chicago, 1961]). It is a common tendency in Greek poetry to emphasize qualities such as brightness or sheen rather than hue, as C. Irwin, *Colour-Terms in Greek Poetry* (Toronto, 1974), was among the first to emphasize. See also Steiner 2001, pp. 97–101; Gage 1993, pp. 11–27; and many of the conference papers in L. Cleland, K. Stears, and G. Davies, *Colour in the Ancient Mediterranean World,* BAR International Series 1267 (Oxford, 2004). See C. W. Shelmerdine, "Shining and Fragrant Cloth in Homeric Epic," in Carter and Morris 1995, pp. 99–107, for a discussion of the highly desirable qualities of shininess and fragrance in Aegean elite textiles and the larger implications of her argument.

73. Quintus Smyrnaeus, *The Fall of Troy* 5.623–25, trans. A. S. Way, Loeb Classical Library 19 (London, 1913).

74. Pausanias, *Description of Greece* 5.12.7–8, trans. W. H. S. Jones and H. A. Ormerod, Loeb Classical Library 188 (Cambridge, MA, 1966): "Of the statues set up in the round buildings, the amber one represents Augustus, the Roman emperor.... This amber of which the statue of Augustus is made, when found in the sands ... of the Eridanus, is very rare and precious to men for many reasons." What better material for the divine princeps?

75. For the Roman preference for a reddish cast in yellow, see Gage 1993, p. 272, n. 74. On the affinity of red and gold in Egypt, see Wilkinson 1994, pp. 106–7. For the Classical world, see Gage 1993, p. 26. The *flammeum,* the most characteristic element of the Roman bridal costume, and the veil of the Flaminica Dialis were deep yellow (*luteum*), the same color as lightning, according to Pliny (*Natural History* 21.22). See Sebesta and Bonfante 1994, esp. chaps. by L. La Follette, "The Costume of the Roman Bride" (pp. 54–64), and by J. L. Sebesta, "Symbolism in the Costume of the Roman Woman" (pp. 46–53), and "*Tunica Ralla, Tunica Spissa*: The Colors and Textiles of Roman Costume" (pp. 65–76). Some amber is similar in color to egg yolks (said to be the color of the *flammeum*). As noted in "Amber Medicine, Amber Amulets" below, amber is attested as a gift for Roman brides.

    For the Egyptians, pure gold, its pigment cognate, yellow, and the color red were the colors of the sun; gold was symbolic of that which was eternal and imperishable. The flesh and bones of the gods were held to be of gold, and thus that was the natural material for their images (Wilkinson 1994, pp. 106–9, 116). E. A. Waarska, in *The Quest for Immortality: Treasures of Ancient Egypt,* exh. cat., ed. E. Hornung and B. Bryan (Washington, DC, 2002), p. 105, no. 20, says that gold represented purity, and bedecking a mummy with such a material was thought to ensure a successful afterlife for its owner.

76. Gage 1993, p. 26, with bibl.

77. J. André, *Étude sur les termes de couleur dans la langue latine* (Paris, 1949), p. 155, discusses the many instances of gold referred to as red in Rome (as cited in D. Janes, *Gold and God in Late Antiquity* [Cambridge, 1998]). For further discussion of the poetic and symbolic vocabulary for the different colors of gold, see P. R. S. Moorey, *Ancient Mesopotamian Materials and Techniques: The Archaeological Evidence* (Oxford, 1994), p. 218.

78. Plantzos 1999, p. 36: "The shape in which a stone was going to be cut was also sometimes determined by its colour."

79. Falernian wine, a product of Campania, was among the most prized in ancient Rome and, as Pliny writes, the second-best wine produced in Italy (Pliny, *Natural History* 14.8.62). On Falernian wine and its golden, red, and dark red colors, see, for example, P. McGovern, S. Fleming, and S. Katz, eds., *The Origins and Ancient History of Wine* (London, 1996); and T. Unwin, *Wine and the Vine: An Historical Geography of Viticulture and the Wine Trade* (London, 1991). See also *The Wine of Dionysus: Banquets of Gods and Men in Basilicata,* exh. cat. (Rome, 2000). While wine is associated with Dionysos (and the Egyptian Bes), honey is associated with the Olympians Zeus and Artemis.

80. See Plantzos 1999, pp. 36, 89. For the use of garnets, hematite, and other red stones for martial subjects, see n. 223.

forehead gleamed the golden horns that cast their brilliance along the banks … and amber dripped from his hair."[95] Why, as Frederick Ahl asks, is the Swan a friend of the sun's child? The answer to this question explains in part why amber was important in ancient Italy, and why the long-necked birds are represented early and often in the "solar" material. The swan was a cult bird in northern Europe during the height of Celtic power, in the Urnfield and Hallstatt phases of European prehistory. "The evidence strongly suggests that this bird was especially associated with the solar cults that were widespread in Europe, and that can be traced from the Bronze Age, into the Iron Age."[96]

The constellation of Eridanus "wets the clear southern skin in its tortuous course and with starry stream flows beneath Orion's dread sword": so writes Claudian in his panegyric of A.D. 404. Here, too, amber's place in Greek myths suggests that it was viewed as an ancient material, something belonging to a great age of the distant past. But it also had a practical life outside myth—by Pliny's time, amber was very common in Rome, and a great number of amber objects were used as jewelry, incense, pharmaceuticals, and furnishings for the dead. Nonetheless, amber's mythological significance would have had a powerful effect on the way the material was seen and employed in everyday life.

Of course, as soon as one begins to delve deeper into the relationship between the myths and the reality of amber, it becomes difficult to distinguish which is which. Myths about amber's role in the mourning of the dead and the actual funerary use of amber, for instance, both have a direct correlation to the fact that amber can sometimes act as a tomb itself.

The connection among amber, tombs, and funerary customs is brought out in a unique Etruscan amber, the bow of a fibula, in the Metropolitan Museum of Art, the so-called Morgan Amber, possibly the most beautiful of all surviving pre-Roman carved amber objects (figure 24).[97] The bow is carved into a complex grouping: a draped and shod female wearing a pointed hat holds the base of a small vase in her right hand and touches it with her left. A young, beardless man, with flowing hair, long garment, and bare feet, supports himself on his left arm. Nestled between them is a long-necked bird, presumably a swan. At the foot of the couch is an attendant. The amber apparently depicts a ceremonial banquet, but is the couple mortal or divine? Are the figures Aphrodite and Adonis (Etruscan: Turan and Atunis) and the bird the goddess's swan? Or is this an elite couple? If so, is the swan a symbol or a part of the event? The iconography of the reclining couple and the ceremonial banquet had spread earlier from the Ancient Near East to Greece and Etruria. Significant Archaic Etruscan sculpted and painted depictions are extant. If this is a funerary object and the subjects divine, rich mythological implications are possible. If the subjects are mortal, the pin could have functioned in some manner as a "substitution" for the deceased.

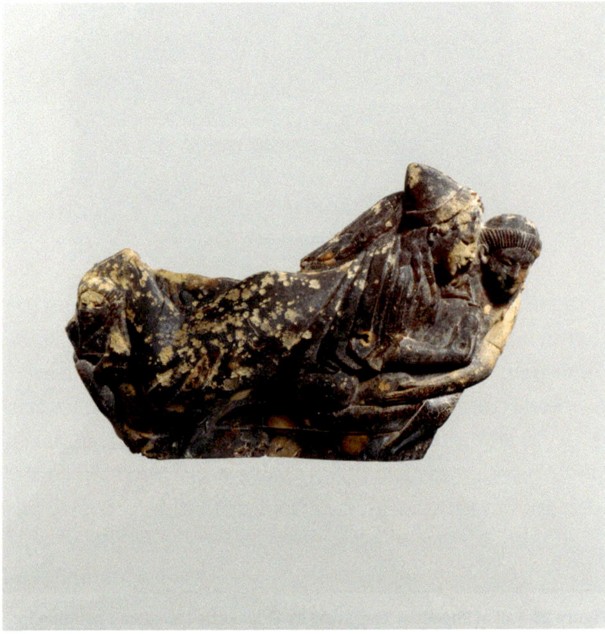

Figure 24 Bow of a Fibula (Safety Pin) with Reclining Figures, Attendant, and Bird, Etruscan, ca. 500 B.C. Amber, L: 14 cm (5½ in.). New York, Metropolitan Museum of Art, 17.190.2067. Gift of J. Pierpont Morgan, 1917. © The Metropolitan Museum of Art / Art Resource, NY.

As Jean-René Jannot writes about the Etruscan-depicted dead:

> *Was [a wall painting, an effigy sarcophagus] considered the physical envelope for that which does not die, the* hinthial *[soul, or shade]? None of these monuments were made to be seen.... Was the deceased, through his material image, believed to be living in the funerary chamber, which has become a house, or in the trench where offerings of food were set out for him?*[98]

Certainly, inclusions in amber—life visibly preserved for eternity—would not have been ignored when preparing amber for funerary purposes.

The insects and flora in amber, which Aristotle and later Pliny and Tacitus point to as proof of amber's origin as earth-born, as tree resin,[99] are apt metaphors for entombment and for the ultimate functions of the funeral ritual: to honor the deceased with precious gifts and to make permanent the memory of their lives. Three of Martial's epigrams are devoted to this correlation:

> *Shut in Phaethon's drop, a bee both hides and shines, so that she seems imprisoned in her own nectar. She has a worthy reward for all her sufferings. One might believe that she herself willed so to die.*
>
> *As an ant was wandering in Phaethonic shade, a drop of amber enfolded the tiny creature. So that she was despised but lately, while life remained, and now has been made precious by her death.*
>
> *While a viper crawled along the weeping branches of the Heliads, a drop of amber flowed onto the creature in its path. As it marveled to find itself stuck fast in the viscous fluid, it stiffened, bound of a sudden by congealed ice. Be not proud, Cleopatra, of your royal sepulchre, if a viper lies in a nobler tomb.*[100]

It is very unlikely that a swift, small snake could be entombed in such a fashion, but it is also only fair to allow Martial a degree of poetic license, given Cleopatra's traditional association with the asp. A more intriguing possibility remains, however: that Martial was describing something he had actually seen or heard about—an early instance of amber forgery.[101]

NOTES

83. Herodotus, *Histories* 3.115.

84. Ibid. 4.32–36. See J. Bouzek, "Xoana," *Oxford Journal of Archaeology* 19, no. 1 (2000): 111; J. Bouzek, *Greece, Anatolia and Europe: Cultural Interrelations during the Early Iron Age* (Jonsered, Sweden, 1997), pp. 35–38; Mastrocinque 1991, pp. 41–45, with reference to J. Tréheux, "La réalité historique des offrandes hyperboréennes de Délos," in *Studies Presented to D. M. Robinson* (St. Louis, 1953), pp. 758–59; and F. M. Ahl, "Amber, Avallon, and Apollo's Singing Swan," *American Journal of Philology* 103 (1982): 373–411. Hughes-Brock 1985, p. 260, points to C. W. Beck, G. C. Southard, and A. A. Adams, "Analysis and Provenience of Minoan and Mycenaean Amber, II. Tiryns," *Greek, Roman, and Byzantine Studies* 9 (1968): 5–19, and the connection of the "Tiryns" type of gold and amber beads to the offerings. See also Fuscagni 1982, pp. 110–11. Callimachus (*Hymn* 4.283–84) differs in this, believing that the offerings are wheat. On this, see C. T. Seltman, "The Offerings of the Hyperboreans," *Classical Quarterly* 22 (1928): 155–59.

85. Ahl 1982 (n. 84, above), p. 378.

86. Apollonius, *Argonautica* 4.611–18.

87. For the passage about the funeral of Achilles (*The Fall of Troy* 3.683–85), see Quintus of Smyrna, *The Trojan Epic: Posthomerica*, trans. and ed. A. James (Baltimore, 2006). For the funeral of Ajax (*The Fall of Troy* 5.625–30), the translation is by A. S. Way (see n. 73, above).

88. This Phaethon is not the only Phaethon of Greek myth; see, for example, J. Diggle, *Euripides' Phaethon* (Cambridge, 1970).

89. Pliny, *Natural History* 37.11.

90. Ovid, *Metamorphoses* 1.750–2.380. See the extensive discussion of Euripides' *Hippolytus* and *Phaethon* in Diggle 1970 (n. 88, above).

91. Diodorus Siculus, *Library of History* 5.23–24.

92. The story may have had particular relevance in Italy (especially in Etruria), a land famous for its fierce boars.

93. *On Marvellous Things Heard* 81–82. A. Spekke, *The Ancient Amber Routes and the Geographical Discovery of the Eastern Baltic* (Chicago, 1957), appears to have been the first to draw attention to this story in relation to amber. See also Grilli 1975 (in n. 52, above); Hughes-Brock 1985; and Mastrocinque 1991, pp. 32–34.

94. See Mastrocinque 1991, pp. 16–22; Dopp 1997; and Geerlings 1996 for further discussion of the planetary and celestial aspects of Phaethon.

95. *Claudian*, vol. 2, "Panegyric on the Sixth Consulship of the Emperor Honorius," trans. M. Platnauer, Loeb Classical Library 136 (Cambridge, MA, 1922). Four amber pendants from Italy, each in the form of a bull-bodied man, may represent this river god.

96. A. Ross, *Pagan Celtic Britain: Studies in Iconography and Tradition* (London, 1967), p. 234, quoted in Ahl 1982 (n. 84, above), p. 390.

97. Metropolitan Museum of Art 17.190.2067, Gift of J. Pierpont Morgan, 1917: *Art of the Classical World* 2007, pp. 284–85, 471, no. 326; Richter 1940, p. 31, figs. 97–98; Kredel 1923–24; and Albizzatti 1919. Richter cites two other ambers with similar subjects, a fragmentary work in the Metropolitan Museum (23.160.96) and an example once in the Stroganoff Collection (Pollak and Muñoz 1912, vol. 1, p. 78, pl. XLVII.1).

98. Jannot 2005, p. 58.

99. Aristotle, *Meteorology* 4.10; Pliny specifies ants, gnats, and lizards, the first two signifying similar-appearing and excellent specimens of amber; Tacitus, *Germania* 45.

100. Martial, *Epigrams* 4.32, 4.59, 6.15, in vol. 2, ed. and trans. D. R. S. Bailey, Loeb Classical Library 95 (London and Cambridge, MA, 1993). See P. A. Watson, "Martial's Snake in Amber: Ekphrasis or Poetic Fantasy?," *Latomus* 61 (2001): 938–43. Was this snake in amber a forgery?

101. See Ross 1998, pp. 6–9 ("fake amber"); Grimaldi 1996, pp. 133–41 ("processed amber, imitations, and forgeries"); D. Grimaldi, A. Shedrinsky, A. Ross, and N. S. Baer, "Forgeries of Fossils in 'Amber': History, Identification, and Case Studies,"

Curator 37 (1994): 251–74; and A. M. Shedrinsky, D. A. Grimaldi, J. J. Boon, and N. S. Baer, "Application of Pyrolysis Gas Chromatography and Pyrolysis Gas Chromatography/Mass Spectrometry to the Unmasking of Amber Forgeries," *Journal of Analytical and Applied Pyrolysis* 25 (1993): 77–95.

Amber and Forgery

If a piece of amber could be guaranteed to have been acquired from an exotic location, such as the distant north, the mythical Eridanus, or the Elektrides islands, or if it embodied one of its more mystical properties—natural luster, powerful magnetism, or particularly impressive inclusions—it would likely have had greater worth as a magical or medicinal item, as well as being more valuable as an ornament. Practically speaking, such a piece would have fetched a much higher price than an unprovenanced or poorer-grade one. Then, as now, the impetus for forgery or false provenance would have been commensurate with price. Roger Moorey, addressing the issue of forgery in relation to blue-colored stones in the ancient Near East, writes that "the desire for rare coloured stones was so great that it stimulated the development of artificial gemstones, made first, before about 2000 B.C., of glazed dull stones or of faïence and increasingly thereafter of glass."[102] It is likely that various tree resins (particularly copal, a hard resin much younger than amber) might have been taken for amber—at least at the time of purchase—either through deliberate deception or because of a genuine misunderstanding.

Of course, because some materials used to imitate amber also possessed, to some degree, the qualities for which amber was prized, they may have been valued in their own right, and it is therefore usually impossible to distinguish cases of successful deception from resins that were never intended as impostors. Tutankhamen's tomb, for instance, was found to contain various nonamber resin objects.[103] Were they forgeries intended to be seen as amber or another high-value resin, or were these materials equally valued for their own sake?

Evidence of other amber-related forgeries in antiquity can be found in Pliny, who discusses the use of amber itself to approximate transparent gemstones, notably amethyst. Pliny also describes a technique for softening amber, a necessary step in clarifying it, and one preliminary to amalgamating small pieces of amber into larger ones, as is still done today. Although there is no extant ancient example of such an amber object, it is a compelling explanation for certain larger works referred to in ancient sources, such as the large drinking vessels mentioned by Juvenal and Apuleius,[104] or the statue of Augustus at Olympia described by Pausanias (see "Color and Other Optical Characteristics," above). What we do have as examples of amalgamated amber pieces are segmented amber fibulae and a few carvings with added patches of amber, held together with glue or by adhesion with oil and heat. fibulae sections were joined with reeds, sometimes covered in metal foil. Today, two pieces of amber may be united by coating their surfaces with linseed oil, heating them, and then pressing them together while still hot.

Probably there was no need to conceal that such pieces were joined or amalgamated, as their craftsmanship was just as impressive as their size. That they were composed of pieces rather than carved from one large chunk of amber would have been generally known, since amalgamation techniques were common in Rome for other media, such as large ivory statues, wood marquetry, and glass. The greatest example of joined amber plaques is the famous Amber Room from Tsarskoje Selo, Russia, now reconstructed. "Compressed" or "mosaic" amber (as it is called today) is often darker and less lustrous than natural amber. Given the immense importance attached to amber's natural sheen, artificial coloring applied to a high-value object might have been deceptive in much the same way as an inclusion forgery like Martial's snake. Pliny was aware that good examples of pieces displaying amber's unique qualities, such as inclusions or brilliance, were valued according to the secret knowledge they seemed to encompass as natural wonders, and he implies as much in his discussion of artificial coloring of amber.

Admittedly, we can only speculate about the exact nature and extent of amber forgery in and before Pliny's time, but it was an early part of a continuing interest in making amberlike materials for scientific, manufacturing, and aesthetic (as well as more dubious) ends.[105] In the early modern period, this interest is documented by no less a figure than Leonardo da Vinci, who describes one recipe for making fake amber from egg whites hardened by heating.[106]

In China, the high value placed on amber has resulted in counterfeiting since at least about A.D. 500, the date of Tao Hongjing's book of *materia medica*. There he warns against false amber and recommends "using the electrostatic ability of amber to attract straw as a means of distinguishing amber from imitations."[107]

More recently, significant modern forgeries of ancient amber objects have come to light. These include an "Assyrian" amber statuette of King Ashurnasirpal in Boston[108] and the *Apollo of Fiumicino* (Paris, private collection), made in the early twentieth century, probably by the same carver responsible for the Getty statuette *Seated Divinity* (figure 25).[109]

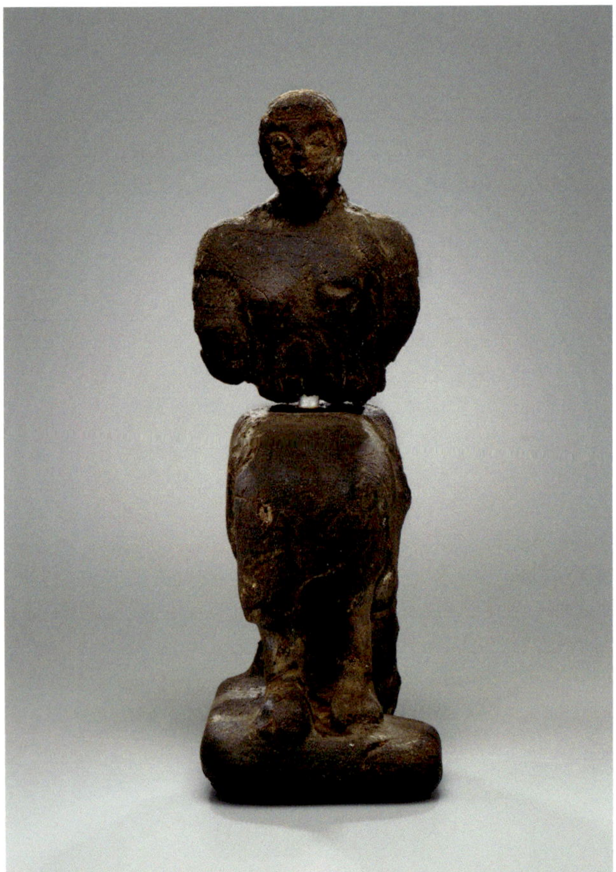

Figure 25 *Seated Divinity* statuette, modern. Amber, H: 28 cm (11 in.), W (of base): 13.5 cm (5 3/10 in.). Los Angeles, J. Paul Getty Museum, 82.AO.51. Gift of Vasek Polak. See cat. no. 57.

Today, an amber counterfeit such as the *Seated Divinity* is made with a mixture of modern materials including synthetic resin and plastics, as well as compressed amber and other resins. The interest today in amber forgery—in fake jewelry and fake specimens—is such that many modern publications and websites are available to help identify and distinguish amber, copal, and the wide range of manufactured-amber imitations.

NOTES

102. P. R. S. Moorey, "Blue Stones in the Ancient Near East: Turquoise and Lapis Lazuli," in Caubet 1999, pp. 175–88.

103. For amber in Tutankhamen's tomb, see S. Hood, "Amber in Egypt," in Beck and Bouzek 1993, pp. 230–35. Sherratt 1995 (in n. 9, above) p. 203, confirms (and refines) Hood's stylistic attribution: he compares the necklace of the late Tumulus culture of central Europe (Reinecke Br C) to a necklace from Barrow 2, Grave 13, at Schwarza, Thuringia. D. Warburton (pers. comm., 2001) pointed out the significance of the feminine necklace in the young king's tomb. Gaslain 2005, pp. 58–60, discusses material associated with amber in Egypt, Mesopotamia, and the ancient Near East, and brings together a critical bibliography. A. Niwinski, "Amber in Ancient Egypt," in *Investigations into Amber: Proceedings of the International Interdisciplinary Symposium, Baltic Amber and Other Fossil Resins, 997 Urbs Gyddanyzc–1997 Gdańsk, 2–6 September 1997,* ed. B. Kosmowska-Ceranowicz and H. Paner (Gdańsk, 1999), pp. 115–19, discusses Egyptian terms for resins and cautions against identifying objects as amber without scientific analysis. M. Cultraro, "L'ambre nel mondo mediterraneo: L'Egeo e le aree di contatto," in *Ambre* 2007, pp. 56–59, also cautions against the identification of amberlike resinous materials without scientific corroboration. For additional views on the existence and use of amber in Egypt, see Serpico 2000; A. Lucas and J. R. Harris, *Ancient Egyptian Materials and Industries* (London, 1962); and E. Daumas, "Quelques notes sur l'ambre jaune dans l'ancien Égypte," *Chronique d'Égypte* 46 (1971): 60.

104. Juvenal (*Satires* 5.38) describes an encrusted amber cup, and Apuleius (*Golden Ass* 2.2.12, 19) speaks of large cups. Strong 1966, p. 34, refers to a fragmentary Roman amber vessel in Rouen: see *Catalogue du Musée de Rouen* (Rouen, 1875), p. 99. Although the large vessels could have been carved from exceptionally large chunks of amber, they instead may have been composed of mosaic amber. Pliny (*Natural History* 37.11) mentions a huge piece weighing thirteen pounds. A piece weighing twenty-seven pounds washed up on the shores of northern Jutland (Grimaldi 1996, p. 50).

105. For discussion of amber imitations, see, for example, Langenheim 2003; and M. Ganzelewski, "Bernstein-Ersatzstoffe und Imitationen," in *Bernstein* 1996, pp. 475–82.

106. Codex Forster 3, fol. 33v. E. Ragazzi, "Historical Amber/How to Make Amber," http://www.ambericawest.com/make_amber/ (accessed July 9, 2011), discusses Leonardo's recipe. Ragazzi cites L. Reti, "Le arti chimiche di Leonardo da Vinci," *La chimica e l'industria* 34 (1952): 721–43, and compares the recipe to an

earlier one (of circa 1424–56) in a British Museum manuscript. He also refers to B. S. Tosatti, *Manoscritto Veneziano: Un manuale di pittura e altre arti—miniatura, incisione, vetri, vetrate e ceramiche—di medicina, farmacopea e alchimia del Quattrocento* (Milan, 1991).

107. Tao Hongjing, *Collection of Commentaries on the Divine Husbandman's Classic of Materia Medica* (reference from Langenheim 2003, p. 279).

108. O. Muscarella, *The Lie Became Great: The Forgery of Ancient Near Eastern Cultures* (Groningen, 2000); and A. T. Olmstead, "Amber Statuette of Ashur-nasir-apal, King of Assyria (885–860 B.C.)," *Bulletin of the Museum of Fine Arts* (Boston) 36 (1938): 78–83.

109. For the Paris statuette, see references for the Getty statuette *Seated Divinity* (82.AO.51, cat. no. 57).

The Ancient Transport of Amber

There is evidence for the movement of amber as early as the Paleolithic era. Rough pieces have been found in ancient dwelling caves in Britain and northern Europe at some distance from amber sources.[110] Early on, amber likely was transported to the Mediterranean via a chain of exchange—there was no defined long-distance amber trade until the mid-second millennium B.C., when it probably was acquired in both raw and more finished forms.[111] It is likely that amber traveled overland to the Mediterranean via the long route between north and south Europe, along the Oder, the Elbe, the Vistula, the Rhine, the Dniester, and other main European rivers.

It also traveled eastward. For a long period it, like tin, was carried by sea through the Gates of Hercules; Phoenicians were likely the main transporters. The Adriatic appears to have been the main destination for amber intended for the markets of the Italian peninsula.[112] Once at the Adriatic, amber must have been moved by water along the Italian coast, finding its way inland along river valleys and mountain passes. It was likely traded from farther west and welcomed along with the Aegean and eastern Mediterranean goods that were transported to the central and western Mediterranean. The existence of raw and worked amber from sites around the Mediterranean and farther afield—on the Iberian peninsula, in Mesopotamia, in Anatolia, at Ugarit on the Syrian coast, and in Egypt—from the Bronze Age onward attests to its widespread value and transmission. Trade in amber was likely a series of short-range transactions from the sources onward, with a few outstanding exceptions. We should imagine seekers traveling to the northern amber deposits to obtain the precious material and learn its secrets. The "knowledge" that accompanies a highly prized substance was as important as the thing itself.

There is no literary evidence for direct trade between Italy and the north until the first century A.D. Pliny the Elder writes of a Roman knight, commissioned to procure amber for a gladiatorial display presented by Nero, who traversed both the trade route and the coasts, bringing back an extraordinary amount of the precious material, which was used to extravagantly decorate the arena. Like the rare animals that were sometimes displayed at such events, amber nourished the idea of exotica from afar—visible affirmation of Rome's domination of the world.[113]

NOTES

110. Unworked pieces have been found in dwelling caves in Europe at the Grotte d'Aurensan in the Hautes-Pyrénées, at Judenes in Austria, at Kostelik and Zitmy in Moravia, at Cioclovina in Romania, and at Gough's cave near Cheddar, Somerset, England, all of which are far from natural sources of fossil resin. An upsurge in the quantity of amber in the archaeological record is observed in the Early Neolithic. White 1992, p. 549, has shown that there is a source-to-distance gradient for Aurignacian personal ornaments and that they are frequently manufactured from exotic materials. Shennan 1993, pp. 62–66, discusses amber's value in light of its acquisition by political-religious elites living far from amber sources. Citing Helms 1988, Shennan summarizes:

> *The spatially distant material, because of its strangeness, has great power, and experience of it can increase the power and prestige of those who acquire that experience.... The ultimate goal of those seeking such goods (shields or shell or stones or holy incense [or amber]) may well be directed towards obtaining (maintaining) access to material manifestations of the power and potency that imbues their cosmos, thereby continuing their close association and inclusion with the dynamics of the universe of which they are an integral part.... Many exchanged items have inherent magical or religious significance as "power-charged" treasures acquired from extraordinary realms outside their own heartland.*

111. Not all students of the material agree that it was traded in both finished and unfinished forms.

112. In Pliny's day, he relates (*Natural History* 37.11) that amber was previously "conveyed by the Germans mainly into Pannonia. From there it was first brought into prominence by the Veneti, known to the Greeks as the Enetoi, who are close neighbors of the Pannonians and live around the Adriatic."

113. J. Kolendo, *A la recherche de l’ambre baltique: L’expédition d’un chevalier romain sous Néron* (Warsaw, 1981).

Literary Sources on the Use of Amber

The archaeological record hints at a variety of uses of amber throughout the ages that are sometimes complemented by the surviving literature, but often are not. Certainly those uses that were by nature magical or tied up with mystery religions are unlikely to have been referred to other than obliquely in any mainstream literature, although they were extensive and widely acknowledged from very early on, through the Classical era and well into the Middle Ages. In addition, as helpful as the archaeological record is in elucidating the use of amber in mourning and burial contexts, it is less so when it comes to the everyday employment of amber as documented in the literature. Its use among the very wealthy ranged from girls' playthings to decorative items such as the amber-encrusted goblet that Juvenal mentions in a satire to sculpture such as the imposing statue of Augustus that Pausanias describes to items for magical and religious purposes—amulets, incense, fumigators, and burnt offerings (which by definition do not leave any physical trace). Martial writes of the pleasant odor amber gives off when it is handled by girls, of "amber nuggets polished by hand," and compares kisses to "well-worn amber."[114] In a letter to Marcus Aurelius, Fronto speaks scathingly of those writers (Seneca and Lucan) who "rub up one and the same thought oftener than girls their perfumed amber."[115] These analogies provide some explanation for the wear on many pre-Roman amber beads; magical use explains it further (figure 26).

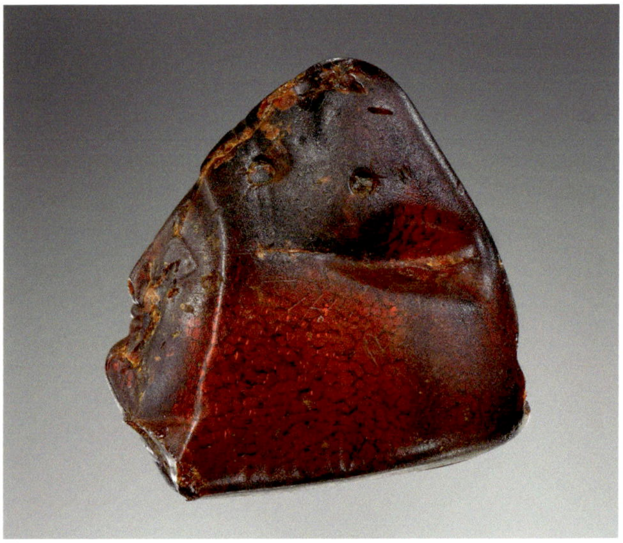

Figure 26 *Female Head in Profile* pendant, Etruscan, 525–480 B.C. Amber, H: 5.7 cm (2 ¼ in.), W: 5.6 cm (2 ⅕ in.), D: 3 cm (1 ⅕ in.). Los Angeles, J. Paul Getty Museum, 77.AO.81.4. Gift of Gordon McLendon. See cat. no. 14.

Pliny (as usual) has a long list of possible uses: amber is carved into figurines (figure 27),[116] fashioned into truffle-cutting knives,[117] made into artificial gems,[118] and in Syria used for spindle whorls. He also describes amber drinking cups, arms, and decorations of the arena (uses that would have been appreciated by men as well as by women), although he prefaces these examples with denunciatory comments at the beginning of Book 37: "The next place among luxuries [after myrrhine and rock crystal], although as yet fancied only by women, is held by amber. All three enjoy the same prestige as precious stones ... but not even luxury has yet succeeded in inventing a justification for using amber."

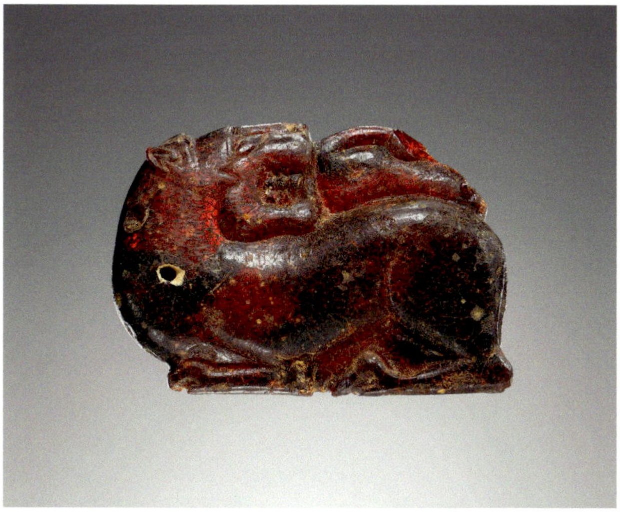

Figure 27 *Lion with Bird* pendant, Etruscan, 600–550 B.C. Amber, H: 4.2 cm (1 ³⁄₅ in.), W: 6 cm (2 ³⁄₈ in.), D: 1.5 cm (³⁄₅ in.). Los Angeles, J. Paul Getty Museum, 77.AO.81.2. Gift of Gordon McLendon. See cat. no. 5.

Amber is also often burned; as Tacitus says: "If you make an experiment of burning amber by the application of fire, it kindles, like a torch, emitting a fragrant flame, and in a little time, taking the tenacious nature of pitch or resin."[119] Pliny observes that "amber chippings steeped in oil burn brighter and longer than the pith of flax."[120] This suggests that amber may have had a practical use as interior lighting. Pliny also cites evidence that the northern Guiones used amber instead of wood as fuel and refers to what must have been a very common use of amber as incense, suggesting that in India, "amber was, to its inhabitants, found to be more agreeable even than frankincense."[121]

Such burning may seem a rather wasteful use of a precious material, but it was essential for offerings, for communication between the human and the divine, and even for feeding the gods, as in Egypt.[122] As Joan Todd has pointed out, "From the earliest recorded times burnt offerings and specifically incense are considered the most sacred gifts of all. The burning of amber would not have been considered a destructive act, but rather an elevated use of the material."[123]

Amber burned as incense was of great consequence in rituals involving solar deities before and during the Classical era, since both amber and incense were symbolic of the sun in the ancient world.[124] Incense, which emitted a fragrant smoke when scattered on lighted coals (in either a stationary or a movable burner or censer), was a regular element in Babylonian religious ceremonies.[125] The thousands of incense burners found in sanctuaries and graves throughout Greece and Etruria attest to the great importance of burning fragrant gums and spices. One of the oldest Etruscan tombs at Cerveteri, opened in the nineteenth century, was found to include "bits of amber and other oriental gums placed around the corpse," as George Dennis recounts. A morsel carried off and later ignited by the excavator "caused so powerful an odour as to be insupportable."[126]

"Incense 'offerings' were a normal part of sacrificial rituals and the use of incense was often called for in magical rituals."[127] In China, a nineteenth-century traveler records, chippings and amber dust left over from cutting figured pieces were used for varnish or incense. "The burning of the odiferous amber is the highest mark of respect possible to pay a stranger or distinguished guest, and the more they burn the more marked is their expression of esteem."[128]

In ancient medical practice, incense, resins, wood shavings, and other odoriferous materials (usually plants) or aromatics were used as a form of fumigation, either alone or in compounds. It is also likely that amber incense was used in divination: omens were read in the plumes and short curls of smoke formed by burning amber (figure 28).[129]

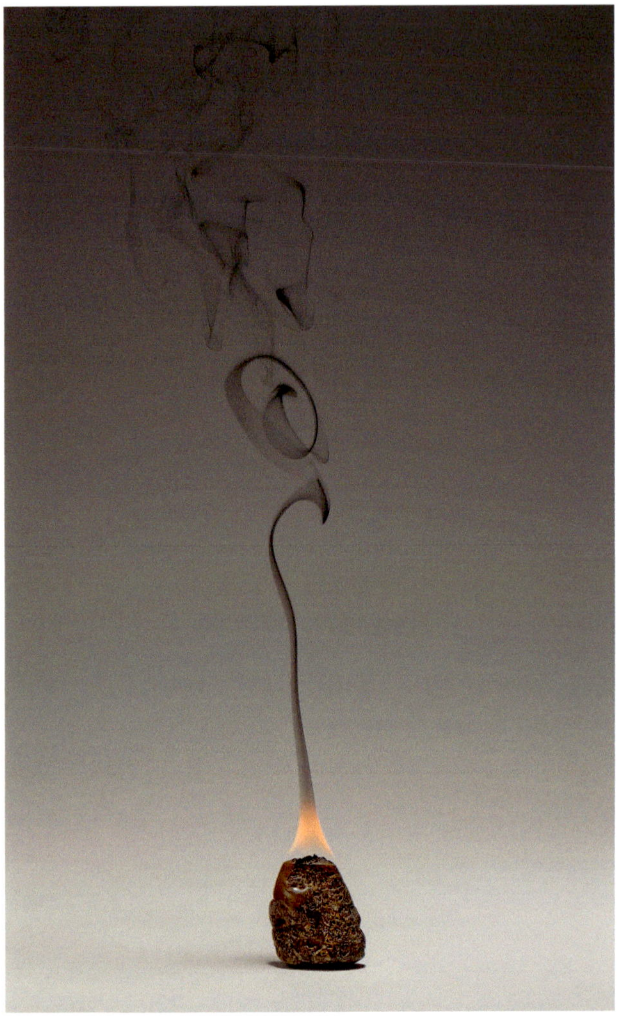

Figure 28 Piece of burning Baltic amber, producing its distinctive flame color and characteristic smoke. Length of amber before burning: 3 cm (1 1/8 in.). Private collection. Photograph © Lee B. Ewing.

Amber incense may have been ground into a powder and mixed with other aromatics, or nitrates, to keep it burning. In Rome, as Karen Polinger Foster shows, "incense was shaped into cones, balls, discs, pyramids, obelisks, granules, and pellets" as it had been in Egypt and the Near East.[130] But the Romans apparently did not follow the Egyptian practice of using figurative incense blocks in forms such as birds or recumbent calves, which clearly suggests a religious element to the burning of incense.[131] A few pieces of unworked amber found in Etruscan graves might be construed as evidence of amber used for fumigation or as unburnt incense.[132] And it may be that the very same amber objects considered then and now as ornament and amulet (for example, birds or recumbent calves) might also have been valued for their potential as light energy or incense.

NOTES

114. Martial, *Epigrams* 5.37.11. Martial compares the kisses of Diadumenus to "well-worn amber" in 3.65 and those of another (an unnamed youth) to "amber thaw'd in a virgin's hand" in 11.8. Juvenal, *Satires* 6.573, makes fun of a woman who clutches "a well-worn calendar in her hands as if it were a ball of clammy amber." Translations by Faris Malik: http://people.well.com/user/aquarius/martial.htm.

115. Fronto, "On Speeches," in *Correspondence*, vol. 1, trans. C. R. Haines, Loeb Classical Library 112 (Cambridge, MA, 1919).

116. "Its rating among luxuries is so high that a human figurine, however small, is more expensive than a number of human beings, alive and in good health." Here, in *Natural History* 37.12, Pliny may refer to simple carvings such as the actors in the British Museum (Strong 1966, nos. 109–13), but it is more likely that he cites Roman masterworks such as the Dionysos group from Esch, the Netherlands: see, for example, A. Zadoks-Josephus Jitta, "Dionysos in Amber," *Bulletin antieke beschaving* 37 (1962): 61–66. Or might Pliny be referring to household Penates of amber, as documented in the House of a Priest at Pompeii?

117. Pliny, *Natural History* 22.47.99. Strong 1966, p. 12, declares such a use "an idiotic affectation," but it may reflect the high regard in which amber was held.

118. Pliny, *Natural History* 37.12, states that "amber plays an important part also in the making of artificial transparent gems, particularly artificial amethysts, although … it can be dyed any color."

119. Tacitus, *Germania* 45.

120. However, Philemon is cited by Pliny (*Natural History* 37.11) as saying that amber does not yield a flame. Strong 1966, p. 24, citing A. Bonarelli, "Le ambre nelle tombe picene," *Rendiconti dell'Istituto Marchigiano di Scienze, Lettere ed Arti* 3 (1927), and Marconi 1933, col. 409, says that "it is recorded that before any regular excavations took place at Belmonte Piceno, the villagers used amber found in ancient tombs as fuel on their fires." Strong adds (but without references), "The same practice is recorded in the Perugia district."

121. Barfod 1996, p. 453. Amber's combustibility (and its corresponding application of being burnt) is suggested by at least two of its ancient names: *sualiternicum* and *thium. Thium* is derived from the old Italic *thyem,* or *thyon.* Ritters, cited by F. Eckstein and J. H. Waszink, was the first to connect *thium* with incense. Still today, amber is an important ingredient of incense in India and many other places in the world and is advertised globally, as a Web search can demonstrate.

On the ancient use of resins in incense, see Langenheim 2003, chap. 8. A. L. D'Agata, "Incense and Perfumes in the Late Bronze Age Aegean," in Avanzini 1997, p. 85, notes that the ultimate origin of the Greek term for incense "can be traced back to the Mycenaean *tuwo* (pl. *tuwea*), which in the Late Bronze Age seems to have been used as a general term for aromatics, and cannot be in any way connected with frankincense." D'Agata presents evidence that "other resins were known in the Aegean [during] the Mycenaean period, and probably also in Minoan Crete." Nearly a ton of terebinth resin and a large group of worked Baltic amber beads were among the cargo of the late-fourteenth-century shipwreck at Uluburun off the Lycian coast (Turkey). See C. Pulak, "Who Were the Mycenaeans Aboard the Uluburun Ship?," in *Emporia: Aegeans in the Central and Eastern Mediterranean: Proceedings of the 10th International Aegean Conference, Italian School of Archaeology, Athens, 14–18 April 2004* (*Aegaeum* 25 [2005]), ed. R. Laffineur and E. Greco, pp. 295–312; and C. Pulak, "The Cargo of the Ulu Burun Ship and the Evidence for Trade with the Aegean and Beyond," in *Italy and Cyprus in Antiquity, 1500–450 B.C.: Proceedings of an International Symposium Held at the Italian Academy for Advanced Studies in America at Columbia University, November 16–18, 2000,* ed. L. Bonfante and V. Karageorghis (Nicosia, 2001), pp. 22–25, 37–39. The *Murex opercula* found on the Uluburun ship is today an ingredient of incense in many parts of the Arab world; see G. F. Bass, "Prolegomena to a Study of Maritime Traffic in Raw Materials to the Aegean during the Fourteenth and Thirteenth Centuries B.C.," in *TEXNH: Craftsmen, Craftswomen, and Craftsmanship in the Aegean Bronze Age; Proceedings of the 6th International Aegean Conference, Philadelphia, Temple University, 18–21 April 1996* (*Aegaeum* 16 [1997]), ed. R. Laffineur and P. Betancourt, p. 163 (with reference to C. Pulak, "1994 Excavation at Uluburun: The Final Campaign," *Institute of Nautical Archaeology Quarterly* 21, no. 4 [1994]: 11.)

On incense in the Greek world, see W. W. Mueller, RE suppl. 15 (1978), s.v. "Weihrauch," pp. 702ff. A. Testa, *Candelabri e thymiateria in Vaticano* (Rome, 1989); and L. Ambrosini, *Thymiateria etruschi in bronzo: Di età tardo classica, alto e medio ellenistica* (Rome, 2002), concentrate on frankincense and myrrh as incense ingredients. C. Zaccagnino, *Il thymiaterion nel mondo greco: Analisi delle fonti, tipologia, impieghi* (Rome, 1998); and C. Zaccagnino, "L'incenso e gli incensieri nel mondo greco," in Avanzini 1997, pp. 100–20, offer a fuller discussion of incense, but no mention is made of amber. However, other ancient authors do describe additional substances burned as incense, as Mueller says. See Aristotle (*Meteorology* 4.10), where he lists in one breath "amber, myrrh, frankincense, and all the substances called 'tears,'" and Theophrastus, *On Odours* 12–13, where he differentiates among myrrh, frankincense, and "anything that is burnt as incense." G. Banti, "Names of Aromata in Semitic and Cushitic Languages," in Avanzini 1997, p. 169, underlines the difficulty in "singling out the gum resins of frankincense and myrrh with respect to other aromata ... particularly in the most ancient literary sources and in the reports by the earliest European travellers." Burnt amber has a delicious odor. From all of the evidence in the ancient sources, archaeological evidence, and the widespread use of amber in incense throughout the world today, it is hard to believe that amber was *not* used as incense (or an ingredient thereof), in fumigation, and/or in sacrifice.

122. Black and Green 1992, p. 109.

123. J. M. Todd, "Baltic Amber in the Ancient Near East: A Preliminary Investigation," *Journal of Baltic Studies* 16 (1985): 292.

124. Shennan 1993 (in n. 110, above), p. 66; Bouzek 1993, p. 141. As Shennan summarizes: "Amber is a prehistoric exemplar of Mary Helms' [Helms 1988] 'political religious exotic experience.' Northern amber thus mirrored southern myrrh as a mystic import to the Mediterranean (and was, on occasion, used in the same way)." Archaeological and linguistic evidence shows that the use of amber as a "gemstone" occurred in Greece and Etruria at the same time in the eighth and seventh centuries, alongside other "well-documented Near Eastern practices such as incense-burning, purificatory rituals, hepatoscopy, and the use of foundation deposits in temples": Faraone 1992, pp. 26–27. See also W. Burkert, "Itinerant Diviners and Magicians: A Neglected Element in Cultural Contact," in *The Greek Renaissance of the Eighth Century B.C.: Tradition and Innovation,* ed. R. Hägg, *Acta Instituti Anthenensis Regni Susiae* 30 (1983): 115–19; and W. Burkert, "'A Seer, or a Healer': Magic and Medicine from East to West," in Burkert 1992, pp. 41–87.

125. Black and Green 1992, p. 109.

126. For amber and other resins surrounding the corpse in the Grotta della Sedia, Banditaccia Necropolis, Cerveteri, see G. Dennis, *The Cities and Cemeteries of Etruria,* vol. 2 (London, 1848), p. 59, n. 4, with reference to P. E. Visconti and A. Torlonia, *Antichi monumenti sepolcrali scoperto nel ducato di Ceri, negli scavi eseguiti d'ordine di Sua Eccellenza il signor D. Alessandro Torlonia signore del Luogo dichiarati dal cav. P. E. Visconti* (Rome, 1836), pp. 29–32.

127. Black and Green 1992, p. 109. Burning and offering incense as a means of communication between the earthly and divine spheres is first attested in the Pyramid Texts of the third millennium and remained a central cult act in Egyptian temples erected by Greek and Roman rulers. In Mesopotamia, as B. Böck, "'When You Perform the Ritual of "Rubbing"': On Medicine and Magic in Ancient Mesopotamia," *Journal of Near Eastern Studies* 62, no. 1 (2003): 10, describes, "the burning of incense plays an important role in magical and latreutic cult because of its association with purity and impurity. Fumigation is part of the veneration of gods and, accordingly, the burning of sweet-smelling fumigants accompanies sacrifice, prayers, as well as intercessions."

128. E. A. Smith, "Concerning Amber," *American Naturalist* 14, no. 3 (March 1880): 106.

129. This practice is documented in Old Babylonian times; see Black and Green 1992, p. 109. See K. Polinger Foster, "Dionysos and Vesuvius in the Villa of the Mysteries," AntK 44 (2001): 43, n. 39

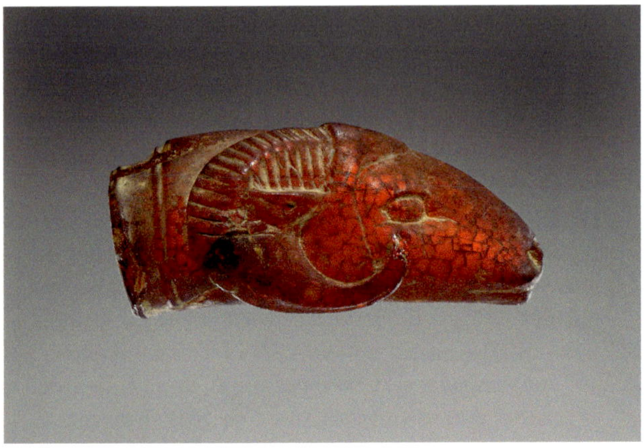

Figure 29 *Ram's Head* pendant, Italic, 500–400 B.C. Amber, L: 3.6 cm (1 2/5 in.), W: 1.9 cm (3/4 in.), D: 1.5 cm (3/5 in.). Los Angeles, J. Paul Getty Museum, 77.AO.81.15. Gift of Gordon McLendon. See cat. no. 45.

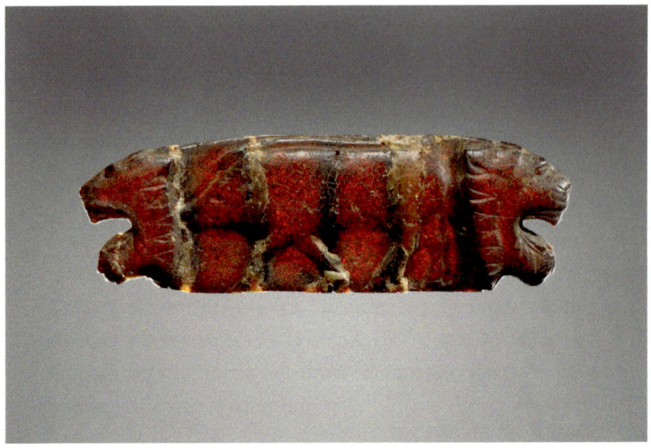

Figure 30 *Addorsed Lions' Heads with Boar in Relief* plaque, Etruscan, 500–480 B.C. Amber, H: 3.6 cm (1 2/5 in.), W: 8.2 cm (3 1/5 in.), D: 1.2 cm (1/2 in.). Los Angeles, J. Paul Getty Museum, 77.AO.83. Gift of Gordon McLendon. See cat. no. 38.

Almost any jewelry object could have had some apotropaic function—and, as Geraldine Pinch remarks in her book on Egyptian magic, it is hardly an exaggeration to say that most Egyptian jewelry had amuletic value. How conscious wearers were of their ornaments' symbolism is a more difficult question to answer.[143]

The same is evidently true for amber objects of adornment. In life, amulets were worn as charms to bring good luck, health, protection, or love, to avert danger, or to cure disease. Figured or inscribed amulets often would have had a sympathetic function;[144] a figure of a boar, such as the Getty plaque *Addorsed Lions' Heads with Boar in Relief* (figure 30), might have brought luck in a hunt, safeguarded the wearer from the boar he was hunting, or even channeled the powers of Herakles or Meleager. Situations of potential crisis, such as a hunt, a dangerous journey, or childbirth, warranted temporary amulets.[145] More permanent amulets, in the form of jewelry, could have provided protection during childhood, throughout an individual's life, and during the fraught voyage to the afterworld, the dangerous realm of spirits and demons. Indeed, amber and amber amulets were important elements in the mourning ritual as permanent tears and as grave gifts.[146]

The impact of the Aegean, the Near East, and Egypt (where women and children wore the majority of amulets) on native Italian customs during the first millennium B.C., a period of contact and acculturation, is evidenced by the amulets' subjects. New images, spells, amulets, deities, and aspects of deities replaced, perfected, or married with the old. Although only a portion of the extant figured ambers can be associated with religious cults, the use of amulets was certainly bound up with secret knowledge of sources of power—the province of skilled practitioners such as magicians, priests, "wise women," healers, and midwives.[147] Practitioners of magic might exert an influence on all levels of society. Theophrastus maintains that Pericles, on his sickbed, was induced by the women of his household to wear an amulet—entirely against his better judgment. The story, whether apocryphal or not, is further evidence for widespread use of amulets among the elite, as well as the lower classes.[148] It is also interesting for its indications about the role of women in promoting such use.[149]

Amulets were especially valuable to women for controlling or increasing fertility, protecting the unborn, helping to ensure safe childbirth, and safeguarding their children. Protective gynecological amulets must have been among the earliest of all amulets. Such items in Italy and the Greek world were age-old, the lore passing from generation to generation, no doubt affected by contact with new populations, practitioners, and magical practices.

One seventh-century B.C. plain pendant in the Getty collection (figure 31) is inscribed with two images, on one side a fish and on the other something resembling the Egyptian symbol of a papyrus clump, or a pool with lotus flowers. This piece is one of forty-three beads from the same parure, its original findspot now unknown. Who scratched the signs? How were they understood? Was the mere presence of Egyptian, or Egyptianlike, writing enough to make the amber more efficacious?

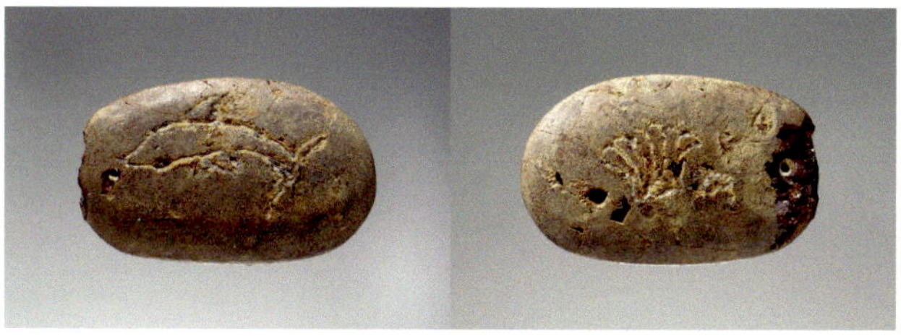

Figure 31 Pendant inscribed with two Egyptianizing hieroglyphs, 7th century B.C. Amber, H: 3.8 cm (1 ½ in.), W: 2.2 cm (⅞ in.), D: 0.8 cm (3/10 in.). Los Angeles, J. Paul Getty Museum, 82.AO.161.285.

The serious dangers of disease for young children and the considerable risks for women in childbirth and early motherhood gave rise to a belief that the dead were jealous of new life, and the need for magical protection of women and children was a compelling one.[150] For a pregnant woman, amber's property of encapsulating living things may have made it an especially powerful *similia similibus* amulet, a "pregnant stone."[151] Resin also heals damage and wounds in trees; could it extend such properties to people wearing it?

The bulla, a lens-or bubble-shaped container, is perhaps the best known of all ancient amulet types. Known in Rome as *Etruscum aurum*, it combined two magical functions: it enclosed amuletic substances, and it symbolized the sun in material, in form, and in its powers.[152] The shape derives from age-old disk amulets of the sun. The bulla was given to high-born boys. The ancient sources relate that the king Tarquinius Priscus was the first to present his son with a gold amulet after the son had killed an enemy in battle, and from that time onward the sons of cavalrymen wore amulets. Ancient sculpture shows that Etruscan boys wore the bulla, and Roman writers recount that it was worn by magistrates, triumphant generals, and even domestic animals. It should be noted that bullae were made not only of gold, but also of other bright metals such as bronze, as is evidenced by bronze bullae of various forms found in Latin and Etruscan graves dating as early as the eighth century B.C.

In fourth-century pre-Roman art, the single bulla and strings of bullae, not only lens-shaped but also pouch-shaped pendants, were worn by elite personages, some recognizable divinities and heroes. Dionysos wears a single bulla on the Praenestine "Cista Napoleon" in the Louvre.[153] On an Etruscan red-figure krater in Florence, an Argonaut wears strings of bullae on his arms, while a companion ties on yet another (figure 32).[154] On a sarcophagus from the Tomb of the Triclinium at Tarquinia, a reclining woman wearing a necklace of bullae, holding a thyrsus and kantharos and keeping a fawn by her side, is clearly a devotee or maybe a priestess of Dionysos/Pacha/Fufluns. On Etruscan mirrors, Aplu, Fufluns, Tinia, Epiur and Maris, young Hercle, Thetis and Alcumene, Athena, and Turan wear bullae.[155] Votive images of women, girls, and boys, and effigies of deceased men, women, and babies, are often shown with a bulla or bullae.[156] A mid-fourth-century B.C. mirror in New York shows Peleus wearing an armlet with bulla-shaped pendants on her left arm and Calaina (Galene), a Nereid, holding a circlet with similar pendants in her left hand (figure 33).[157]

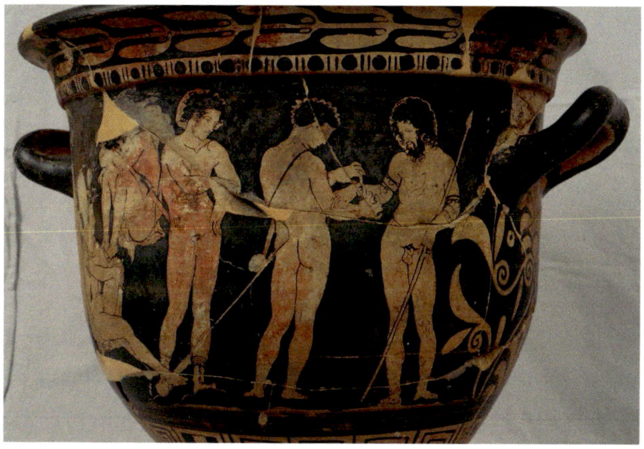

Figure 32 Red-figure crater attributed to the Argonaut Group (detail), Etruscan, early 4th century B.C. Florence, Museo Archeologico Nazionale, 4026. Photo: Nicolo Orsi Battaglini / IKONA.

the Mediterranean littoral, and in the Near East. Such groupings are documented as early as the Early Dynastic period (third millennium B.C.) at Ur.[169] Mesopotamian texts specifically refer to figured amulets in the context of protection and healing, amulets that were to be either carried and worn by the living or placed on various parts of the deceased's body. Strings of amulets are documented as hanging in houses in the ancient Near East. In Greek, Cypriot, and Etruscan art, babies and children (and some Greek young women) are depicted wearing amulets tied onto a long cord worn diagonally across the body. This tradition may well be the ancestor of the Roman *crepundia*. As Demetrius Waarsenburg argues, the *crepundia* (charms strung together and used as rattles for children) can be connected to these assemblages of amulets, implying that they originally had a more profound significance.[170]

Although nearly all figured amber pendants excavated in Italy were found in funerary contexts, many of them had "lives" and an owner or owners (not necessarily the deceased) before they became part of the mourning ritual. Interments could contain both old and new pieces. Some may have been heirlooms, already venerable and powerful, made so by provenance, status, or accrued potency.

Some beads and pendants show signs of use—of handling, of pulling on the suspension perforations, of rubbing. Was the rubbing done to enliven the electromagnetic properties of the amber? To release its fragrance? For the tactile sensation? To activate amber's divine associations? For medicinal and magical purposes? To enact the magic of the amulet's imagery?

The blurred features of some figured ambers must be due to handling in the course of amuletic use. Several examples from controlled excavations seem to confirm this. A female head from a grave at Latronico retains sharp groovings in the hair and crisp delineations in the diadem, but has smoothed facial features (its tiny chips are likely from modern times). It has a standard perforation through the top of the pendant but also a secondary perforation through the temple area, front to back, which has been elongated by gravity and pull, very like the holes on heirloom Tibetan or African amber beads. The Herakles and satyrs' heads from a woman's grave, Tomb 106 at Braida di Vaglio, which may be at least a generation older than the burial, are salient examples of nonuniform use wear. The face of the Herakles pendant is especially worn.[171] Some figured ambers from another of the Braida di Vaglio tombs, Tomb 102, that of a little girl, are clearly worn on the prominent surfaces of the face. The features of one of the frontal female faces is nearly worn off, and three of the rams' heads, as well as the pendant in the form of a dormant feline, show evidence of use wear. This is in contrast to the comparatively fresh surface of other ambers from the tomb, including the recumbent sphinx (which is also at least a generation older than the burial).

The woman's Tomb 48 at Melfi-Pisciolo included at least five figured pendants, but only one female head in profile shows considerable surface wear. It contrasts with the male subject, a crisply detailed winged nude youth in a Phrygian hat with a shield at his side and sword in his hand.[172] A large pendant of Eos carrying off a youth, perhaps Kephalos, from a burial of circa 350 B.C. at Tricarico–Serra del Cedro, is an extreme example of face-rubbing: the youth's face is nearly lost.[173] Female heads from a documented find at Valle Pega (Spina) and rams' heads from excavated tombs at Bologna show well the contrast between the better-preserved tops of heads and the more abraded faces.[174] A number of the Getty female and rams' heads illustrate similar patterns of wear. Many other carved amber objects from burials throughout Italy (and Serbia) bear signs of wear: pulling troughs at the suspension hole, as in a head of a satyr from Palestrina (figure 36), handled or rubbed surfaces, and repairs, such as the drilling of replacement perforations or securing broken pieces in mounts.[175]

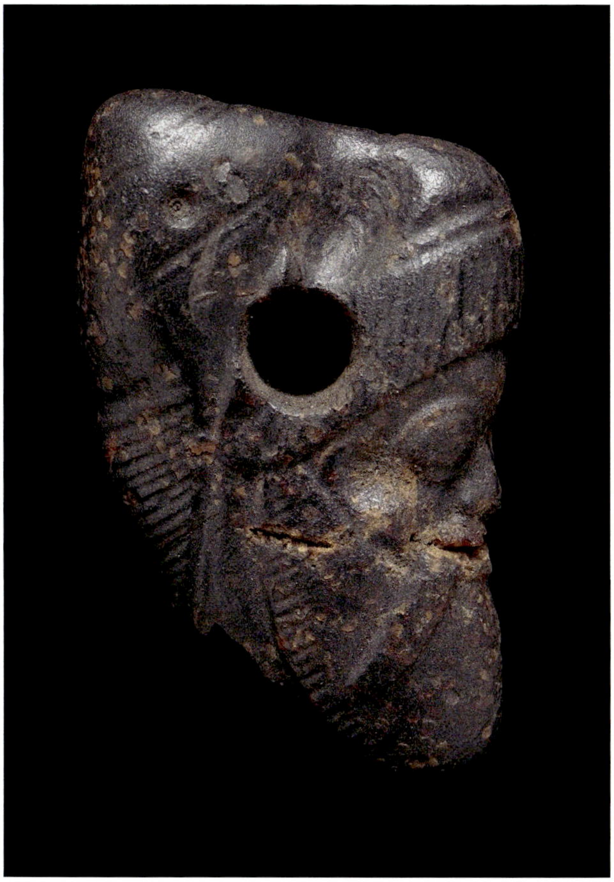

Figure 36 *Dancing Figure or Head of Satyr*, Etruscan or Italic, early 5th century B.C. Amber, legacy dimension: 7.5 x 4.7 cm (2 15/16 x 1 7/8 in.). Museum of Fine Arts, Boston. Gift of Miss C. Wissmann, 02.253. Photograph © 2011, Museum of Fine Arts, Boston.

The sometimes disfiguring large drilled holes in the faces deserve special comment. Why and when were they bored? Raw amber pieces are sometimes found with large round holes in their center, the result of resin forming around a branch or twig (now disintegrated). If a piece of amber was purposely perforated before it was made into an object, the act might have occurred anywhere between the Baltic and Italy, and at any time, for it is likely that amber moved south in both worked and unworked form from earliest times. On a practical level, the holes may have been drilled into the amber to better protect it when it was suspended from a pin, or, once the piece was cored, it would have been suitable for wearing on a pin. The smoothed prominent surfaces of the Getty pendant *Winged Female Head in Profile* (figure 37), the multiple through-bores, the abrasion troughs in the suspension perforations at the top, and the central hole all indicate that this pendant must have been used over a period of time before it was finally interred in a grave.[176] How and by whom amber pendants were used during life is a subject for speculation. Pliny's account is one useful source of information, and the few surviving Archaic and Classical illustrations of people (and divinities) wearing figured elements and amulets around their necks and limbs are valuable evidence for figured pendant usage outside the grave context.

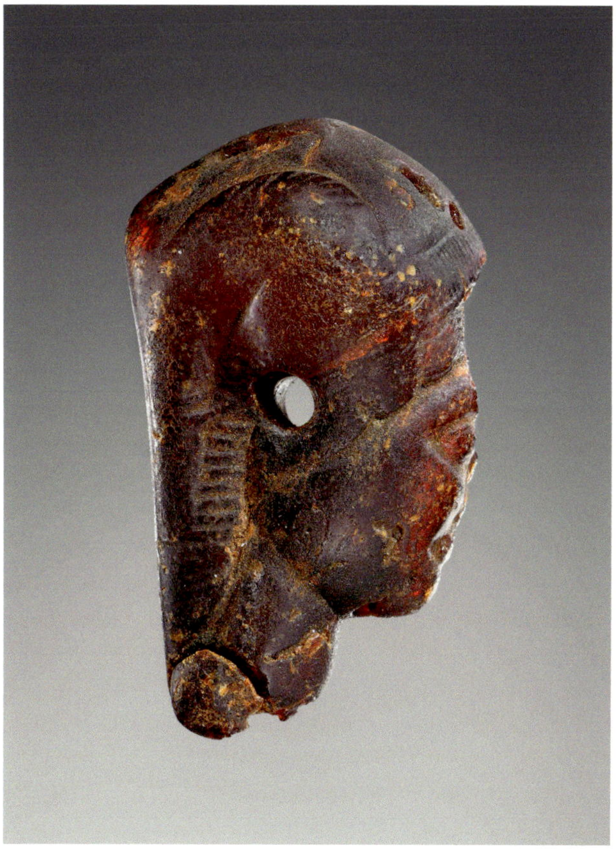

Figure 37 *Winged Female Head in Profile*, Etruscan, 525-480 B.C. Amber, H: 7.9 cm (3 1/8 in.), W: 4.9 cm (1 9/10 in.), D: 2.5 cm (1 in.). Los Angeles, J. Paul Getty Museum, 76.AO.85.2. Gift of Gordon McLendon. See cat. no. 15.

Paintings or sculptures of figures wearing a string with a single amulet or a group of them (as opposed to necklaces designed with repeating elements) are uncommon in Archaic and Classical art from Italy, but the depictions that do survive depict bullae-wearing men, women, and children, horses, and even ravens. Human figures of both sexes wear them around the neck and on the upper arms. The single ornaments include gorgon masks and the heads of animals, such as fawns, lions, and rams. A number of terracottas of seated goddesses from Greek sanctuaries in Magna Graecia, for example, wear strings of figured elements, among them bulls' heads.[177] On Greek vases, on Cypriot terracotta sculptures of temple boys, and on Laconian bronze images of partly clothed young women are seen cross-torso carriers bearing various kinds of amulets: crescents, boar tusks, circlets, and other shapes. Women wearing a single lotus-blossom pendant are represented in terracottas, bronzes, and plastic vases of the late sixth and fifth centuries. Pomegranates and simple flowers are also not unusual.

All amulet wearers depicted on Etruscan fourth-century mirrors are elite subjects, and most are identified as divinities and heroes. Two examples are important for amber pendants, especially because of the material's association with Apollo/Aplu and Dionysos/Fufluns. On many fourth-century Etruscan mirrors, Aplu wears pendants around his neck or on his upper arm. On a mid-fourth-century Etruscan bronze mirror in Naples, the infant Dionysos/Fufluns is already adorned with a ribbon of amulets during his birth from Tinia's thigh. Fufluns as a youth, now with a necklace of amulets but otherwise unadorned, is kissed by his mother, Semele, on another in Berlin.[178]

Key illustrations of animal pendants in use are painted in the Tarquinian Tomb of Hunting and Fishing (circa 510 B.C.) (figure 38).[179] On the back wall of the main chamber, the male banqueter wears a necklace of three (possibly amber) rams' heads almost identical to the Getty amber rams' heads (figure 39). In the first room of the tomb (figure 40), simple carriers with ram's- and lion's-head pendants, similar to those in the Getty (figure 41), hang from branches. This room of the tomb may depict the grove of Apollo or a Dionysian setting.

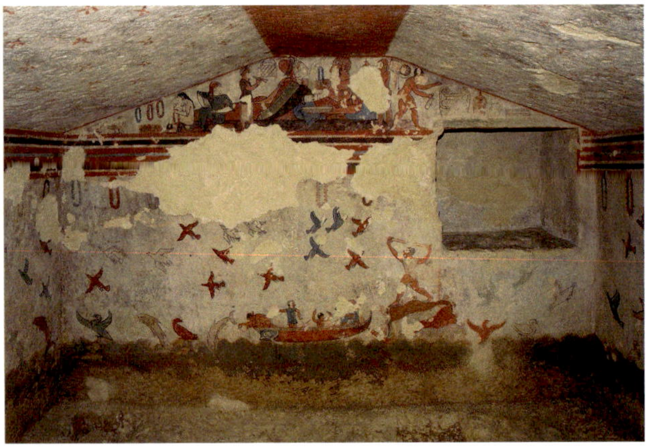

Figure 38 *Reclining Couple with an Attendant*, back wall of the Tomb of Hunting and Fishing, Tarquinia, Etruscan, ca. 510 B.C. Fresco. By permission of La Soprintendenza per i Beni Archeologici dell'Etruria Meridionale, Roma / IKONA.

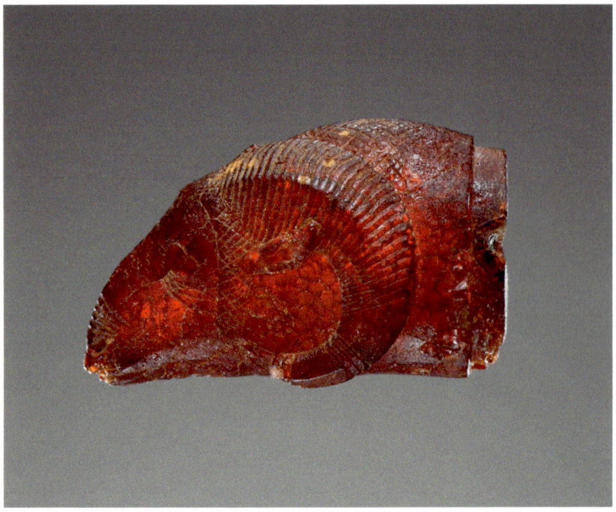

Figure 39 *Ram's Head* pendant, Etruscan, 525–480 B.C. Amber, L: 3.6 cm (1 2/5 in.), W: 2 cm (4/5 in.), D: 1.8 cm (7/10 in.). Los Angeles, J. Paul Getty Museum, 76.AO.82. Gift of Gordon McLendon. See cat. no. 39.

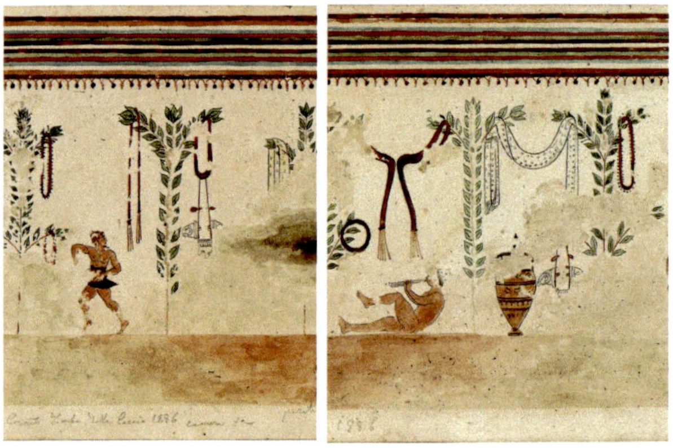

Figure 40 Figured amulet necklaces in the antechamber of the Tomb of Hunting and Fishing, Tarquinia, Etruscan, ca. 510 B.C. Fresco. Details from nineteenth-century watercolor painting by G. Mariani. From Steingräber 2006, p. 96.

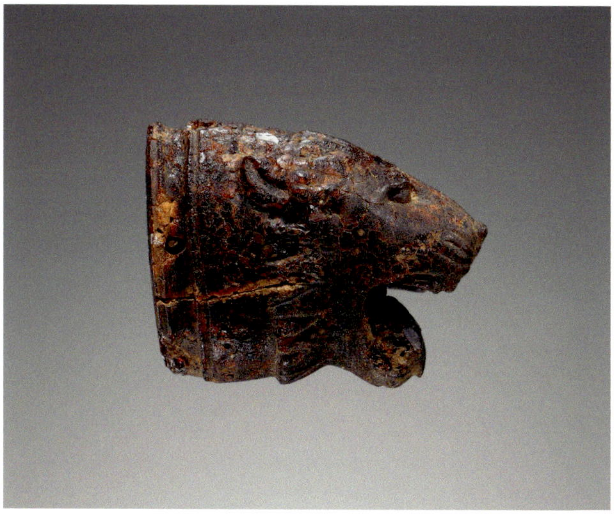

Figure 41 *Lion's Head* pendant, Etruscan, 550–500 B.C. Amber, H: 2.8 cm (1 1/10 in.), W: 2.2 cm (9/10 in.), D: 3.8 cm (1 1/2 in.). Los Angeles, J. Paul Getty Museum, 76.AO.80. Gift of Gordon McLendon. See cat. no. 33.

NOTES

133. Pliny, *Natural History* 37.11. Negroni Catacchio 1989, p. 659, linking a modern custom with this report by Pliny, notes that in many regions of Italy in relatively recent times, it was popular to present amber necklaces to young women as their first precious object and as a *portafortuna.* Negroni Catacchio also cites an eloquent passage in Ovid's retelling of the Phaethon story. See *The Metamorphoses of Ovid,* a verse translation by A. Mandelbaum (New York, San Diego, and London, 1993), p. 51, lines 365–66: "The stream's clear waters bear that amber off, and it will then adorn young wives in Rome." The gift of amber necklaces to immortal brides is also described in Nonnus, *Dionysiaca* 38.99, 40.400. In nineteenth-century Poland, following folk tradition, brides wore amber necklaces, usually of three strings, during their wedding, necklaces that may have been passed down from generation to generation. At least one of the beads would have had an inclusion, as I. Łapcik notes in "The Gold of the Baltic Sea: Amber in Art and Culture," in *Languages and Cultures of the Baltic Region: Collection of Papers, International Conference of Young Scholars,* vol. 2, ed. Y. Khramov and T. Khramova (Riga, 2007): http://www.sta-edu.lv/conf2007 (accessed November 24, 2009).

134. Compare, for example, this Egyptian text: "The infant is protected by the gods, the child's name, the milk he sucks, the clothes he wears, the age in which he lives, the amulets made for him and placed around his neck." F. Lexa, *La magie dans l'Égypte antique, de l'Ancien Empire jusqu'à l'époque copte,* vol. 2 (Paris, 1925), pp. 32–33.

135. Caesarius, *Sermons* 13.5, 14.4. See also Dickie 2001, pp. 304–5. Dickie suggests that the amber amulet "may well have had writing on it, or a magical symbol."

136. For a selection, see Strong 1966, nos. 119–23 (including ring pendants). Crescent-shaped pendants have a long history in the Mediterranean. See A. Zadoks-Josephus Jitta and A. M. Gerhartl-Witteveen, *Roman Bronze Lunulae from the Netherlands* (Leiden, 1977); and H. Wrede, "Lunulae in Halsschmuck," in *Wandlungen: Studien zur antiken und neueren Kunst, Ernst Homann-Wedeking gewidmet* (Munich, 1975), pp. 243–54. A *lunula* could be a single pendant on a carrier or one of many pendants in an ornament. The necklaces of the Archaic Sicilian terracotta Athana Lindia type wear complex pectorals, and the *lunulae* can have either upturned or downturned ends: M. Albertocchi, *Athana Lindia: Le statuette siceliote con pettorali di età arcaica e classica, Rivista di Archeologia*, suppl. 28 (Rome, 2004).

137. Strong 1966, p.12.

138. Kotansky 1991, p. 113: "The use of unengraved materials as amulets continues unabated into the Roman period side by side with talismans and phylacteries that carried texts.... Magical texts (often containing just symbols or very short spells) ... often [are] inscribed on small, semiprecious stones that are then set into rings and necklaces or otherwise simply carried in an individual's clothing." Kotansky provides an excellent list of sources for gemstones and magic, but singles out Philipp 1986 (n. 7, above).

139. Bonner 1954, p. 151, in reference to A. Bertholet, "Die Macht der Schrift in Glauben und Aberglauben," *Abhandlungen der Deutschen Akademie der Wissenschaften zu Berlin 1948,* no. 1 (Berlin, 1949).

140. Kotansky 1991, p. 107: "Simple uninscribed amulets are difficult, if not impossible, to identify; even when they carry some tell-tale symbol or design they remain silent about their specific purpose or the source of their efficacy. Those, however, that are inscribed with texts (no matter how brief) provide information about the ancient medical and religious contexts of their use."

The tradition of tying on amulets and using knots in magic is attested in Egypt and the ancient Near East as early as the third millennium. A. Livingstone, "The Magic of Time," in *Mesopotamian Magic: Textual, Historical and Interpretive Perspectives,* ed. T. Abusch and K. van der Toorn (Groningen, 1999), pp. 131–37, calls for further ancient Near Eastern–area studies of "stones, their individual characters, and the tying on of amulets." The action of tying was one part of the magic, the substance another, and the spell or charm said over the amulet still another. Thus, the magical rite included the actions that accompanied the words, while the objects or ingredients used in the rite were equally important; see Pinch 1994, p. 76. The stone's role actively implemented the communication between suppliant and superior; see Winter 1999, p. 51. In a similar vein, Gordon 2002 (in n. 7, above), p. 83, confirms: "The spells in the magical papyri generally contain two elements, the preparation of materia magica and an accompanying incantation, whose function is either to activate the inherent properties of the material, or to invoke a named divinity and his or her metamorphs. Although the balance between these elements is variable, we may call this the tacit or implicit model of good practice, a model whose appropriateness was learned by practitioners in the course of their training." J. Borghouts, *Ancient Egyptian Magical Texts* (Leiden, 1978), p. ix, emphasizes that "spells are the verbalized core matter of the rite."

141. Many objects excavated from Italian tombs (of as early as the eighth century B.C.) are generally considered amulets. These include flints, fossilized shark teeth, shells of various species, bears' claws and teeth, boars' tusks, faïence figures of Bes, and "Phoenician" glass masks. Many are commonly described as jewelry or by an equivalent word, but rarely as amulets.

142. T. G. H. James, "Ancient Egyptian Seals," in Collon 1997, p. 39. See also Ritner 1993; Andrews 1994, esp. pp. 100–106; and Wilkinson 1994, pp. 82–95.

143. Pinch 1994, p. 105.

144. The word *sympathetic* is used in the sense of "sympathetic magic." As is written in one surviving Egyptian medical papyrus, "still in some circumstances magic is needed to attract the sun's influence": J. F. Borghouts, "The Magical Text of Papyrus Leiden I 348," *Oudheidkundige Mededelingen uit's Rijksmuseum van Oudheden te Leiden* 51 (1971): 165–67.

145. Pinch 1994, p. 105.

146. Golden tears of amber might have been thought to be everlasting tears of mourning. For the ancients familiar with the Phaethon or Meleager myth, the tears may have called up the weeping of the Heliades or the Meleagrides. Amber objects are found on the body, unassociated in the tomb, on top of cremated ashes, and, in rare cases, outside the container within the grave complex. H. Horsnaes, *The Cultural Development in North-western Lucania, c. 600–273 B.C.* (Rome, 2002), p. 85, reminds us that "personal 'gifts' and ritual objects may have had plural functions (indeed, one object would often belong to more than one of these categories): the practical function in the rituals taking place during the burial, the display of wealth/status for the community attending the burial, or the needs of the deceased in his/her afterlife."

147. Dickie 2001.

148. Kotansky 1991; Dickie 2001, p. 93, nn. 54–56.

149. Dickie 2001, p. 93.

150. See V. Dasen, ed., *Naissance et petite enfance dans l'Antiquité: Actes du colloque de Fribourg, 28 novembre–1er décembre 2001*, Orbis Biblicus et Orientalis 203 (Fribourg, 1994), provides a bibliography of the critical texts and secondary literature on amulets and spells of protection (against dangers of unspecified origin) for the pregnant woman, the fetus, parturition, and the newborn. See also V. Dasen, "Amulettes d'enfants dans le monde grec et romain," *Latomus* 62 (2003): 275–89. Bonner 1950 was my introduction to the subject of amulets in connection with women, birth, and children, followed by A. A. Barb, "Diva Matrix," *Journal of the Warburg and Courtauld Institute* 16 (1953): 193–238; and Pinch 1994. See also Johnston 1995; and J. J. Aubert, "Threatened Wombs: Aspects of Ancient Uterine Magic," *Greek, Roman and Byzantine Studies* 30, no. 3 (1989): 421–49. Added to the ancient evidence, overviews such as J. Musacchio, *The Art and Ritual of Childbirth in Renaissance Italy* (New Haven, 1999), and systematic analyses such as G. Bellucci, *Catalogue descriptif d'une collection d'amulettes italiennes, envoyée à l'Exposition universelle de Paris, 1889* (Perugia, 1889; repr., 1980), and G. Bellucci, *Il feticismo primitivo in Italia, e le sue forme di adattamento,* 2nd ed. (Perugia, 1919), show the long duration of charms and amulets in Italy. Many uterine amulets are for quieting the womb, while others are to still or retain a "wandering womb."

151. Magical stones that protect pregnant women are listed in most ancient lapidaries. See n. 68.

152. Juvenal (*Satires* 5.163–65) calls the bulla the *Etruscum aurum,* and some Roman writers (Pliny, *Natural History* 33.4; Festus, *De significatione verborum* 26.25; Plutarch, *Vita Romulus* 25) refer to the bulla as a specifically Etruscan ornament. The importance of the bulla for high-born Etruscan boys is evidenced by the third-century B.C. bronze statuettes *Putto Carrara* and *Putto Graziani* in the Museo Gregoriano Etrusco in the Vatican: Cagianelli and Sannibale 1999, pp. 110–34, with nos. 2–3.

P. G. Warden, "Bullae, Roman Custom, and Italic Tradition," *Opuscula Romana* 14, no. 6 (1983): 69–75, outlines the amuletic custom of the bulla, drawing attention to one from Campovalano that contains three small stones and to elaborate figured bullae displaying, for example, apotropaic devices, Bes, or the gorgoneion. J. Sebesta, "The Costume of the Roman Woman," in Sebesta and Bonfante 1994, p. 47, notes the apotropaic nature of both the bulla and the band of the *toga praetexta.* Macrobius (1.6.8–14), discussing a bulla worn by a triumphant general, says it enclosed curatives (*remedia*) that were believed to be strong against *invidia. Invidia* is one of the words used to describe the dangers amulets were intended to prevent or act against. See M. Dickie, "The Fathers of the Church and the Evil Eye," in *Byzantine Magic,* ed. H. Maguire (Washington, DC, 1995), pp. 9–27 (with essential bibl.), where he shows that the term *evil eye* as such was hardly used in Classical antiquity and the Christian world:

> *The terms most often used are, by Greek speakers, φθόνος and βασκανία, and by speakers of Latin,* invidia *and* fascinatio *or* fascinus. *What men feared under these headings was not a single object with a secure and fixed identity but a complex of objects with shifting identities, and identities that coalesce.... The more or less constant factor in this constellation of fears was envy: men were afraid lest their good fortune would draw envy on their heads. The mighty feared it would come from their fellow men, demons, the gods, fortune, the fates, and a malign supernatural power they called simply φθόνος or* invidia. *(p. 12)*

See also J. Russell, "The Archaeological Context of Magic in the Early Byzantine World," in Maguire 1995 (see above), pp. 37–38.

The same word (*invidia*) was used in nineteenth-century Italy for the same purposes, as revealed in Bellucci 1889 and Bellucci 1919 (in n. 150, above).

Archaeological evidence for Roman domestic animals with bullae is to be found in the bronze bullae-ornamented horse tack buried at Populonia: Warden 1983 (see above), p. 70, with reference to A. Minto, *Populonia* (Florence, 1943), pp. 185–86, pl. 49.5. R. D. De Puma called my attention to the many bulla-wearing animals in Etruscan art, including the terracotta horses from the Temple of the Queen's Altar, Tarquinia, and the ravens on Etruscan mirrors. Exempla of human bulla wearers are on the stone sarcophagus from the Tomb of the Sarcophagi, Banditaccia Cemetery, Cerveteri (Museo Gregoriano Etrusco). Round bullae are worn by the deceased male on the lid and by a woman and both horses on the box front: B. Nogara, *Guide du Musée de sculpture du Vatican I: Musée et Galeries Pontificaux* (Vatican City, 1933), p. 412; and R. Herbig, *Die jüngeretruskischen Steinssarkophage: Die antiken Sarkophagenreliefs* (Berlin, 1952), p. 46, no. 83, pls. 1–2.

A. Coen, "Bulle auree dal Piceno nel Museo Archeologico Nazionale delle Marche," *Prospettiva* 89–90 (1998): 94, has best articulated the difference between the wearing of multiple bullae by various personages and the wearing of the single bulla by boys. The bulla was offered up to the Lares on the day of Liberalia at puberty, thus connecting the boy to Liber and the sphere of Dionysian activity. Coen hypothesizes that the gold bullae buried with high-status individuals, women particularly, connote a particular status and were worn in view of the "religious salvation" and heroization of the subjects represented on the bullae. Coen notes that bullae are frequently found in graves with *coronae aureae*, perhaps also Dionysian. Figured gold bullae (dating to as early as the sixth century B.C., but mainly of the fourth) usually are worn in multiples; they include obvious Dionysian subjects as well as age-old aversion devices, the gorgoneion being a notable example. If the bulla-wearing Dionysos on the Praenestine "Cista Napoleon" is also Liber, the image may be a link to the tradition of boys dedicating their bullae to Liber at puberty. See n. 156, below.

A subject still deserving closer study is the relationship between the large figured amber pendants (found mainly along the Adriatic and in the Basilicata) and the pictorial gold bullae and pectorals of the fifth and fourth centuries B.C. (found mainly in Etruria, Latium, and Picenum). Both are made from materials with solar connotations and figured with apotropaic, heroic, and divine subjects, especially ones associated with rebirth and most particularly with Dionysos.

In addition to the bibliography above, see Bonfante 2003, pp. 143–44, n. 95; Dickie 2001; Haynes 2000, p. 282; Cagianelli and Sannibale 1999, pp. 117–18, 133; R. E. A. Palmer, "Locket Gold, Lizard Green," in *Etruscan Italy: Etruscan Influences on the Civilizations of Italy from Antiquity to the Modern Era,* ed. J. F. Hall (Provo, UT, 1996), pp. 117–27; Waarsenburg 1995, p. 409, nn. 1050–52; S. Stone, "The Toga," in Sebesta and Bonfante 1994, pp. 20, 41, n. 37; A. Stout, "Jewelry as a Symbol of Status," in Sebesta and Bonfante 1994, pp. 76–77; H. R. Goette, "Die Bulla," *Bonner Jahrbücher* 186 (1986): 133–64; F. Roncalli in *Santuari d'Etruria,* ed. G. Colonna (Milan, 1985), pp. 37–38; H. Gabelmann, "Römische Kinder in Toga Praetexta," Jdi 100 (1985): 497–541; M. Torelli, *La storia degli Etruschi* (Rome and Bari, 1984), pp. 23–25; Cristofani and Martelli 1983, p. 11; and A. Andrén, "Oreficerie e plastica etrusche," *Opuscula Archaeologica* 5 (1948): 94–99.

The largest and most "canonically" apotropaic of all amber pendants may be that excavated from a woman's tomb (Tomb 94) at Belmonte Piceno: Rocco 1999, p. 62, nn. 161, 343, 473, fig. 27; Negroni Catacchio 1989, pp. 679–80, pl. 9a; Marconi 1933, cols. 421–23, pls. 29.4–5; and I. Dall'Osso, *Guida illustrata del Museo Nazionale di Ancona* (Ancona, 1915), pp. 42, 65ff., fig. 127. The large, lens-shaped amber has a relief gorgoneion in its center and seven feline and human heads carved around its edge. The drilled holes on its periphery could have been used to attach additional small pendants.

An Egyptian text describes how a solar amulet such as a bulla or an amber (or both) might work: "The hand and seal of the sun god are the mother's protection. Each morning and evening, she recites the magic spells over an amulet that she hangs around her child's neck. She prays to the rising sun. She implores him to take away the dead who would like to steal her child. She does not give her child to the thief from the kingdom of the dead": Borghouts 1978 (in n. 140, above).

153. G. Bordenache Battaglia with A. Emiliozzi, *Le ciste prenestine, I: Corpus*, vol. 1 (Rome, 1979), pp. 181–82, n. 59.

154. Florence, Museo Archeologico Nazionale 4026. Could his bullae be of amber, considering the Argonauts' destination of the northern lands, the ancient association between this voyage and amber, and amber's safeguarding and buoyant properties?

155. For examples of bulla wearers (including demons) on Etruscan mirrors, see ES 2, pl. 166; ES 3, pl. 257; ES 4, p. 30, pl. 298; and ES 5, p. 60. See also LIMC 3 (1986), s.v. "Fufluns" (M. Cristofani), p. 532, n. 11; L. B. van der Meer, *Interpretatio etrusca: Greek Myths on Etruscan Mirrors* (Amsterdam, 1995), pp. 93–95, figs. 38, 42, 56, 60, 122, 125; LIMC 1 (1981), s.v. "Amatutunia" (G. Colonna), p. 586, n. 1; and LIMC 1 (1984), s.v. "Ares/Laran" (E. Simon), p. 502, n. 19. Two other named bulla-shaped pendant wearers are Peleus (armband) and Calaina (holding a circlet), who are depicted on Metropolitan Museum of Art 09.221.16, Rogers Fund, 1909: G. Bonfante, "Note on the Margin of a Recent Book: Calaina," *Etruscan Studies* 6 (1999): 8–9; and *Corpus Speculorum Etruscorum* 3, no. 14.

156. The extraordinary series of fourth-century B.C. terracotta votive figures from Lavinio are richly ornamented with figural bullae of various forms: *Enea del Lazio: Archeologia e mito,* exh. cat. (Rome, 1981). An extraordinary sarcophagus-lid figure with

similar bullae (circa 400–350 B.C.) was found at Cerveteri: Cristofani and Martelli 1983, pl. VIII.

157. The jewelry represented on the New York mirror (see n. 155, above) is compared by R. Nicholls to that on a mirror with Amphiaraos in the Fitzwilliam Museum: *Corpus Speculorum Etruscorum Great Britain*, no. 2.8. Nicholls also discusses the significance of the armlet in Etruscan art.

158. Bulla-shaped amber pendants (the commonest form of pendant) are documented in the seventh-century Foundation Deposit at Ephesus and in women's graves in Etruria and southern Italy from the eighth century onward. Unfortunately, many of the known bulla-shaped amber pendants are without secure provenance. The largest amber bulla known to me comes from Belmonte Piceno Tomb 94, a grave typed as female by I. Dall'Osso (cited by Rocco 1999, p. 107, n. 473). The bulla was found in a woman's tomb with iron armor and arms, parts of a cart, bronze torques, bracelets, fibulae of various kinds (including ones with amber segments and one with bronze bullae pendants), and other amber objects. Rocco 1999, p. 86, no. 143, discusses an ivory cylinder from the same tomb.

159. See n. 75, above.

160. Amber might have been especially effective in magically attracting the sun, due to its inherent magnetic property and because of amber's "sympathetic" brilliance and color: like would attract like. The verb "to fix" in reference to amulets is borrowed from the Hermetic writings in reference to talismans. See D. Pingree, "Some of the Sources of the *Ghāyat al-hakīm*," *Journal of the Warburg and Courtauld Institutes* 43 (1980): 1–15, quoted by E. Reiner, "Magic Figurines, Amulets, and Talismans," in *Monsters and Demons in the Ancient and Medieval Worlds: Papers Presented in Honor of Edith Porada*, ed. A. Farkas et al. (Mainz, 1987), p. 27.

161. Pingree 1980 (in n. 160, above), p. 3. Plantzos 1999, p. 110, notes that "the ability of a lens—*hyalos*—to attract the rays of the sun ([Aristophanes,] *Clouds* 760–75)" was common knowledge.

162. Kotansky 1991, p. 108, with reference to P. W. Schienerl, "Der Ursprung und die Entwicklung von Amulett behältnissen in der antiken Welt," *Antike Welt* 15 (1984): 45–54, esp. 50–54.

163. Plato, *Republic* 426b1–2.

164. Amulets of clay, stone, ivory, bone, and other materials are among the earliest surviving sculpted objects from Italy. The early Neolithic and Chalcolithic clay heads and figurines from cultic caves include nude and partially dressed figures and heads with necks, but no isolated faces. See K. Holmes and R. Whitehouse, "Anthropomorphic Figurines and the Construction of Gender in Neolithic and Copper Age Italy," in *Gender and Italian Archaeology*, ed. R. Whitehouse (London, 1998), pp. 95–126.

165. Melfi, Museo Archeologico Nazionale del Melfese "Massimo Pallottino" 118680–81 (the female heads) and 118678–79 (the satyr heads) from Tomb 48, Ripacandida: Bottini 1987, pp. 9–12, figs. 13–15, pl. III.

166. Bottini 1993 p. 65; and Bottini 1987, p. 10, n. 39.

167. See n. 155, above, for the Etruscan mirror with Calaina (Galene) in New York. A "so-called 'chaplet' or string of beads is carried as an attribute by a goddess who appears on the palace sculpture of King Assurnasirpal II of Assyria, and on Neo-Assyrian seals, the goddess carrying the chaplet is sometimes Ishtar (Inana)": Black and Green 1992, pp. 51–52.

168. See n. 150, above.

169. Goff 1963 (in n. 7, above), pp. 162–211. For the Sumerian material, see, for example, the beads and amulet group from the tomb of Queen Puabi, discussed by H. Pittman in *Treasures from the Royal Tombs of Ur,* exh. cat., ed. R. L. Zettler and L. Horne (Philadelphia, 1998), pp. 95–96, no. 33 (with critical comparanda).

170. For a recent discussion of *crepundia* and Roman amber, see M. Lista, "L'ambra dei Romani in Plinio: Dal moralismo alla *devotio,*" in *Ambre* 2007, pp. 254–59. Waarsenburg 1995, pp. 458–59, n. 1299 (with bibl.), notes that "although by Imperial times, crepundia had become restricted to protective charms for children, Apuleius (*Apologia* 56.3) confirms that they had a religious significance *(sacrorum crepundia).*" See also V. Dasen, "Protéger l'enfant: Amulettes et *crepundia,*" in *Maternité et petite enfance dans l'Antiquité romaine,* exh. cat., ed. D. Gourevitch, A. Moirin, and N. Rouquet (Bourges, 2003), pp. 149–51.

171. Potenza, Museo Archeologico Nazionale 96684 (satyr) and 96685 (Herakles, identified in all publications as "maenad"), from Tomb 106, Braida di Vaglio: *Magie d'ambra* 2005, ill. p. 117; Bottini and Setari 2003, p. 66, nos. 310 (Herakles) and 311 (satyr), fig. 37. S. J. Schwarz confirmed my identification of the head as a Cypriot-type Herakles (pers. comm., September 22, 2006); see LIMC suppl. 2009, vol. 1, add. 2, s.v. "Hercle" (S. J. Schwarz), pp. 247–48.

172. Melfi, Museo Archeologico Nazionale del Melfese "Massimo Pallottino" 51436–40, from Tomb 48, Melfi-Pisciolo. The frontal female heads, inv. 51436–37, are each drilled with numerous stopped bores. Inv. 51436 even has bores in the cheek and chin. For the female heads from this tomb, see Bottini 1993; Bottini 1987; and *Popoli anellenici* 1971, p. 125, pl. LIII. The two other pendants, female heads, inv. 51439–40, are in poor condition.

173. Eos and Kephalos (identified by A. Bottini), Matera, Museo Nazionale "Domenico Ridola" 169680, from Tricarico-Serra del Cedro, Tomb 60, middle of the fourth century B.C.: *Magie d'ambra* 2005, ill. p. 128. This pendant is likely older than the burial. The intact woman's Tomb 952 from Lavello-Casino, dating to the middle of the fifth century B.C. (Melfi, Museo Archeologico Nazionale del Melfese "Massimo Pallottino"), included three large amber pendants suspended in the groin

area and several necklaces composed of glass-paste eye beads, bone pendants, and amber beads and pendants. Although the necklace ambers are in poor condition, two hitherto unidentified pendants (a ram's head and a siren; no known inv. nos.) show evidence of pulling wear at the suspension holes: *Ornamenti e lusso* 2000, p. 57; *Treasures* 1998; and "La tomba 952 di Forentum" (undated pamphlet, Melfi museum, above).

174. For the amber ram's head from Adria, see *Due donne* 1993. For the Bolognese (Certosa) material, see A. Zannoni, *Gli scavi della Certosa di Bologna* (Bologna, 1876); and G. Muffatti, "Paste vitree, alabastri, oggetti in osso, avorio e ambra," StEtr 35 (1967): pl. 77a. For other ambers from the area, including recent and previously unpublished older finds, see L. Malnati, "L'ambra in Emilia Romagna durante l'età del Ferro: I luoghi della redistribuzione e della produzione," in *Ambre* 2007, esp. pp. 122–29, 152–59.

175. The female head from Tomb 90 at Latronico–Colle dei Greci is Policoro, Museo Nazionale 216349: *Ambre* 2007, p. 239. E. Brizio, "Verucchio, scoperta di sepolchri tipo Villanova," NSc 10 (1898): 373, reported that an amber ring from Tomb 11 at Verucchio was repaired in antiquity with "sewing stitches."

176. Amber pendants are not alone in showing signs of use wear—from touching, rubbing, kissing, or other kinds of abrasion as the objects came into contact with the body or clothing. See Ritner 1993 on kissing, spitting, and other acts in Egyptian ritual magic. Ritual washing may also have been a cause of the uneven wear. The Africanist Zoë Strother (pers. comm., August 2005) recounts her interview with a Central Pende man who described how he washed his ivory pendant in river sand to keep it white. Compare the ivory mask in the Tervuren Collection (7676): *Masterpieces from Central Africa: Royal Museum for Central Africa, Tervuren,* exh. cat., ed. G. Verswijver et al. (New York, 1996). For more examples of beads and pendants in amber and other materials that show evidence of use, see R. K. Liu, *Collectible Beads* (Vista, CA, 1995), pp. 35–37 and passim.

177. Some classes of amulet wearers deserving closer study include the Laconian acrobats and dancers; babies and toddlers; Cypriot temple boys; and certain female divinities. Among the last are seated divinities from Sicily (Gela, the extraurban sanctuary of Predio Sola; Selinus, the Malophoros Sanctuary) and southern Italy (Metaponto, San Biagio). The amulets worn by youngsters and athletic young women (on mirror supports) include many time-honored fertility subjects: the crescent moon, lotus blossoms, lotus flowers, and the sun.

178. For examples of these two gods adorned with pendants, see L. Bonfante, "Fufluns Pacha: The Etruscan Dionysus," in *Masks of Dionysus,* ed. T. H. Carpenter and C. Faraone (Ithaca, NY, 1993), pp. 224–31, figs. 21, 24. The Naples mirror is Museo Archeologico Nazionale ES, pl. 82; the Berlin mirror is Antikenmuseum Fr. 36, ES, pl. 83.

179. For the Tarquinian Tomb of Hunting and Fishing, see, most recently, Steingräber 2006. On p. 95, he notes the importance of Dionysian elements in the tomb. Haynes 2000, p. 229, interprets the tomb as Dionysian; compare Simon 1998, who reiterates her belief that its plants are laurel and signify it as the grove of Apollo. Brown 1960, p. 106, was the first to make the connection between the painted images and excavated gold animal-head pendants.

Early Iron Age and the Orientalizing Period

After about 1200 B.C., amber was much scarcer throughout the Mediterranean until about the mid-eighth century B.C., when it begins to reemerge appreciably in archaeological contexts. For the most part, it was at the end of the eighth and especially during the seventh centuries when amber was most popular in Greece and peninsular Italy. This is not to leave out a few extraordinary tenth-to-early-eighth-century exceptions, notably at sites in Italy, in Latium, at Castel di Decima, and, most recently, in the Roman Forum and in the Basilicata, in the area between the Agri and the Sinni, where, in the graves of elite women, remarkable amber parures were discovered. This is the case with a girdle with interspersed bird-shaped beads from the Enotrian Tomb 83 at Latronico.[188] On the whole, amber-embellished objects were buried in both male and female graves, but figured amber is almost exclusively found in those of women and children.[189]

Carved *figured* ambers of eighth-to-seventh-century date are characteristically small (on average, roughly fingertip size), suggesting that these works, mainly pendants, were carved from small pieces. None are composites, that is, works made from almost imperceptibly joined pieces, as is characteristic of contemporary fibulae from Etruria, Campania, and the mid-Adriatic. Among the earliest figured finds are those from the eighth-century necropolis at Veio Quattro Fontanili. They include a standing ithyphallic male, monkeys,[190] a horse, a duck, and a human lower leg and foot, as well as both scarabs and scaraboids, some of which have intaglio horses engraved on their flat side.[191] All of these are amuletic subjects of great antiquity, and truly Orientalizing.[192] A cinerary urn buried in the First Circle of the Interrupted Stones at Vetulonia (of circa 730–720 B.C.) contained a number of high-status objects, including an amber scarab, thus indicating an object interred after cremation.[193] The scarab may well have been an import, like the accompanying glass beads and bronze Phoenician bowl, although the urn also contained locally produced objects. A number of female graves in and around Magna Graecia each contained but one small waterbird, which may be related to the Egyptian duck amulet, a symbol of regeneration; it may also be related to the duck symbol of northern Europe. Since the Bronze Age, the duck, a multivalent symbol both guardian and apotropaic, was believed to connect the chthonic and other worlds.[194]

In Greece, worked amber was buried in foundation and votive deposits as well as, more rarely, in graves. A pair of Geometric-date tombs (possibly of priestesses or princesses) at Eleusis offer critical evidence of amber in the burial of women of the highest rank.[195] The rich tombs include sumptuous grave gifts, among them necklaces of gold, amber, and faïence, and amber-inlaid ivory furnishings. The presence of glowing *elektron* bears witness to the lavish and exceptional occasion of the entire funeral process.

Both figured and nonfigured ambers have been excavated at sanctuaries dedicated to a limited number of divinities, mainly female. These include objects from the sanctuaries of Artemis (Ephesus), Artemis Orthia (Sparta), Hera Limenaia (Perachora), and Apollo Daphnephoros (Etretria). Intaglios were found at Perachora, and two animals at Aetos (Ithaca). The earliest date to the decades around 700 B.C. and represent birds at rest and couchant animals, and they, like the contemporary Italian objects, are generally quite small. At Ephesus, the foundation deposit was buried circa 700 near the cellar of the temple of Artemis. Anton Bammer has suggested that the ambers (and accompanying ivory objects) are the remains of a pectoral worn by an early statue of the goddess.[196] Other figured Greek works of this period include the by-now traditional subjects of figured amber: crouching monkeys, recumbent lions, human heads, birds, and other species.[197]

The seventh-century B.C. ambers from Italy are almost exclusively mortuary and more extensive in number, type, and size than the contemporary Greek examples. As is characteristic of all art from the Orientalizing period, they take on a character different from the eighth-century material, although birds, especially ducks, retain their popular status, as they do in other figurative arts in Italy. At some sites, figured amber is found in combination with faïence amulets of Egyptian fertility and protective subjects.[198] The primary seventh-century finds have come from Etruria, Campania, and Latium; Etruria Padana and elsewhere in the mid-Adriatic; and from the Basilicata. Recent discoveries in southern Italy and at the Adriatic site of Verucchio (near Rimini) have greatly modified the picture of amber importation and use. One rare figured subject from the extraordinary amber-rich graves at Verucchio is a fibula decoration of addorsed ducks.[199]

Figured ambers excavated at southern Etruscan sites include the ubiquitous monkeys and a number of standing "nude" females, their arms in various poses associated with fertility.[200] An exceptional example, dating to the first half of the seventh century, is the elaborate grouping of amber pendants and beads (possibly a collar) found on top of the cremation layer in a tomb at Vetulonia.[201] Little else accompanied the strings of amber: the figured pendants include a fish,[202] a scaraboid, seven monkeys, and eleven standing female figures dressed only in collars and armlets, with legs apart, the vulva exposed, and hands placed on the lower abdomen. The most important pendant represents an enthroned female giving birth, the infant's head appearing between her legs.[203] This tiny amber is the strongest evidence to date for a direct link between amber and childbirth.

Many other types of figured amber from the second half of the seventh century correspond to standard Egyptian amuletic iconography. Among the most popular are the dwarf deities, such as Bes and Pataikos-Ptah—the most common Egyptian protective genies.[204] Bes was known to protect sleepers and women in childbirth and safeguarded the young mother and her children. Both figures have solar associations; the Pataikos-Ptah figure, part adult and part infant, symbolized the infant sun. Almost without exception, the images on early amber carvings were reiterations of Egyptian-sourced solar and rebirth symbols.

The main focus in this catalogue is amber in the form of figural subjects, but the many beads and pendants of this period in botanic or shell forms are also important, since they, too, served a similar role via a metonymic process. Amber cowrie pendants, common in Italy from the seventh to the fifth centuries B.C., were potent subjects of fertility and childbirth, since the mature cowrie shell was thought to resemble the vulva. The extraordinary Getty *Cowrie Shell / Hare* pendant (figure 42), for instance, combines two subjects: fertility and regeneration. Scarab-cowrie combinations, such as that represented by a ninth-century B.C. amber from Tursi (Basilicata), do the same. In Egypt, both real cowries and imitations in gold and other materials were strung together to make girdles and worn in the pelvic region.[205]

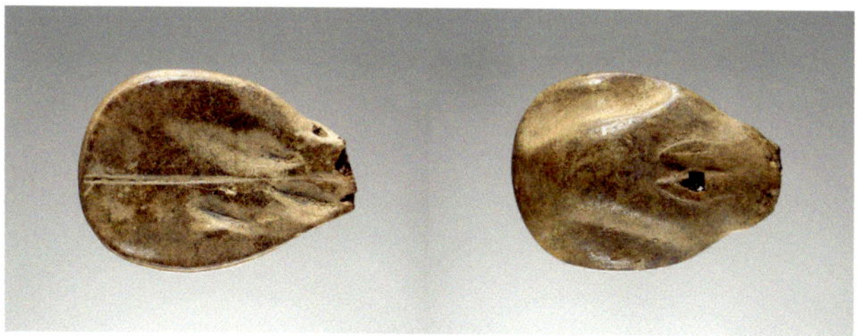

Figure 42 a & b. *Cowrie Shell / Hare* pendant, Italic or Etruscan, 600–500 B.C. Amber, H: 3.7 cm (1½ in.), W: 2.6 cm (1 in.), D: 1.4 cm (½ in.). Los Angeles, J. Paul Getty Museum, 79.AO.75.28. Gift of Stanley Silverman. a) front; b) back. See cat. no. 30.

The most important surviving ensemble of the seventh century from Italy is that of a high-ranking woman buried at Latin Satricum (Tomb VI).[206] The grave, dated to circa 650/640 B.C., contained a flint (actually a Neolithic obsidian scraper)[207] and more than five hundred amber objects—fibulae, spindles, nonfigured beads and pendants, and numerous figured objects. The medley of stylistic and iconographic connections of the objects is typical of the period and place, but the burial is without parallel: it is the largest single burial with amber from ancient Italy. The figured pieces include nude females and males (some doubled and addorsed), fantastic creatures,[208] and fish, and some of the pendants were carved from large amber blanks. Some pendants are unique, others variants on or copies of Egyptian subjects: fish, Bes, and *patakoi*. The unworked pieces of amber, here and in other tombs, may also have served as fumigants, unburnt incense, or apotropaics.[209] This grave's goods and the many contemporary large amber fibulae of the mid-Adriatic of these decades speak to new sources (geographic or cultural) of or new access to big pieces of jewelry-grade amber.

NOTES

188. S. Bianco in *Magie d'ambra* 2005, pp. 94–96, ill. p. 99.

189. This is theorized on the basis of a small percentage of excavations or published accounts; the number of unpublished graves and deposits with amber objects and the amount of pre-Roman amber in non-source-country museums and collections (from old or unreported finds and uncontrolled

excavations) is unfortunately very high. The exceptions are critical (such as the male Tomb 43 at Melfi-Pisciolo).

190. Orientalizing Greek and Etruscan images of nonhuman primates are generically referred to as "monkeys" in the literature, although some may represent baboons, especially the hamadryas baboon (*Papio hamadryas*), as well as a long-tailed monkey (*Cercopithecus*) and the green monkey, or vervet (*Cercopithecus aethiops*). The prototypes of the eighth-to-seventh-century amber pendants from Italy (Etruscan, Latin, Faliscan, Picene) are Egyptian in invention, but they also may have derived from Phoenician examples and could be related to northern Mesopotamian, northern and western Syrian, Old Babylonian, and Anatolian types and symbolism. In Egypt, amulets in the form of monkeys and baboons are first known in the Old Kingdom, made of steatite and faïence, then of amethyst and carnelian in the Middle Kingdom, and in a wider variety of materials from the New Kingdom onward. The green monkey is most often the subject of Egyptian and Phoenician simian amulets: its humanlike features, the females' motherly love, its cleverness and ability to mimic, and its greenish color (symbolic of freshness and regeneration) account for its popularity. It participates at the side of the dwarf as an emissary of Ra, the sun-god, in magical invocations for successful parturition and thus has a solar aspect (Andrews 1994, p. 66). In Egyptian glazed-composition faïence maternity amulets, where it is joined with Bes, the green monkey takes on the role of nurse for the newborn and is connected to music and dance, as associated with birthing. For the monkey and maternity, see also Bulté 1991, pp. 99–102. Monkey representations in the Levant seem to carry several connotations, of both Near Eastern and Egyptian origin, including veneration, eroticism, good luck, and best wishes. In erotic scenes on Old Babylonian terracottas, simian dancers often keep company with dwarfs. As S. Schroer and J. Eggler, "Monkey," in *Iconography of Deities and Demons in the Ancient Near East,* http://www.religionswissenschaft.uzh.ch/idd/prepublication.php (accessed November 12, 2009), p. 1, note for Mesopotamian and Elamite art, "Just like in Egypt, there is a proximity between the monkeys and the Nude Goddess. This may be due to their playful nature, but also their excitability ... leading to their association with sex and eroticism."

Amber and glazed-composition amulets of monkeys might work in various direct and indirect forms of magic: to ensure love and sexual fulfillment; to provide sexual aid in this world and the next, to aid in rebirth and rejuvenation, to assist in the care of newborns, and to inject humor (a potent aversion technique). On the nonhuman primate in Egyptian art generally, see Andrews 1994, pp. 66–67; and A. Kozloff, ed., *Animals in Ancient Art from the Leo Mildenberg Collection* (Cleveland, 1981), pp. 67–69, nos. 54–56. For a wide range of opinions about "monkeys" in Etruscan art, see Waarsenburg 1995, p. 415–16, and esp. 445–50. See also Bonfante 2003, pp. 138, 141; Negroni Catacchio 1999, pp. 280–82; Waarsenburg 1996; F.-W. von Hase, "Die golden Prunkfibel aus Vulci, Ponte Sodo," *Jahrbuch des Römisches-Germanischen Zentralmuseums Mainz* 31 (1984): 269–75; J. Szilágyi, RA 1972: fasc. 1:111–26; and D. Rebuffat Emmanuel, "Singes de Maurétanie Tingitane et d'Italie—Réflexions sur une analogie iconographique," StEtr 35 (1967): 633–44. For an Etrusco-Corinthian *aryballos* in the form of an "ape," see B. A. Kathman in Kozloff 1981, pp. 95–96, no. 95.

For the monkey in the Minoan world, see N. Marinatos, "An Offering of Saffron to the Minoan Goddess of Nature: The Role of the Monkey and the Importance of Saffron," in *Gifts to the Gods: Proceedings of the Uppsala Symposium 1985*, Boreas 15, ed. T. Linders and G. Nordquist (Uppsala, 1987), pp. 123–32, who argues convincingly for a religious function for monkeys and interprets various Minoan roles for them: as adorants, as intermediaries between humans and the goddess of nature, as her servants, and as guardians. Marinatos draws parallels with Egyptian and Anatolian images of squatting monkeys (nn. 10, 17) and suggests the images' possible entry into Crete in the Middle Bronze Age, but points also to Mesopotamian examples of the squatting posture. Both Egyptian and Near Eastern prototypes are proposed, with reference to R. D. Barnet, "Monkey Business," *Journal of Near Eastern Studies* 5 (1973): 1–10; and C. Mendelson, "More Monkey Business," *Anatolian Studies* 33 (1983): 81–83. F.-W. von Hase 1984 (above) proposes Phoenicians as intermediaries in the transition of the motif to Italy. For a view on the possible permutation of the "monkey" type into human imagery in early Greece, see S. Langdon, "From Monkey to Man: The Evolution of a Geometric Sculptural Type," AJA 94 (1990): 407–24.

To be added to this discussion are the simianlike "emaciated humans" of the Old Babylonian period, the clay plaques of the goddess Nintu, and the separate statuette images in the same form. D. Parayre, "Les âges de la vie dans le répertoire figuratif oriental," *KtèMA* 22 (1997): 67, identifies the figures as representing premature or deformed fetuses. See her figs. 10a (stamped relief possibly from Tell Asmar, Louvre) and 10b (bronze statuette, Cincinnati Art Museum). Parayre suggests that the fetus images may be figural transpositions of the *šumma izbu* series, listing the precautions to take in the case of premature, nonviable, or monstrous births. If the amber pendants represented such fetuses rather than monkeys or baboons, they would be extraordinary "like banishes like" amulets. Alternatively, if the amber monkeys are identified with the Minoan interpretation of the type (following Marinatos), they may be associated with the local nature goddesses in Crete, as in Mesopotamia.

191. For Italian finds of eighth- and seventh-century date, Waarsenburg 1995 is the most complete compendium of objects and earlier bibl., including Massaro 1943. The Iron Age Greek amber finds are listed in Strong 1966, pp. 21–24 (with earlier bibl.). The horse imagery, which appears early and remains until the fourth century B.C., deserves closer study. Although the horse has good connotations throughout the ancient world (the Egyptian hieroglyph for "beautiful," *nefer,* is a prancing horse), it had both positive and negative aspects in

Greece. "The horse was strongly associated with Poseidon, a dark and marginal god, a god of the frightening sea and destructive earthquake. According to myth and cultic tradition, Medusa and Erinys (or Demeter-Erinys) each assumed the shape of a mare to become the consort of Poseidon, and subsequently bore him the foals Pegasus and Areion.... From Homer onwards, [Erinys and Medusa] represent the grim, horrific and threatening aspects of the chthonic world": Johnston 1995, pp. 375–76, nn. 36–38. An amber horse may have worked as a danger-averting object.

192. On the Orientalizing phenomenon in Italy, see D. Ridgway, "The Orientalizing Phenomenon in Campania: Sources and Manifestations," in Prayon and Röllig 2000, which takes the phenomenon far beyond Campania. Ridgway's term *medley* is useful in describing sources of Orientalizing art. Also apt is his assessment of the term *Phoenician:* "We cannot simply call the orientalia (and Orientals) in question 'Phoenician' *e basta*." The term encompasses considerable diversity; as coined by the Greeks, it was used to describe Bronze Age Canaanites, Iron Age Phoenicians, and Punic Carthaginians. See also I. J. Winter, "Homer's Phoenicians: History, Ethnography, or Literary Trope? (A Perspective on Early Orientalism)," in Carter and Morris 1995, pp. 247–72. Compare Lapatin 2001, p. 38, n. 3, who concludes that the terms *Phoenician* and *North Syrian* are useful and readily understood stylistic labels, despite their inaccuracies and problems.

193. Poggio alla Guardia Necropolis, Tomb 7. Haynes 2000, p. 15, cites the burial as indicating early connections with the Near East.

194. Waarsenburg 1995, p. 428. The birds are waterfowl, often ducks, represented as if afloat. See S. Bianco (with bibl.) in *Magie d'ambra* 2005; and Franchi Dell'Orto 1999. An eighth-century necklace of bulla-shaped bronze pendants inset with convex pieces of amber and with sleeping ducks above and below (mirrored compositionally) is an important early Italian object that associates amber, the sun, and ducks.

195. For a recent consideration of the pair of tombs, see J. B. Connelly, *Portrait of a Priestess: Women and Ritual in Ancient Greece* (Princeton, 2007), p. 224. For the larger discussion of precious materials and grave gifts in death, ceremony, and burial, sources consulted include C. Sourvinou-Inwood, *"Reading" Greek Death: To the End of the Classical Period* (Oxford, 1995); S. Campbell and A. Green, eds., *The Archaeology of Death in the Ancient Near East* (Oxford, 1995); M. Parker Pearson, *The Archaeology of Death and Burial* (Gloustershire, 1999); and D. Bolger, *Gender in Ancient Cyprus* (Lanham, MA, 2003).

196. A. Bammer, "Kosmologische Aspekte der Artemisionfunde," in *Der Kosmos der Artemis von Ephesos*, Sonderschriften des Österreichischen Archäologischen Instituts 37, ed. U. Muss (Vienna, 2001), pp. 11–26.

197. Also mentioned by Mastrocinque 1991, p. 68.

198. This is noted by Waarsenburg 1995; and Mastrocinque 1991, p. 78.

199. Verucchio (Rimini), Lippi Necropolis, Tomb 27, inv. 11392: P. von Eles, entry for no. 395, in Bartoloni et al. 2000, p. 295; *Verucchio* 1994, p. 161, n. 553, pl. LXI. See also Franchi Dell'Orto 1999, pp. 91–92.

200. Nude and partially clothed humans (with primary and secondary sex characteristics exposed) were potent signs of sexuality, both promoting fecundity and controlling conception, but such pieces also would have encompassed powerful apotropaic, guardian, and positive-attraction forces. For "fertility" gestures, see P. Demargne, *La Crète dédalique: Études sur les origines d'une renaissance* (Paris, 1947), pp. 38–39; Haynes 1985, p. 253, no. 21; Waarsenburg 1995, pp. 433–34, (with additional bibl. and pertinent comparanda, including ivory and *bucchero* caryatid supports of ritual vessels). For the relevant caryatids, see H. Salskov Roberts, "Some Observations on Etruscan Bowls with Supports in the Shape of Caryatids or Adorned by Reliefs," *Acta Hyperborea* 1 (1988): 69–80. Demargne, on the basis of the Cretan material, distinguishes nine types of pudical gestures (and their predecessors). For this Orientalizing material, the gestures may be read as they may have been in Egypt: the pose or gesture is a "still." As Wilkinson 1994 explains, a figure's gesture may be the visual recording of the most characteristic movement within a *sequence* of movements. The image thus registers the most memorable or significant movement or gesture in a sequence.

201. Florence, Museo Archeologico Nazionale 7815.

202. This fish pendant is close to the Egyptian *lates* amulet type, an emblem of the goddess Neith, one of the four great protectresses of the dead.

203. Haynes 2000, p. 100, queries the identity of the figure between the legs of the seated woman—is it a child or a monkey? It must represent a birthing scene, the throne a birthing chair, the head that of an infant human. For the tiny birthing amber, see also Waarsenburg 1995, p. 429; von Hase 1984 (in n. 190, above), p. 274; Massaro 1943, p. 46; I. Falchi, *Vetulonia e la sua necropoli antichissima* (Florence, 1891), p. 101, pl. 7.4; and L. Pernier, "Vetulonia: Il circolo del monile d'argento e il circolo dei lebeti di bronzo," NSc 22 (1913): 425–37.

204. Bes was closely associated with Hathor, as was the related dwarf-god Pataikos-Ptah. Although dwarf figures were associated with a number of gods, they were commonly linked with Bes, often called simply "the dwarf." V. Dasen, "Pataikos," *Iconography of Deities and Demons* (in n. 190, above), p. 1, summarizes: "The term *pataikos* is first used by Herodotus (*Historiae* 3.37) to describe representations of the god Ptah in the form of a dwarf equated with Hephaistos," and "it remains unclear whether [pataikoi] depict various forms of one and the same god, or a group of dwarf gods, as with Bes." Connected with solar and rejuvenating symbolism, they were regarded as a solar hypostasis, embodying the morning form of the sun-

god, newly born and old at the same time. "Their association with the continuing process of creation may have motivated their identification with Ptah in his capacity as a creator god and likewise with Horus, Khnosu, Osiris, and other youthful and regenerative gods." In respect to protection, Pataikos-Ptah seems to have been concerned with both the living and the dead; it aimed to guard the family, especially pregnant women and small children, against unpredictable negative forces. As prescribed in magical spells, *pataikoi* could be worn around the neck as helpers during delivery. *Pataikoi* are often discovered in burials, where they had a strong afterlife symbolism; see Dasen, above (with refs.). For Bes, see esp. M. Malaise, "Bes et les croyances solaires," in *Studies in Egyptology Presented to Miriam Lichtheim,* ed. S. Israelit-Groll (Jerusalem, 1990), pp. 680–729. See also V. Dasen, *Dwarfs in Ancient Egypt and Greece* (Oxford, 1993), pp. 54–83; LIMC 3 (1986), s.v. "Bes" (A. Hermary), pp. 98–112; and Pinch 1994. For the Egyptian and imitation Egyptian amulets of Bes figures and *pataikoi,* see also H. Győry, "To the Interpretation of Pataikos Standing on Crocodiles," *Bulletin du Musée Hongrois des Beaux-Arts* 94 (2001): 27–40; and Andrews 1994, p. 39. Hathor, the "goddess of sexuality, fertility, and childbirth, was also a funerary goddess who presided over the necropolis; she helped women give birth in this world but also facilitated the rebirth of the deceased into the afterlife": G. Robins, "Dress, Undress, and the Representation of Fertility and Potency in New Kingdom Egyptian Art," in Kampen 1996, p. 28. For the dwarf amulet as a health amulet of Hathor, see also G. Pinch, *Votive Offerings to Hathor* (Oxford, 1993). These dwarf images may have functioned not only to protect the state of birthing, but also to control fertility and birth spacing—equally critical issues for young families—for the protection of the mother's health and that of her young. On birth spacing, see n. 15.

205. On cowries, cowroids, and cowrie-shell imitations in Egypt, see Pinch 1994, p. 107; Andrews 1994, p. 42; and R. E. Freed in *Quest for Immortality* 2002 (in n. 75, above), p. 102, no. 17. For a discussion of the cowrie in amber, see 79.AO.75.28 (cat. no. 30).

206. For the find, see the exhaustive treatment in Waarsenburg 1995; and Waarsenburg 1996.

207. Waarsenburg 1995, pp. 410–11, nn. 1058–64: the "flint" likely originated on the nearby island of Ponza and is thus one of several secondarily reused in the Iron Age. Obsidian "flints" are found in central Italy in tombs dating from the ninth to the seventh centuries and in several Latin votive deposits, including in Satricum. A tomb from Terni yielded a Neolithic flint wrapped in an embossed bronze sheet medallion with a representation of Bes. Waarsenburg suggests that the "flint" from Tomb VI would have been known in antiquity as a *ceraunium,* or lightning stone. P. Tamburini in *Antichità dall'Umbria a New York,* exh. cat. (Perugia, 1991), p. 276, discusses such lightning stones and cites A. Cherici, "Keraunia," ArchCl 41 (1989): 372, n. 37. Tamburini points to the ancient belief "in the heavenly origin of prehistoric flintstones found by chance on the ground ... [and] their relation to the thunderbolt" and "to their simple apotropaic function." Still in early-twentieth-century Italy, Neolithic flints are recorded as important amulets to protect against lightning, and to protect people, animals, houses, and land against natural disasters, as G. Bellucci (in n. 150, above) shows. In Etruria, both Menerva and Tinia could hurl thunderbolts, and as such they may have had oracular faculties, as suggested by G. Camporeale, "La manubia di Menerva," in *Agathos daimōn: Mythes et cultes; Études d'iconographie en l'honneur de Lilly Kahil* (Athens, 2000), pp. 77–86. Waarsenburg 1995, p. 411, notes that "a functional and semantic relationship seems to have existed also between Eileithuia, lightning and the Elysium.... An entry in [the *Suda*] states that *Eilusion*—normally the afterlife world—was also used to denote a place hit by lightning." Was the flint a special amulet of protection against lightning?

A carved amber in New York likely represents a thunderbolt (a perfect marriage of subject and material). Metropolitan Museum of Art 1992.11.22, Purchase, Renée E. and Robert A. Belfer Philanthropic Fund, Patti Cadby Birch and the Joseph Rosen Foundation Inc., and Harris Brisbane Dick Fund, 1992: *The Metropolitan Museum Annual Report* (1991–92), p. 37; C. A. Picón, "Carved Ambers," *Recent Acquisitions: A Selection, 1991–1992: The Metropolitan Museum of Art Bulletin* 50, no. 2 (Fall 1992): 10; *Art of the Classical World* 2007, pp. 295, 473, no. 339.

The bracelet pendant worn by the male figure on the Etruscan stone sarcophagus of a couple from Vulci, now in Boston (Museum of Fine Arts 86.145), appears to be either a shark's tooth or a "flint."

208. The most frequent form of demons is that of a hybrid or monster, and the demonic "frequently serves as a classificatory marker that is part of a larger system of boundaries used to express or reinforce a society's values": Johnston 1995, p. 362. "The demon is situated between two taxa that are considered mutually exclusive: the hybrid nature of demons, noted by Smith, is a form of this": Johnston 1995, p. 363. Johnston cites J. Z. Smith, "Towards Interpreting Demonic Powers in Hellenistic and Roman Antiquity," *Augsteig und Niedergang der römischen Welt* 2.16.1 (1978): 425–39, who therein develops the precepts of M. Douglas, *Purity and Danger: An Analysis of Concepts of Pollution and Taboo* (London, 1966).

209. See Waarsenburg 1995, p. 430, on the unworked pieces in the Archaic votive deposit. See n. 126 above for reference to amber and resins in a tomb at Cerveteri.

The Archaic and Afterward

The most important reference to amber from around 600 B.C. may be only apocryphal. It concerns the early Greek philosopher Thales of Miletos, the first to recognize amber's magnetism, which he argued was proof of a soul or life, even in inanimate objects. Did he observe this property at home when watching women spinning Miletos's famous wool with an amber spindle and distaff?

After about 600 B.C., the record shows a change in amber use. Individual pieces and long strings of worked amber became much rarer throughout the Mediterranean, perhaps owing to relative scarcity or to fluctuations in trade or even its interruption (perhaps by the Celts). Thus, amber finds from the next decades take on a particular importance. Most are very small pieces used for inlay, in multimedia fibulae, in small ivory and bone boxes, or in furnishings. Four composite ivory or bone and amber figured objects dating to the first half of the sixth century are of considerable iconographic importance: a pair of plaques from the Picene territory, from Tomb 83, that of an elite woman, at Belmonte Piceno; and a pair from a late Hallstattian Celtic tomb of an elite woman at Asperg, Germany. The two Picene plaques each represent a winged female figure flanked by two smaller female figures. The winged female is represented in the schema of *Potnia Theron* (Mistress of the Animals) or perhaps another (now unknown) divinity of protection and fertility. The carving is on all sides; the faces (now lost) were inlays of amber. Giulia Rocco attributes the reliefs to Picenum, noting the Greco-Orientalizing character of the figures and their relationship to portrayals of Artemis in the Laconian world.[210] The Halstattian furniture plaques with amber-faced sphinxes are generally believed to be Laconian.[211]

The figured ambers of the sixth to fourth centuries B.C. range in size from the tiny (20 mm) to the very large (250 mm) and are formed in a range of subjects, some traditional and some new to the material. They are mainly pendants and fibulae bow decorations. Not only is there a wide distribution of finds on the peninsula, but many of the individual pieces are of exceptional size. This is the third great flourishing of archaeologically evidenced amber importation in the Mediterranean-rim area before the time of the Flavian emperors.

Most large sixth-to-fourth-century figured works demonstrate a new respect for the original shape of the raw material in its naturally occurring forms—rods, drops, or sheets—and figural subjects accommodate the ancient resin's form. Italian figured ambers of the eighth and seventh centuries continued ancient traditions, but new kinds of amuletic figuration developed during the sixth century B.C. in response to changing local and contemporary magical, medicinal, and sacral needs. The multifarious seventh-century fertility and hunting divinities began to be replaced by Olympian subjects and hitherto unknown faunal and fabulous subjects. Rams, lions, and boars (figure 43) take the place of frogs, monkeys, dogs,[212] and sphinxes. Sirens now proliferate, and dancers appear. Pendants in the form of detached heads, of either specific female divinities or other figures, are among the few traditional subjects that retain their important place right through to the beginning of the fourth century B.C. Yet despite the change in iconography, the *categories* of appropriate subjects do not appreciably change: they are still the protective and regenerative subjects of tradition, the subjects that could enhance or focus the powerful properties of amber.

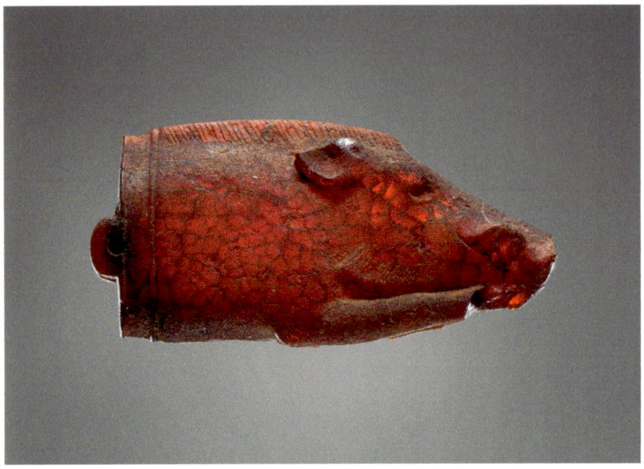

Figure 43 *Foreparts of a Recumbent Boar* pendant, Etruscan, 525–480 B.C. Amber, L: 5 cm (1 9/10 in.), D: 1.3 cm (1/2 in.), H: 2.4 cm (9/10 in.). Los Angeles, J. Paul Getty Museum, 76.AO.84. Gift of Gordon McLendon. See cat. no. 37.

A number of exceptional (unprovenanced) ambers can be dated to the sixth century based on their style and iconography. Among them are the Getty *Divinity Holding Hares* group (figure 44), the Getty *Ship with Figures* pendant (see figure 6), a two-figure group in London, *Satyr and Maenad* (see figure 17, which is perhaps instead a dancing male and female),[213] and a group of four pendant figures, possibly from Ascoli Piceno, now in Philadelphia: two crouching nude males and an addorsed pair of draped female figures.[214] A recumbent lion, found in a circa 560–550 B.C. tomb at Taranto, is a rare example of a piece from a Greek colonial city.[215] These are "contemporary" works for their time, but they also evince artistic connections to older central Italic and Etruscan art, to the eastern Mediterranean, and to East Greek and Peloponnesian art. This wide range of influences might suggest simple explanations: itinerant carvers with a rich artistic vocabulary or a workshop in the ambient of a great crossroads. While both may be accurate, this line of thought underemphasizes the magical aspects of the imagery. Alongside such evocative and wide-ranging explanations should be considered the fact that the figured ambers were made as amulets, or objects following a "prototype" or recipe, or modeled according to tradition and prescription, which required the practitioner to absorb various symbol systems and modes of representation. There must have been persistent types, and a long-lived oral tradition behind them. Because precision of execution is essential to efficacy, "magical practices have little potential for modification, change, and interpretation and thus tend to be slower to change than most other aspects of culture."[216] What Jaś Elsner queries from the starting point of a large-scale offering at Delphi is relevant here:

> *In what sense is an image identical with the deity or activity it represents? The magical and theological properties of images, as well as the way the offering of the Orneatai could actually substitute as a ritual, hint at a much more dynamic interpenetration of image and referent, representation and prototype, than we usually allow for in discussions of mimesis.... Here ... the context of the image asserts the actual* presence *of its prototype.*[217]

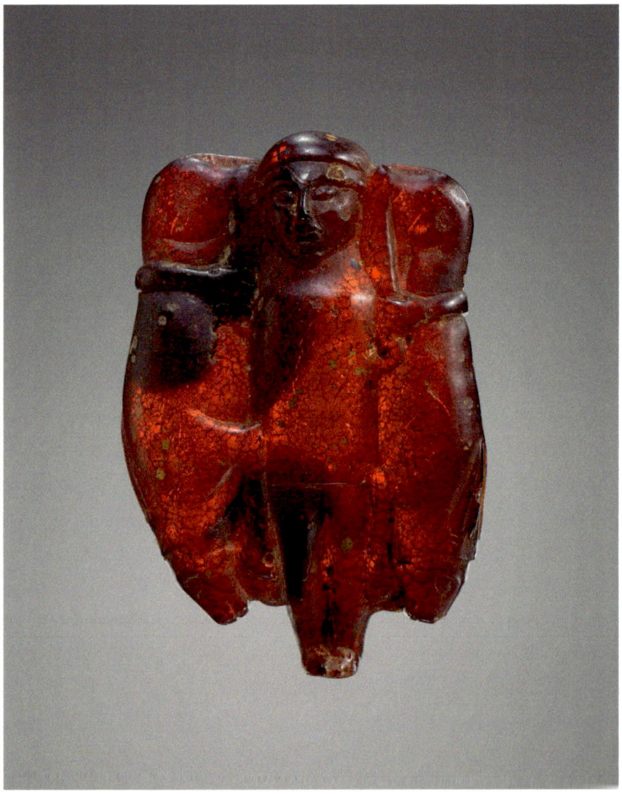

Figure 44 *Divinity Holding Hares* pendant, Etruscan, 600–550 B.C. Amber, H: 9.7 cm (3 4/5 in.), W: 6.4 cm (2 1/2 in.), D: 2.4 cm (9/10 in.). Los Angeles, J. Paul Getty Museum, 77.AO.82. Gift of Gordon McLendon. See cat. no. 4.

In contrast to these "international style" works are a scattering of amber carvings, markedly Ionian in style, that date to the second half of the sixth century B.C. Where they were carved is not known for sure, but some have old provenances: Falconara, in the mid-Adriatic, for the amber in New York; Sala Consilina, for the flying-figure ambers in the Petit Palais, Paris; Armento, for the London kouros. Another tiny kouros in Paris,[218] two of the Getty *Heads of a Female Divinity or Sphinx* (see figures 18 and 45), and the Getty *Kore* (figure 46) and her animal companions, the ram and boar pendants (see figures 29, 39, and 43), if from the Italian peninsula, would be additional evidence of the presence of Ionians (or Ionian models).

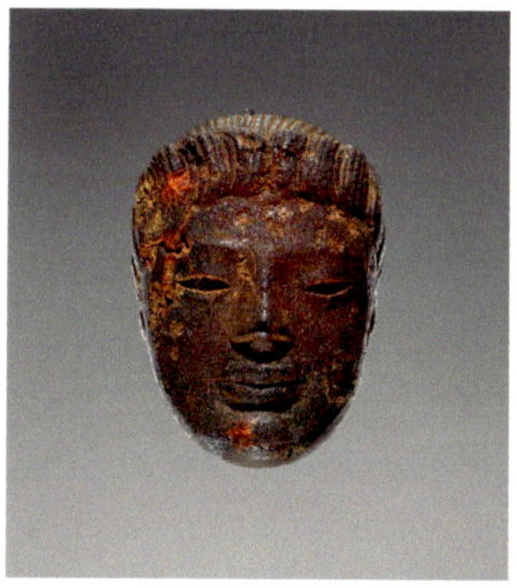

Figure 45 *Head of a Female Divinity or Sphinx* pendant, Etruscan, 550–520 B.C. Amber, H: 3.4 cm (1 3/10 in.), W: 2.4 cm (9/10 in.), D: 1.6 cm (3/5 in.). Los Angeles, J. Paul Getty Museum, 76.AO.79. Gift of Gordon McLendon. See cat. no. 11.

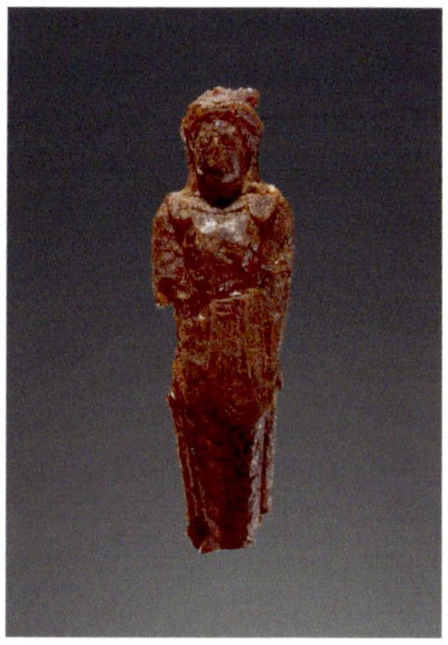

Figure 46 *Standing Female Figure (Kore)* pendant, Etruscan, 525–500 B.C. Amber, H: 6.7 cm (2 5/8 in.), W: 2 cm (7/10 in.), D: 0.9 cm (3/8 in.). Los Angeles, J. Paul Getty Museum, 76.AO.77. Gift of Gordon McLendon. See cat. no. 8.

Three burials, rich in amber objects that date to the end of the sixth century B.C., demonstrate the tradition (extending back to the Bronze Age) of burying strings of ambers in elite females' tombs: at Sala Consilina, with three necklaces totaling a minimum of 114 pieces; in Tomb 102 at Braida di Vaglio, with nearly 300; and at the "princely" tombs at Novi Pazar (Serbia), with over 8,000 individual amber beads, pendants, and related objects. In the Braida di Vaglio tomb, the skeleton is that of a young girl.[219]

Fifth-century finds are more concentrated outside coastal sites in Latium, Etruria, and Magna Graecia. They are dispersed at the fringes of Etruria and the mid-Adriatic area and in Campania and the Basilicata. A very large number of surviving figured ambers, mainly pendants, can be dated by context or style to the period of about the mid-fifth to the early fourth century B.C. They range in subject from the now-traditional rams' heads, female figures, detached heads and faces, and satyrs to whole animals and mythological creatures in repose to more innovative images. The new subjects reflect the plurality of cultural and commercial relations established among Greeks, Etruscans, and other indigenous peoples, and many show the incorporation of new ways of attracting the good, averting the dangerous, or picturing the voyage to the afterworld and its guides. New to amber, but already established by this date in vase and wall painting, bronzework, and gold, all of which have come from graves, are action figures: Dionysian revelers vintaging or dancing,[220] a charioteer, a swaying Danaid, and figures in flight, sirens especially.[221] Athena, with lionskin, shield, and lance, is in movement: the Pyrrhic dance?[222] Aggressive subjects, of rape, imminent or active combat, or triumph over death, emerge: Eos and Kephalos (or Tithonos), Herakles killing the Nemean lion, Ajax, or Achilles lying in wait.[223] Only in a few cases, such as these, can heroes and divinities be surely identified.

The style and iconography of the ambers of this period come out of the artistic traditions of Greece (including Magna Graecia), Etruria, and other Italic areas. Some heads have old-fashioned "divine" hairstyles and large, severe faces, conjuring up Near Eastern divinities. Most wear old-fashioned Etruscan dress, the significance of which deserves more attention. Generally speaking, the Archaic style has a secure hold throughout the fifth century B.C. and well into the fourth. Some works are very like other sculptural works and compare well with the corpora of ivories, bronzes, and terracottas. Others, significantly, are old-fashioned in style: many have the oversized eyes of much earlier art, kept alive in the millennia-old schemata of divine and heroic representations of the Near East; some are remarkably like Hittite sculptures. The huge eyes signify the figure's identity and the characteristic keenness of sight of the supernatural. Wide-open and exaggerated, the eyes of the amber heads project a dazzling gaze, emphasizing the efficacy of their role as *apotropaia*, or devices of protection and danger aversion (figure 47).[224]

both pre-Roman and Roman-period ambers. The medieval treatise states: "Rub it with both hands on a hard sandstone moistened with water until it takes on the shape you want to give it; then on another stone of the same kind, which is finer and smoother, until it becomes completely smooth."[251]

Theophilus then suggests the use of a flat abrasive surface to sand the nodule. Evidence of this is found on the remarkably well-preserved flat inside of the Getty pendant *Head of a Female Divinity or Sphinx* (see figure 18).

The amber pieces must have been further abraded, carved, graved, and polished into the desired subjects, perhaps refined with engraving (and, more rarely, drilling). Sketching was likely done with a sharp scriber of metal, stone, or flint.[252] Pliny refers to "Ostacias" (possibly flint), which is "so hard that other gemstones are engraved with it."[253] Engraving required a rotating instrument, such as a bow drill, the standard tool of a gem engraver.

The narrow-bore suspension perforations, usually transverse, were drilled with particular attention to how the pendant would hang or would attach to a carrier. The narrowness of the borings suggests that the ambers would have been suspended from plant filaments, such as linen, or silk. Many larger pendants have multiple long borings, again usually transverse, signifying that more than a simple filament was needed for the suspension, or that the pendant was part of a complex beaded apron, neck ornament, or girdle or was sewn directly onto clothing. The Getty's large *Ship with Figures* (see figure 6) and the Getty *Kourotrophoi* group (see figures 35 and 53), to name three examples, have multiple perforations and would have required more than one carrier, and a system of knots. The circa 600 B.C. multipiece pendant in Trieste[254] and the circa 500 B.C. composite pendants from Novi Pazar[255] were made possible by complicated stringing/knotting systems. The through-borings are all visible in the transparent amber.

Not only would the stringing have secured the pendants, but both the knots and the action of tying the knots were critical to amuletic usage. In magical practice, tying a knot implies hindering negative actions. Demons and their corresponding diseases were believed to be caught by knots, bands, threads, strings, and amulets. Knots thus could actively play a protective or benevolent role. The pendant-amulets would have been tied on, attached, or suspended as an essential aspect of their efficacy, as we learn from ancient literary sources.[256] The large frontal holes of some figured works are secondary to the transverse perforations and, as discussed below, could be used to attach works to pins or even to a piece of furniture.

The final stage of the work was probably to polish the surface, likely with oil and a fine abrasive or cloth. To bring out the brilliance of the stone, Theophilus instructs: "Finally, put the tile rubbings, moistened with spittle, on a goatskin free of dirt and grease, which is stretched on a wooden frame and secured below with nails, and carefully rub the crystal on this until it sparkles all over." For amber, such polishing and rubbing would bring up the luster and the fragrance, releasing the amber's ambrosial perfume; if that were not enough, the piece could have been rubbed with perfumed oils. We might imagine how this would have added to amber's attraction and mystery, especially if it were in a divine image. The delicious odor might have "[matched] the emanation of fragrance that forms so regular a part of divine epiphanies."[257] As a divine characteristic, fragrance was itself imbued with the power of everlasting life.

All of the surviving pre-Roman figured ambers (at the Getty Museum and elsewhere) reveal an understanding of the morphological and structural characteristics of the ancient resin. Compositions tend to be compact, without projecting parts; the potential points of weakness are minimized in the designs. In human figures, legs and feet are close together, arms and hands are attached to bodies, and necks are short; animals may have their legs tucked beneath themselves, heads reverted, and tails curled around their haunches. The best-preserved works retain signs of surface burnishing, which once enhanced the optical qualities of the amber: its transparency, brilliance, luster, and color.

The earliest figured amber objects from Greece and Italy, dating to the eighth to seventh centuries, are small, from about 48 mm, and frequently imitate small-scale sculptural objects in other media, including ornaments and amulets. The Orientalizing amber carvings are comparable to works in ivory, bone, wood, faïence, precious metals, gemstones, and bronze. Many appear to be direct translations into amber. Examples are the Egyptian and Egyptianizing scarabs, scaraboids, monkeys, dwarves, and other time-honored amuletic subjects. In these small works, there is no evidence of the amber's natural shape, and little wastage. Some excess may have been used to make tiny beads or inlay, as flux in goldsmithing,[258] or as incense or medicine. A number of pendants in the Getty collection, all dating to about the last third of the sixth century B.C., correspond closely to objects of Ionian Greek (or Ionian-influenced) art. Among the finest examples are the two *Heads of a Female Divinity or Sphinx* (see figures 18 and 45), the *Kore* (see figure 46), and some of the rams' and lions' heads.

A different approach to the material emerged at the beginning of the sixth century B.C. The natural form of the amber nodule is preserved, even enhanced, by the design. Some objects, such as the Getty *Hippocamp* (see figure 52), suggest that the lumpy nodule of amber may even have dictated the subject. The subjects of the multifigure pendants are distorted as they wrap around the exceptionally large amber pieces. In order to comprehend the entire subject, the pendant needs to be physically turned in every direction. In any one view, the figures are deformed, but as the pendant is turned, the shapes shift. The Boston *Dancing Youth* (figure 55) and the British Museum *Satyr and Maenad* (see figure 17) are excellent illustrations of this approach. The compositions are illogical, the scale of the figures is skewed, parts are missing, the heads and bodies are twisted or wrapped around the amber in an anatomically impossible manner. Is this because of the sanctity of the whole piece of amber? Are the figures deformed as part of the magic, the shapes shifting as the object is turned over? Might this shape shifting—a common demonic talent—be part of the attraction?[259]

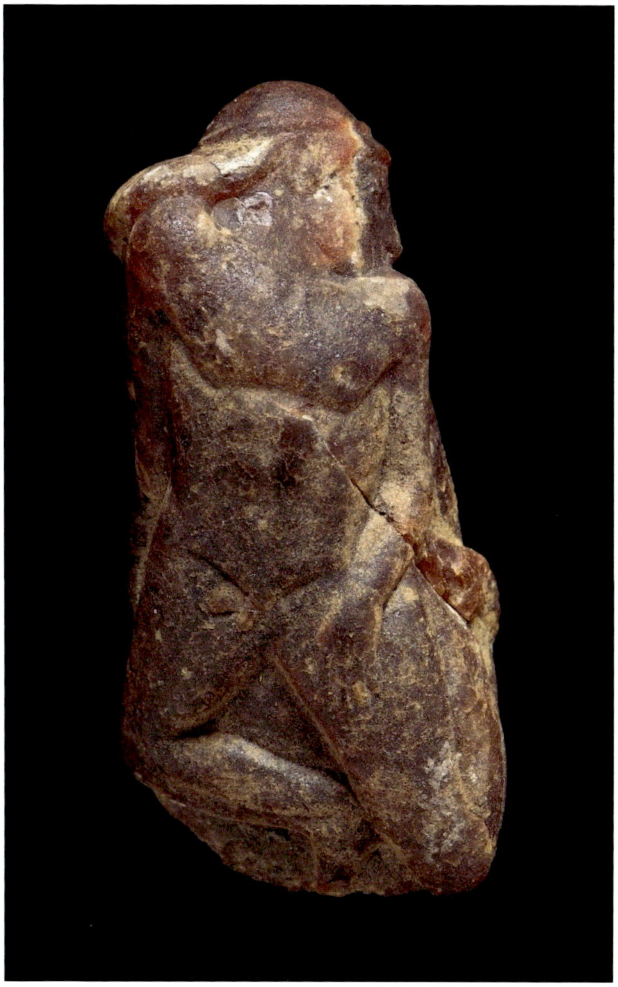

Figure 55 *Dancing Figure*. Etruscan or Italic, early 5th century B.C. Amber, 7.2 x 3.7 cm (2 13/16 x 1 7/16 in.). Museum of Fine Arts, Boston. Gift of Miss C. Wissmann, 02.254. Photograph © 2011 Museum of Fine Arts, Boston.

A variant of this approach is seen in a number of animal subjects, best exemplified by the Getty *Lion* (see figure 54). In this work, there must have been no appreciable wastage. The natural form of the amber blank is obvious and the subject embellishes, rather than conceals, the idiosyncrasies of the raw material. In such cases, the outline, depth, and undulations of the surface are incorporated into the design, with the result that animals and anthropomorphic figures are compacted, splayed, or contorted.[260] A few anthropomorphic pendants are worked in the round, but many have flattish, plain backs. Since the amber was transparent, the carving would have been visible from any angle, an extraordinary sight especially if the piece was figured on all sides. The reverse of the *Lion* allows it to be seen from below, a view only chthonic beings might have. These are extraordinary sculptural objects; in the ancient world, perhaps only in-the-round rock crystal carvings are comparable.[261]

und Mühlacker (Kr. Vaihingen)," *Hallstattforschungen in Nordwürttemberg* (Stuttgart, 1970), p. 21, fig. 9, pls. 10–11, 61–62, 68–69. Amber was also inset into gold and silver, as the Latin tombs (especially Tomb 102) from Castel di Decima show: see, for example, M. R. Di Mino and M. Bertinetti, eds., *Archeologia a Roma: La materia e la tecnica nell'arte antica* (Rome, 1990).

Inset amber eyes are found on the (possibly seventh-century Ionian) ivory lion staff heads (chance finds) from Vasilkov near Smêla: Boardman 1980, p. 259, fig. 301; and E. H. Minns, *Scythians and Greeks: A Survey of the Ancient History and Archaeology on the North Coast of the Euxine from the Danube to the Caucasus* (Cambridge, 1913; repr., New York, 1971), pp. 78, 193, fig. 85. A number of Greek headpieces for horses (*prometopidia*) from southern Italy have eyes of ivory inset with amber irises; compare Getty 83.AC.7.1. Votive eyes of ivory and amber were excavated at the Syracusan Athenaion: see Strong 1966, pp. 22–23. For a possibly Etruscan seventh-century ivory bed inlaid with amber, see G. Caputo, "Quinto Fiorentino: Avori applicativi incastonati d'ambra," StEtr 56 (1989–90): 49ff.; and A. Mastrocinque, "Avori intarsiati in ambra da Quinto Fiorentino," BdA 10 (1991): 3–11. For an East Greek or Lydian *kline* from a later sixth-century B.C. grave in the Athenian Kerameikos cemetery, see U. Knigge, *Der Kerameikos von Athen: Führung durch Ausgrabungen und Geschichte* (Athens, 1988), p. 101. Rocco 1999 compares the Hallstattian examples from Asperg, Hundersingen, and Römerhügel to the Orientalizing bone and ivory objects from Italy; D. Marzoli, in Bartoloni et al. 2000, pp. 397–98, no. 587, compares them to the furnishings from Etruscan tombs. See also A. Naso, "Egeo, Piceno, ed Europa central in period arcaico," in *L'Adriatico, i Greci e L'Europa: Actes du colloque (Venice-Adria 2000),* ed. L. Braccesi, L. Malnati, and F. Raviola (Padua, 2001), pp. 87–110. In the Byzantine *Suda,* under *elektron,* it is noted: "ancient beds used to have their feet set with dark precious stones and amber." See "Elektra," trans. A. Ippolito, March 16, 2006, *Suda on Line*, http://www.stoa.org/sol/ (accessed November 27, 2009). Such elaborate objects correspond well to the literary descriptions of earlier Near Eastern furniture, marvelous works worthy of the gods' attention: see, for example, Winter 2000, p. 29, who cites a text of Ashurnasirpal I (1049–1030 B.C.) in which an ornate bed of precious wood, gold, and precious stones, made for the inner chamber of the temple of the goddess Ishtar, is described as "shining like the rays of the sun (god)."

247. Massaro 1943, pp. 36ff., no. 27/a, records that the bored concentric eyes of female pendants from the Circolo dei Monili preserved traces of silver inlay (reference from Waarsenburg 1995, p. 429, n. 1123).

248. See A. Hermary, "Un petit kouros en bois de Marseille (fouilles de la Bourse)," RA 1997: 227–41, n. 14, figs. 5a–d (inv. H 34), who dates the Marsailles kouros "third-quarter to end of the seventh century." K. A. Neugebauer, *Antiken im deutschen Privatbesitz* (Berlin, 1938), no. 255, dates the pair of ivory kouroi in a German private collection to circa 500 B.C. Two late-seventh-century bone kouros pendants were excavated at the sanctuary of Artemis Orthia at Sparta: see Marangou 1969, pp. 163–64, nos. 109–10, figs. 138a–c.

Much Egyptian wooden and ivory (or bone inlaid) furniture, the Orientalizing wooden throne from Verruchio (*Verucchio* 1994), and the furniture from Gordion are exempla of the technical and stylistic similarities between ivory- and woodworking. See, for example, O. Krzyszkowska and R. Morkot, "Ivory and Related Materials," pp. 320–31, and R. Gale, P. Gasson, N. Hepper, and G. Killen, "Wood," pp. 334–71, in Nicholson and Shaw 2000; G. Herrmann, "Ivory Carving of First Millennium Workshops: Traditions and Diffusion," in *Images as Media: Sources for the Cultural History of the Near East and the Eastern Mediterranean (1st Millennium BCE)*, ed. C. Uehlinger (Fribourg, 2000), pp. 267–82; E. Simpson and K. Spirydowicz, *Gordion ahşap eserler / Gordion Wooden Furniture* (Ankara, 1999); G. Herrmann, ed., *The Furniture of Western Asia: Ancient and Modern* (Mainz, 1996); R. A. Stucky, "Achämenidische Hölzer und Elfenbeine aus Ägypten und Vorderasien im Louvre," AntK 28 (1985): 7–32; and O. W. Muscarella, *The Catalogue of the Ivories from Hasanlu, Iran* (Philadelphia, 1980), who writes, "That the same artisans who carved the ivories also worked with wood and bone is attested at Hasanlu [which date prior to 800 B.C.] ... and this situation ... fits into a general pattern known from other Near Eastern sites." Rocco 1999 frequently refers to the relevant hard materials in understanding the Picene bone and ivory material. As noted in n. 246 above, both Rocco 1999 and Russo 2005 draw significant connections among amber, bone, and ivory carvings.

249. Delphi Museum 10413–14, circa 550 B.C. See Lapatin 2001, no. 33, for illustrations and bibl. (note especially the photographs of the heads during restoration). Attention to detail (*akribeia*) was much praised by ancient critics, records Lapatin 2001, p. 135, with reference to R. Meiggs, *Trees and Timber in the Ancient Mediterranean World* (Oxford, 1982), pp. 302–5.

250. Forming holes from both ends toward the center prevents "blowout"—a technique already in evidence in the earliest bead- and pendant-making. Modern craftspeople recommend placing amber underwater when making perforations to avoid shattering the material or cracking the holes.

251. Theophilus, Book 95, *The Various Arts,* trans. C. R. Dodwell (London, 1961), pp. 168–69. G. Kornbluth, *Engraved Gems of the Carolingian Empire* (University Park, PA, 1995), pp. 9–10, provides the useful model of using Theophilus.

252. The sketching might have been done in a manner similar to that which Theophilus, Book 98 (see n. 251, above), p. 166, recommends for carving a prepared piece of bone. Chalk is spread as the ground for drawing figures with lead. Theophilus advises scoring "the outlines with a sharp tracer so that they are quite clear."

253. Pliny, *Natural History* 37.15, 37.65.

254. Trieste, Civico Museo di Storia ed Arte 9795. Pendant-pectoral from Santa Lucia di Tolmino / Most na Soči, Tomb 3070, end of the seventh or beginning of the sixth century B.C.: *Ambre* 2007, p. 120, fig. III.8.

255. For the most recent discussion of this composite jewelry, see Palavestra and Krstić 2006, pp. 94–115.

256. Kotansky 1991, pp. 107–8. Kotansky, p. 124, n. 6, recommends that "the verb περιάπτειν should be regularly translated cognately, viz. 'to wear/attach/suspend a περίαπτον,' or the equivalent."

257. Steiner 2001, 101. On the ambrosial fragrance of the gods, see also Lapatin 2001, p. 55; Richardson 1974 (in n. 82, above), p. 252; and Shelmerdine 1995 (in n. 72, above).

258. T. Follett, "Amber in Goldworking," *Archaeology* 38, no. 2 (1985): 64–65; but compare G. Nestler and E. Formigli, *Granulazione Etrusca: Un antica tecnica orafa* (Siena, 1994).

259. Johnston 1995, p. 363. This may push the concept of shape shifting, but such a concept is relevant for the magical aspects of some amber pendants. The appearance of shape shifting could be conceived as an attestation of the artisan's skill in making what were perhaps to be considered *daidala.*

260. Because of this, each amber object is unique. Figures contorted, splayed, or wrapped around planes are seen in ancient Near Eastern animal representations as early as the fourth millennium, and some Mycenaean ivories and Middle Assyrian alabaster vessels suggest that such figure manipulation was well established much earlier. Of the art made in or imported into pre-Roman Italy, contorted and splayed figures are found in ivory work, scaraboids, plastic vases, some bronzes (especially utilitarian items such as feet or handles), and gold objects of adornment. An early example is the ivory lion group from the Barberini Tomb (Museo Nazionale Etrusco di Villa Giulia, Rome), thought by Brown 1960, p. 5, to be Syrian work. The resonance in Etruria for wrapping figures (divine, heroic, and demonic subjects most especially) around planes may reflect several generations of contact with art from the Orient. Eastern Greece seems to have been a direct source not only for the large-scale stone carving of the Cortona altar (see n. 262, below), but also for later, small-scale bronzework, such as the Vulcian naked youth riding the winged lion of an incense burner's foot (circa 450 B.C., from Olympia: Olympia Museum B 1001) and the (possibly Orvietan) bronze tripod feet with representations of Peleus wrestling Thetis and Perseus decapitating Medusa (circa 470–460 B.C., provenance unknown: Florence, Museo Archeologico Nazionale 710–11). The dating and localization of the bronzes follow Haynes 1985, nos. 118–19.

261. Perhaps only Chinese amber carvers and Japanese *inro*- and *netsuke*-makers have exploited the material and figural form to the same degree. D. G. Mitten (review of Strong 1966, AJA 71, no. 3 [July 1967]: 323) was the first to point out the visual relationship: "Many of these strange lump-sculptures look almost as if they had been intended as hand-pieces, a sort of *netsuke* of the late archaic Italic world."

262. Tumulus II of Melone del Sodo at Cortona: P. Zamarchi Grassi, "Il tumulo II del Sodi di Cortona (Arezzo)," in Bartoloni et al. 2000, pp. 141–42, no. 109.

263. On binding in magic, see Gager 1992; Faraone 1992; and Faraone 1991.

264. The insertion of materials into an amulet or "talismanic statue" is not uncommon in ritual and magical practice. The amber bullae from Satricum have a large vertical piercing unrelated to the suspension perforation, which Waarsenburg 1995, pp. 409–10, takes to be meant for the insertion of a charm. He relates the amber specimens to the original idea of the bulla as a locket. (On the bulla, see n. 152.) There are also vertical borings in the bottle-shaped pendants and the seated monkey of the necklace from Praeneste in London: see Strong 1966, p. 53, no. 23, pl. IX. Were the inclusions in amber conceived as naturally inserted material? Might there have been a preference for specific inclusions, such as a lizard? In Egypt, "the lizard was symbolic of regeneration because of its ability to regrow limbs and tail if they were injured or lost" (Andrews 1994, p. 66).

265. Strong 1966 and others think the plugs might have been made for coloristic effects. It is more likely that they were originally the same color but have suffered from increased oxidation and thus have more rapidly darkened. The original attempt may have been to make the piece appear uniform, as large "tears" of amber.

266. Additional pendants with large secondary holes include a large Eos group and the large frontal head with wings in a New York private collection (Grimaldi 1996, pp. 150–51; and Negroni Catacchio 1999, pp. 289–90); a draped, dancing figure from Oliveto Citra, Aia Sofia district, Tomb 1 (Paestum, Museo Nazionale OC/00082: Mastrocinque 1991, pp. 129, 133, fig. 84; and P. C. Sestieri, "Ambra intagliata da Oliveto Citra," ArchCl 4 [1952]: 16, pl. 14); a winged female figure (perhaps Lasa) (Shefton Museum of Greek Art and Archaeology 286: B. Shefton, *Archeological Reports* [1969–70]: 58–59, figs. 11–12); two other, very different sirens in the Shefton Museum (nos. 298, 596: unpublished); a pair of satyr heads from Palestrina in Boston (Museum of Fine Arts 02.252–53: Mastrocinque 1991, pp. 131–32, figs. 73–74); a head from Tomb 9, Rutigliano-Purgatorio Necropolis, which has a lateral through-bore in the top of the head and is still attached to a silver pin (Taranto, Museo Archeologico Nazionale 138144: *Ornarsi d'ambra: Tombe principesche da Rutigliano,* ed. L. Masiello and A. Damato [Rutigliano, 2004]; Mastrocinque 1991, p. 131, n. 408; and G. Lo Porto in *Locri Epizefirii: Atti del XVI Convegno di studi sulla Magna Grecia* [Naples, 1977], pl. CXV); a winged female head from Tomb 10, Rutigliano-Purgatorio Necropolis, also still attached to its bronze pin (*Ornasi d'ambra* 2004; and Negroni Catacchio 1993, pl. XIII). A satyr head in Milan has a large frontal hole: N. Negroni Catacchio, "Un pendaglio in ambra in forma di

protome maschile," *Notizie dal chiostro del Monastero maggiore: Rassegna di studi del Civico museo archaeologico e del Civico gabinetto numismatico di Milano* 15–18 (1975): 37, 39, pl. XXV. A large, unpublished head of Herakles in a lionskin helmet (art market, Geneva) has a large central hole through the forehead. Two ambers on the London art market, allegedly found together in the Basilicata, a female head and a horse's head, were originally pendants that saw considerable use (there are pulling troughs on the upper edges of the suspension perforations). The two were later bored and attached to a wooden(?) support with silver nails, fragments of which still remain.

The Production of Ancient Figured Amber Objects

As a result of unauthorized archaeological activity since at least the nineteenth century, a great number, perhaps the majority, of sixth-to-fourth-century B.C. figured ambers are undocumented or lack sure provenance. This places greater importance on works with solid documentation for a discussion of culture and meaning. It is often the case that findspot is equated with place of origin, and grave goods are associated with ownership by the deceased or assumed to be direct evidence of daily dress and customs. The existence of high-value objects such as amber and gold in elite graves must be considered in light of their role as ingredients in a larger network of cultural relationships. Amber and gold, incense and precious textiles were internationally recognized prestigious and valuable objects, suitable for exchange, gift giving, and status display. Not all objects were new. They may have been tokens of guest friendship, or heirlooms or funerary gifts from family or clan members or people with some other relationship. Such "antiques" may have been valued for their history, provenance, or established efficacy (sacral, magical, or medicinal). Celebrations of alliances, marriages, and other rituals were likely occasions for the gathering, exchange, special commissioning, and social display of such objects. Some ambers may have been highly prized prestige objects—treasures gained from purchase, plunder, or presentation—and were meant to be circulated within an aristocratic network. Emporia, palaces, or possibly sacred sites might support established as well as itinerant artisans. And the gifting of things, old and new, could not have been a rare occurrence in the pre-Roman period when amber reigned. Travel and travelers (for reasons of commerce, politics, religion, or celebration) meant gatherings of people at sanctuaries and "princely" centers, where high-status objects might be purchased or commissioned, and where jewelry or magic or medicine may have been procured. The "'cultural clearing houses,' the intermediate centers where goods and ideas were received, adapted, mixed—and passed on,"[267] places such as Pithecoussai and Rhodes in earlier centuries, or a city such as Vulci in the sixth century B.C., are important examples to consider. The extent to which the existence of such centers resulted in a web of autonomous secondary routes—along with a whole range of other cultural outcomes[268]—demands our attention, especially with a mythic material such as *elektron*. An indigenous palatial center such as Braida di Serra di Vaglio (Potenza, Basilicata) is an Italic example of a place where the "circulation" of both objects and people, and interchange among foreigners and colonial Greeks and Etruscans and the indigenous population, might be found. Traders and makers of amber objects might include residents as well as itinerants.

It is important to say a word about style: the efficacy of pre-Roman ambers may have been determined in part by the resin's assured provenance (from the north), by its form (it should follow established guidelines or a prescription), and by its appropriate style(s). The very duration of time-honored forms and style—the long life of Egyptian subjects and forms in amber, or the importance of Ionian- and Etruscan-looking ambers deep into the fourth century B.C.—underlines the conservative functions of figured ambers. It was seemingly important that works look as if they were made by, or followed the prescriptions of, Egyptians, Ionians, or Etruscans. This visual resemblance, perhaps a stamp of authenticity, may have assured their potency or "branded" the objects' magic. In this way, the style, "a way of doing things,"[269] is a culturally significant variable. In the case of amuletic ambers, the style can be said to play a critical role in defining the genuineness and efficacy of the objects. In addition, there appear to be prototypes—not only schemata, but actual models—that were followed for centuries. It is possible that certain works were on view for a long period, through public display in ceremonial circumstances or via circulation. If some works were family or clan heirlooms, they may have been valued for one or more reasons, economic, sacral, medical, or magical. To find individual style in a copy of a copy is a challenge indeed.

In a search for the artistic origins of some figured ambers, scholars have tended to look for individual hands, schools, and centers of production. Connoisseurship and archaeological sleuthing have identified master artisans. Much progress also has been made in siting some groups of objects, drawing them around schools or the hands of particular artisans, and there are undeniable stylistic connections between groups of carved ambers.[270] However, there are many reasons to consider paradigms that move beyond individuals, workshops, and centers of manufacture. As touched on above, many students of figured ambers see an undeniable Etruscan connection in the subjects and "art" of these objects. Some emphasize Magna Graecian, Campanian, Lucanian, or other Italic elements. This author has long advocated for the Ionian, and even more specifically the Milesian, aspects of many amber pendants.[271] Other scholars, notably Nuccia Negroni Catacchio, have charted well-stated arguments for several regional centers.[272] Canosa is a good candidate for the fifth century B.C., as Angelo Bottini has argued.[273] Armento is another.[274]

But why (and where in) these centers? Was there a religious site or sacred sanctuary there? A market? A venerable studio? A school of pharmacology? Raw materials and finished products were easily portable, and not only was the use of amber amulets pervasive, but the iconography of some types—the form of a detached head (figure 57), to cite the most numerous—was consistent over time. There is also evidence that carvings of different dates and styles could be buried together, as in the grave of the young girl of Tomb 102 at Braida di Vaglio.[275]

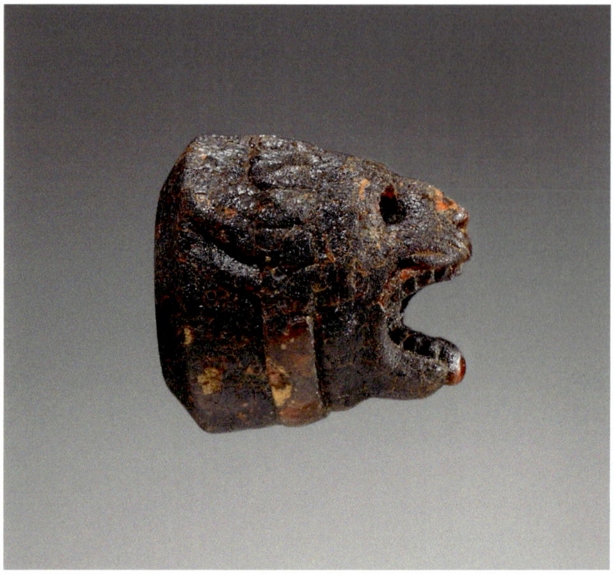

Figure 57 *Lion's Head* spout or finial, Etruscan, 525–480 B.C. Amber, H: 1.9 cm (3/4 in.), W: 1.7 cm (2/3 in.), D: 2 cm (7/10 in.). Los Angeles, J. Paul Getty Museum, 76.AO.81. Gift of Gordon McLendon. See cat. no. 34.

The great potency of amber made it the province of healing specialists, too. Although it is possible that itinerant craftsmen produced the amber carvings of pre-Roman Italy, and that they did so in court settings, as has been proposed,[276] these hypothetical models emphasize the craft and deemphasize the special function of figured amber objects in medicine, magic, and mourning.

The terms *craftsman* and *artisan* imply *métier*, instruction, apprenticeship or training, and the production of *art*. It must be kept in mind that amber is relatively soft and easy to work and was not, of necessity, the exclusive province of skilled artisans. While the Getty pendant *Head of a Female Divinity or Sphinx* (see figure 45) may be equal to the finest of contemporary temple dedications or cult imagery, many figured ambers are art only by modern definition. The material was the force behind its usage, and therefore the workers of amber might well have encompassed pharmacists and religious functionaries, including priests or priestesses, magicians, healers, seers, midwives, and sorceresses.[277] Was an amber object an heirloom, a gift, an exchange object?[278] Or was it produced and/or purchased at a time of crisis? What was most important about these objects was *how well they worked*: as social indicators, as prestige objects, as gifts, as items in transition rituals, as ornamentation, *materia medica*, and amulets. Knowledge of the incantations necessary to accompany them and their specific magical role as amulets was essential. Any analysis of how ambers functioned for the living and the dead needs first to consider who would have possessed such information.

In what activity was an amber involved? This question is especially important when it comes to the most long-lived and geographically widespread amulet types, of which a substantial number (early as well as late) are schematic in manufacture. The sixth-century B.C. female heads from Eretum, for example, are small and schematic, their features formed primarily by abrasion.[279] Such is also the case with a number of crude heads in the Getty collection. Since both the material and the subjects of pre-Roman amber amulets suggest an association with healing, the protection of women, infants, and children, and the aversion of danger, some may have been acquired at the sanctuaries of healing divinities, where old traditions were kept alive or powerful images were on view in special settings or ceremonies. Some pieces may have been spoils, gifts, or dedications.

There is much to be learned about the making of power objects, jewelry, and amulets from Egypt and Mesopotamia, where the literary sources and the archaeological evidence are especially rich, and from the later Greek tradition of inscribed amulets, among the earliest of which were found in the south of Italy. With noninscribed amulets, the situation is more complicated and more open to misinterpretation. Nevertheless, information can be mined from earlier, concurrent, and later traditions. Especially valuable are ancient amulets with writing, which appear frequently in Roman times, as well as ancient handbooks with instructions on the preparation of rites and amulets. These reveal a great deal about the workings of amulets: the stated purpose, the ingredients, the time and place for performance, accompanying gestures, and the incantations themselves. For specific objects, however, we may never know the answer to the question "Was the preparation, inscription, or donning of the amulet conceived or enacted as a ritual act or in a purely perfunctory manner?"[280]

The differing possibilities for who made the amber pendant heads, and in what kind of context, are not necessarily mutually exclusive. A female head pendant excavated at Lavello may be a local product, for it has formal connections with earlier Etruscan art, with the art of (Laconian) Taranto, and with local Italic production, as Maria-Cecilia D'Ercole has shown.[281] Was it carved by a local artisan who offered up key elements of the image in her/his own style? What was the model? How old was it, and where was it seen? Or was it made by an itinerant who had absorbed a large visual vocabulary, sculptural repertoire, or pharmacopoeia—whatever the correct lexicon may be? And according to which traditions, and which kind of "instructions"? Another example might be a group of pendants in the form of frontal female heads in the British Museum, possibly found together at Armento, which some scholars believe are Campanian, or made under Campanian influence, as is Donald Strong's opinion.[282] In each of the two cases, the heads may have been produced at a sanctuary of the divinity represented in the amber, by a local carver as a commissioned good, by an itinerant, for the open market, or even as filled "prescriptions." Relevant here are the critical questions Jean Turfa asks about offerings and exchange in Greek votive tradition: "The large number of terracottas manufactured from the same molds or workshops at sites like Kirrha, the staging port for Delphi, suggests seasonal production or supply from factory to sanctuary, and thus the sanctuary as the 'retail' supplier of votives."[283] These heads, like all amber amulets, were valuable in every sense, and their value may have depended in part on where or by whom they were made. And they were just the sort of thing to have accrued further value by being displayed, worn, or buried at a place distant from their manufacture. A carved amber or group of ambers may have been carried in the pouch of an itinerant artisan, trader, or healer. Before it played a role in a sanctuary or in the rituals of death, the amber may have been traded or gifted elsewhere, to be copied or remembered. Carved ambers may have had many lives and been involved in many activities. Made from a material as old as the earth, formed into deeply significant subjects only to be interred once again, these gems of the ages offer new windows onto the past.

NOTES

267. Ridgway 2000 (in n. 192, above), p. 236.

268. Ibid., with reference to A. Peserico, "L'interazione culturale greco-fenicia," in *Alle soglie della classicità: Il Mediterraneo tra tradizione e innovazione; Studi in onore di Sabatino Moscati*, ed. E. Acquaro, vol. 2 (Pisa and Rome, 1996), pp. 899–924.

269. These ideas were articulated with the help of M. Hegmon, "Technology, Style, and Social Practices: Archaeological Approaches," in *The Archaeology of Social Boundaries*, ed. M. T. Stark (Washington, DC, and London, 1998), pp. 264–79. "A way of doing something" is found on p. 265.

270. See Strong 1966, p. 31. He argues convincingly that if the analogies he put forward are valid, "it leads to the conclusion that the bulk of the better pieces were made under the strong influences of Campanian art of the sixth century." Strong thinks that Lucania was the center of such manufacture but does not rule out centers in Apulia. Others who have published strong arguments about other sites of manufacture are Russo 2005 (in n. 246, above); Bottini and Setari 2003 (with earlier bibl.); Palavestra and Krstić 2006; Palavestra 2003; D'Ercole

1995, pp. 284–85; Mastrocinque 1991, passim; Bottini 1987, pp. 11–12; and La Genière 1961, pp. 87–88.

271. D'Ercole 2008, pp. 52–69, convincingly argues for an Ionian working in Etruria for the Herakles and the Nemean Lion group of circa 530–500 B.C. in Paris (Bibliothèque nationale, Cabinet de Médailles, Fröhner 1146).

272. This has also been done by a number of University of Milan students, noted by Negroni Catacchio 1999.

273. Bottini 1987, p. 12, has suggested several reasons for this but emphasizes the existence of a clientele capable of appreciating and acquiring luxury articles. Might the draw have been a temple, cult, shrine, or healer at Canosa or Armento (see n. 274)?

274. On Armento as a center, see, most recently, A. Bottini, "Le ambre nella Basilicata settentrionale," in *Ambre* 2007, pp. 232–33.

275. Bottini and Setari 2003; A. Bottini (pp. 541–48) and E. Setari (p. 644) in Pugliese Carratelli 1996; Bottini and Setari 1992; Bottini and Setari 1995; Bottini and Setari 1998; and E. Pica in *Treasures* 1998, pp. 224–25, pls. 32–33. See also E. Greco, *Archeologia della Magna Grecia* (Rome, 1992).

276. For the amber from Tomb 102, E. Setari summarizes in Pugliese Carratelli 1996, p. 643: "Native craftsmanship can in no way be excluded, but they were probably part of a palace-based activity, the work of traveling craftsmen with various cultural origins." E. Pica in *Treasures* 1998, p. 224, hypothesizes that the amber objects "came from the shops of itinerant indigenous artisans who reworked both colonial Greek and Etruscan-Campanian models." This idea is elaborated in Bottini and Setari 2003. Bottini 1987, pp. 11–12, proposes a modulated picture: the possibility of a fixed center of production at a major center and the existence of itinerants using acquired models (particularly aristocratic Greek ones) while introducing innovations. The types of drinking vessels in the Braida di Vaglio necropolis indicate the acculturation of Greek rituals of wine consumption alongside native traditions. For a recent note on this tomb, with the wine service as possible evidence of the Dionysian aspect of the burial, see Causey 2007. On the Greek customs of wine drinking and the adoption of the symposium in Italy, see A. Rathje, "The Adoption of the Homeric Banquet in Central Italy in the Orientalizing Period," in *Sympotica: A Symposium on the Symposium,* ed. O. Murray (Oxford, 1990). The earliest representation from Italy of feasting while reclining is the Etruscan symposiast on the lid of a two-handled calyx vessel from Tomb 23 from the necropolis at Tolle, dating to circa 630–620 B.C. See G. Paolucci, ed., *City Archaeological Museum of Thermal Waters: Chianciano Terme* (Siena, 1997), fig. 90; and Haynes 2000, p. 108.

277. "A seer, or a healer of illnesses, or a carpenter who works on wood, or even an inspired singer," named by Eumaios (*Odyssey* 17.381–87), are four kinds of high-ranking strangers, any one of which (theoretically) could have been involved in aspects of amulet construction. For discussion of the passage and the translation see Nagy 1997. See also Burkert 1992, pp. 41–87.

278. Bottini 1987 discusses the figured ambers of two "princely" tombs at Melfi-Pisciolo as being older than their (second half of the fifth century B.C.) contexts.

279. The Eretum pendants are from Tomb XIII: see P. Santoro, "Sequenza culturale della necropoli di Colle del Forno in Sabina," StEtr 51 (1985): 13–37; and Losi et al. 1993, p. 203. Santoro published Tomb XIII as a child's grave (P. Santoro, "La necropolis di Colle del Forno," in *Civiltà arcaica dei Sabini nella valle del Tevere* [Rome, 1973], pp. 39–44), but this is not certain per Losi et al. 1993, p. 209, n. 1.

280. D. Frankfurter, "Narrating Power: The Theory and Practice of the Magical *Historiola* in Ritual Spells," in Meyer and Mirecki 1995, p. 3.

281. D'Ercole 1995.

282. Strong 1966, pp. 67–71, no. 44–3.

283. J. M. Turfa, "Votive Offerings," in De Grummond and Simon 2006, p. 108, n. 37. She cites J.-M. Luce, "Les terres cuites de Kirrha," in *Delphes: Centenaire de la "grande fouille" réalisée par l'École française d'Athènes (1892–1903),* ed. J.-F. Bommelaer (Leiden, 1992), pp. 263–75; and J. Uhlenbrock, "Terracotta Figurines from the Demeter Sanctuary at Cyrene: Models for Trade," in *Cyrenaica in Antiquity*, BAR International Series 236, ed. G. Barker et al. (Oxford, 1985), pp. 297–304.

Catalogue

Orientalizing Group

The first six objects presented here, *Female Holding a Child (Kourotrophos)* (77.AO.84, cat. no. 1), *Female Holding a Child (Kourotrophos) with Bird* (77.AO.85, cat. no. 2), *Addorsed Females* (77.AO.81.1, cat. no. 3), *Divinity Holding Hares* (77.AO.82, cat. no. 4), *Lion with Bird* (77.AO.81.2, cat. no. 5), and *Paired Lions* (77.AO.81.3, cat. no. 6), are similar in style, technique, state of conservation, and size. Subject also relates them. Because of this, and because the six were part of the same donation, it is posited that they come from the same original context.

As is argued below, the six were produced in northern internal Etruria in the first half (or perhaps in the third quarter) of the sixth century B.C. and have stylistic connections to Greek Arcadian and Ionian small bronzes, as well as to contemporary Etruscan votive bronzes, relief work, and *bucchero*. All can be shown to have specific ties to subjects and styles current in the Near East and Cyprus. The North Syrian and "Phoenician" aspects are salient.

These objects would have belonged to an elite person. The size alone of the largest three pendants would have signaled their exceptional value even before craftsmanship transformed the lumps of amber into traditional subjects of great power and status. As ornaments and amulets, the ambers could not but have made a spectacular impression, if only because of the optical characteristics of the rare material and its associations. The imagery enhanced the amber's value. The age-old vocabulary that gave form to these glistening jewels made them good luck–inviting, danger–averting, protective objects. Although there are no close parallels in amber or other media for the individual works or the group, they belong to the vocabulary, iconography, and styles of Etruscan Orientalizing art. The subjects are women, children, and wild fauna—lions, hares, and migrating waterfowl. In format, the six include three heraldic compositions, a squared-up group of a lion with its prey, and two pairs of an adult and child in a side-by-side pose.

All six ambers are better understood when looked at in the context of contemporary and slightly earlier production from Greece, especially from the Peloponnesus and South Ionia, as well as ivories, bronzes, gold, faïence, and shell carvings from the Near East and Cyprus, including Cypro-Phoenician objects, and Orientalizing carved ambers and ivories from Picenum and Latium. This is a range similar to the visual vocabularies of other Orientalizing amber and ivory carvings, as carefully analyzed by A. M. Bisi and G. Rocco (for Picene ivory and bone carvings), P. G. Warden (for four "Picene" ambers in the Museum of Archaeology and Anthropology, University of Philadelphia), and D. J. Waarsenburg (for ambers from Satricum in the Villa Giulia).[1] Similarly rich stylistic and iconographic links are characteristic of some Orientalizing bronze reliefs, Praenestine ivory work, and Felsine stelai, and many of the small finds from Samos, a number of unique carvings in wood and in ivory in particular.[2] Because of the Etruscan aspects of the Italian-provenanced works, it seems most likely that they all were produced on the peninsula for locally based commissioners and purchasers.

The *realia* of the amber figures' thoroughly Etruscan dress is matched by the waterfowl depictions. The bird of 77.AO.85 is a white-fronted goose, and that of the *Lion with Bird* pendant (77.AO.81.2) a mute swan. These species have long histories in the ancient world and its art, and both long-necked waterfowl accrued a rich lore and symbolism. The species are highly distinctive migrants to Italy and elsewhere in Europe, and both are excellent table fare. The hare, too, is good eating.

Each of the six might also have been read as incorporations, or as symbolic, of a female nature divinity. This may be the principal divinity of popular Etruscan religion, who was worshipped in a variety of forms and under different names.[3]

Although the six may have been used during the life of one or more powerful persons, some of their iconography seems to be funerary: the right-hand-on-breast gesture made by the figures of 77.AO.84, 77.AO.85, and 77.AO.81.1, the common mantle of 77.AO.85, and the hares of 77.AO.82 may have held special funerary meaning. Further study may support the identification of the large female figures as divinities with chthonic as well as afterworld aspects; the same may be true of the waterfowl and hares. Individually and as a group, the six remarkable ambers invite questions about their commissioning, making, owning, and burial. At this point, it is feasible to interpret these as the property of one or more political-ceremonial specialists, and to posit that these amulet-ornaments may have served as insignia.[4]

There are only a few documented pre-Roman burials with significant numbers of figured ambers. All such intact, published graves also included numerous other high-status objects, providing not only evidence of the elaborate rituals that accompanied the deceased, but also a glimpse of ideas about the tomb and the afterworld. These comparison burials, which are rich in figured amber, also contain nonfigured amber beads and pendants, plus many other high-value objects, of bronze, precious metals, ivory, or ceramics, among their durable goods. Textiles and other now-perished organic goods that must have accompanied the deceased have left few traces. These six pendants, then, may be evidence of what must have been an extraordinary burial.

NOTES

1. A. M. Bisi, "Due avori piceni di tradizione vicino-orientale," *Studi urbinati di storia di filosofia e letteratura Urbino*, ser. B, 3, 55 (1981–82): 79–83; and A. M. Bisi, "Componenti siro-fenicie negli avori piceni," in *La civiltà nelle Marche: Studi in onore di Giovanni Annibaldi* (Ripatransone, 1992), pp. 128–39, shaped my argument originally. See also Warden 1994; Waarsenburg 1995; and Rocco 1999.

2. Warden 1994 outlines the issue succinctly. Strøm 1971 was instrumental to my study of this group of ambers.

3. H. Nagy, "Divinities in the Context of Sacrifice and Cult on Caeretan Votive Terracottas," in De Puma and Small 1994, p. 221. She refers to A. Pfiffig, *Religio Etrusca* (Graz, 1975), p. 98. Waarsenburg 1996 and Waarsenburg 1995 discuss in detail the representation of the Great Mother in the Orientalizing period, with special attention to Astarte. There is a visible absorption of a female nature divinity's aspects by several female and male divinities, but the concept of a Great Goddess is fraught, as Moorey 2004 points out.

4. For this idea, compare the placing of certain ritual pre-Columbian gold objects in the graves of political-ceremonial specialists by other specialists. For example, in the Costa Rican Guanacaste and Central regions, finds from funerary contexts show that both new and previously used objects were deposited. Some are local, but there is also evidence of interchanges from around the region. The production of sumptuary and ritualistic objects in diverse materials suggests the existence of such specialists, who required the use of insignia associated with those possessions that, at some point, were deposited in graves. The archaeological evidence suggests that during the period of A.D. 300–800, as these societies became more hierarchical, with greater social stratification, there was greater consolidation of experts in political-ritualistic activities, and the number and quality of grave offerings increased and changed. There is also evidence, represented by images in clay, of women carrying out various roles. See, for example, S. K. Lothrop, *Archaeology of the Diquis Delta, Costa Rica*, Papers of the Peabody Museum of Archaeology and Ethnology at Harvard University 51 (Cambridge, MA, 1963).

1. Pendant: Female Holding a Child (*Kourotrophos*)

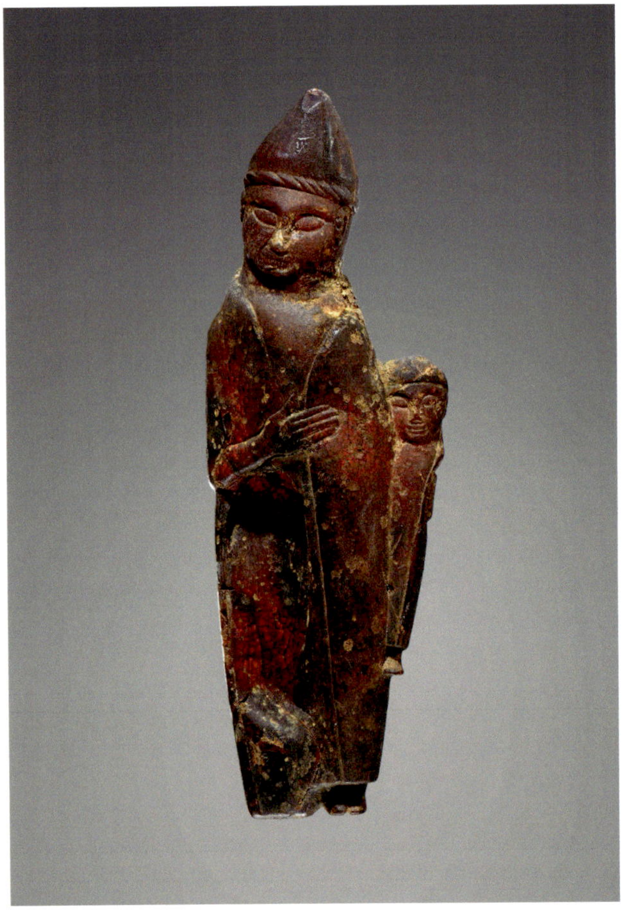

Accession Number	77.AO.84
Culture	Etruscan
Date	600–550 B.C.
Dimensions	Height: 130 mm; width: 45 mm; depth: 18 mm; Diameter of suspension holes: 2.5 mm; Weight: 55.2 g
Subjects	Amulets; Artemis; Birds; Etruscan culture; Funerary use of amber (also Burial); Ionia, Greece (also Ionian, Greek); Kourotrophos

Provenance

–1977, Gordon McLendon (Dallas, TX), donated to the J. Paul Getty Museum, 1977.

Condition

The pendant is intact and in good condition. There is a long, curved fissure in the lower right section of the larger figure's heavy cloak, extending to the base. There are numerous minute chips on the child's head and on the adult's nose, chin, and left side of the neck, and along the cloak's left shoulder. There is an old chip on the tip of the hat. A pattern of minute cracking extends over the surface of the entire piece. There are inclusions at the hem on the right side, at the right elbow, at the top of the child's head, and scattered throughout the adult's body. The pendant's patina varies from yellow-ocher to brown. In ambient light, the amber is reddish brown, and in transmitted light, translucent and ruby-red.

Description

The two figures form a compact composition. The physiognomy, pose, gestures, dress, and relative scale of the figures suggest that the larger figure is a woman and the smaller figure is a child. The woman wears a long, heavy cloak and a conical hat and is shod in close-fitting boots. The raised area at the collarbone suggests the presence of a close-fitting undergarment. Although there is no sign of the undergarment's hems or selvage edges, it is probably a long, close-fitting, unbelted chiton. Bunched cloth at the top of the cloak forms a kind of collar at the back of the neck and around the shoulders. Engraved vertical lines extend from the lower edge of the sleeve slits to the hem. On the left side, the cloak hem falls to the ankles, just above the small feet, and on the right, to ground level. The two front edges of the cloak join below the chest, at the woman's solar plexus. Her open right hand is placed at this junction. Her somewhat bulbous conical hat stands high off her head. The hat's rim is rounded and protruding; it is engraved with short diagonal striations, creating a design resembling cable molding. On the proper left side of the hat, a graved line, interpreted here as a seam, runs from the apex to the rim.

The woman's left forearm emerges from the cloak as it encircles the upper body of the child; her left hand lies flat on the child's upper arm. The upright, frontally and rigidly posed child tilts back toward the body of the woman. From the back, it appears that the child is under the mantle of the adult. The child wears a miniature version of the adult's garments, but the hood/collar section of the mantle is pulled over the head. The mantle fits snugly around the brow, curves behind the right ear, and drapes forward over the shoulder and chest before extending to the ankles. The tiny shod feet are set side by side and jut straight outward. Below them is a spur of amber. The child, too, appears to wear a long chiton.

The woman's head is large, full, and round, and her neck thick, short, and cylindrical. The child's neck is characteristically short, and the head is a smaller, more delicate, and slightly more schematic version of the adult's. No hair is visible on either figure. Both the woman's and the child's eyes are blank, almond-shaped bosses, turned up slightly at the outer canthi (the right eye of each is slanted more sharply upward at the outer canthus than are the left eyes), and are surmounted by an eyebrow ridge that moves smoothly from the temples over the orbits. The eyes of both figures, set between a bulging brow and cheeks, are almost as big in profile as in full face, but the child's eyes are less richly modeled than the woman's, and the line separating the eye from the brow is longer. Both figures' brows are low and flat. The noses are short, smooth, and triangular, and the bar-shaped mouths are formed as horizontal protrusions. The chins are wide. Both figures have round, flat ears with an opening at the position of the tragus. The woman's ears are placed close to the hat rim. The child's right ear, a flatter version of the woman's, is far back on the head.

The tilt of the woman's head, the illogical location of her feet, the scale of the figures, the child's placement, and the concave depression on the lower part of the woman all suggest that the sculptural configuration is dependent on the amber blank's original shape.

In addition to the engraved grooves and lines, some traces of abrasion are visible on the woman's neck and in the depression of the lower right section of the heavy cloak. Two perforations form a V-shaped suspension system, each extending from a hole drilled at the shoulder and meeting just below the position of the woman's right wrist. The amber may have been strung with one carrier forming a loop. Alternatively, if there were two filaments, each could have been knotted at the point below the woman's right wrist. In either case, the figures would have stood upright when suspended.

Discussion

77.AO.84 and 77.AO.85 (cat. no. 2) belong to the category of divinities known as child-carriers, or *kourotrophoi*, and are composed in the side-by-side pose exclusive to heroes and divinities, a schema of great antiquity.[1]

For the style and the forms, the principal amber comparisons for 77.AO.84 are a pendant in London of two standing figures[2] and a group of four Etruscan amber pendants in Philadelphia, perhaps excavated at Ascoli Piceno. One of the latter pendants, MS 2536, a fragmentary standing woman, is the single best parallel for 77.AO.84 in style and physical type.[3]

The physiognomies of the women and children of 77.AO.84, 77.AO.85, and the relevant Philadelphia pendant are characteristic of Orientalizing Etruscan sculpture. They all have long, almond-shaped eyes, named the Blind Eye by Emeline Richardson.[4] In common with the Etruscan votive bronzes that Richardson groups together as Orientalizing Early Etruscan Ladies, the Getty Museum six and their amber comparanda exhibit the same solid, rounded Assyrian forms; in Richardson's words, "the convex surfaces of the 'Assyrian' tradition as well as its convention of a fully draped figure and some of its massive quality."[5] The ambers, like the Ladies, were likely carved in northern Etruria.[6] An excellent comparison for 77.AO.84 in the Ladies group is the small votive bronze of a woman in London (British Museum 1907.3–11.1), which Richardson points out is the earliest shoe-wearer and is closer to the "Assyrian" aesthetic than any other.[7] She dates this figure to the end of the seventh century B.C.

While Volterra may have been the center of such production, specific details of dress and style among these ambers also draw them close to Chiusi. One example is a bronze in Leiden (Rijksmuseum van Oudheden H3 ZZZZ 1), said to have come from Montalcino (in the ancient Chiusine territory).[8] A large bronze from Brolio, which should be considered Chiusine, is another excellent parallel.[9] The bronzes and ambers reveal several similarities, including a geometric structuring of the figures, their proportions—especially the smallness of the hands and feet in relation to the rest of the body—and the form of the faces, fingers, and thumbs.

Each of the constituent parts of the dress worn by the woman of 77.AO.84—long chiton, cloak, boots, and hat—is Etruscan. Larissa Bonfante refers to the mantle similar to those worn by 77.AO.84 and 77.AO.85 as a kind of cape "raincoat."[10] Richardson describes it as a distinctly Etruscan garment, names it the Heavy Cloak, and underlines its dependence on *masculine* Near Eastern models.[11]

The carver of 77.AO.84 indicated the important sartorial details of the Heavy Cloak, which must be based on an understanding of the actual garment. The collar/hood section of the cloak is formed by bunched fabric. The child's cloak is drawn over its head, so no collar is formed. The line that extends from just below the armhole to the hem may represent a seam joining the garment's front and back sections, but more likely it indicates the meeting or overlap of the selvages.

Under the cloak, the large 77.AO.84 figure wears a long chiton. Both male and female figures wear the "the Oriental, clinging, unbelted tunic," as Sybille Haynes describes it.[12] At the Archaic Building Complex at Poggio Civitate, tunic-wearers animate the terracotta frieze plaques,[13] and two—one passenger in the two-wheeled cart (perhaps a woman) and the woman seated on the curved throne in the assembly of seated figures—also wear an enveloping cloak.

The tall hat of 77.AO.84 is distinct in its slightly bulging conical shape, in the manner in which it is worn (toward the front of the head and concealing all hair, front and back), and in its construction. Despite its simple form, the carver articulates specific details of its structure: the egglike bugle, the seam line on the front, the tiny dimple at the apex, and the rounded, upturned brim with diagonal striations. The form of the brim and the seam line indicate that the hat is of leather, skin (fur side inward), or leather lined with fur, and not of felt; the horizontal line likely indicates an upturn, and the diagonal lines, the whipstitching.[14]

It is rare to find representations of Etruscans wearing pointed hats before about the mid-sixth century, but afterward, various kinds of conical hats—originally a male fashion in the Near East—appear everywhere, in innumerable variations of type, size, and even number worn at the same time.[15] Conical hats were a female fashion in Etruria. (They are related to but not identical with the pointed hat worn only by haruspices.) The hat of 77.AO.84 stands out as unusual within the repertoire of Etruscan pointed headgear and dates to the moment just before the fashion took off in Etruria. The combination of its bulbous shape, construction, and placement on the head sets it apart from later-fashionable types. The contemporary Oriental masculine parallels include the hats of "Asian" figures represented in Egyptian relief sculpture, as well as the hats worn by some Cypriot priests and some divinities on Phoenician engraved gems. Antecedents are found in Hittite and other North Syrian sculpture. The bulbous shape is like that of the traditional crown of Upper Egypt. Oriental precedents for *female* figures wearing cone-shaped hats are few but significant, and among them are an image of a Hittite goddess in a pointed hat on a silver rhyton[16] and Near Eastern *lamassu*, the beneficent protective female deities first seen in the Neo-Sumerian period. Contemporary and just slightly earlier Greek parallels for females wearing this distinctive hat type are to be found in Magna Graecia and Laconia. Many of the sixth-century B.C. terracottas of Artemis excavated at the Metapontine spring sanctuary of San Biagio wear nearly identical hats. As Gesche Olbrich has argued, the San Biagio type is closely related to the Tarantine Artemis and Artemis Bendis types of terracottas,[17] and the Artemis of the San Biagio sanctuary is closely related to the Laconian Artemis Orthia, who herself has important connections to the Near East and the Minoan-Mycenaean worlds.

Carved amber figures with pointed hats are few, and in each case, they differ from the type and personage of 77.AO.84. The female figure of the New York "Morgan Amber" (see introduction, figure 24), the bow of a fibula,[18] wears a hat with one seam near the midfront and a large, flat upturn. It is set back on the head, and the front of the hair shows. Two of the wingless flying figures from Sala Consilina in the Petit Palais, Paris, are hatted. The bee-divinity (perhaps Ideo with Zeus) sports a pointed, beanielike hat with six seams and a dog-toothed turnup; the vessel-carrier wears a swirl-wrapped hat with a simple upturn.[19] Other amber figures in pointed hats include numerous female pendants (Etruscan and Italic, late sixth to early fourth centuries), but these usually display small, close-fitting, pointed caps, which are sometimes worn under veils and over garments.

Other important Etruscan parallels for the hat of 77.AO.84 are a unique pair of hats depicted in the Tomb of the Funeral Couch, where they are placed on an ornate bed,[20] and the hats worn by certain bronze figures. Six Middle Archaic bronze draped female figures, four winged and two wingless, were part of a wheeled vehicle (a *carpentum*, or mule-drawn cart) found at Castel San Mariano near Corciano (Perugia).[21] Another example is a Late Archaic bronze, a rare type of votive kore, from Volterra.[22] The figure (she must be a divinity) on the apex of a large bronze kyathos handle from Bisenzio, of the late sixth century B.C., holds a small raptor.[23] She and the *Potnia Theron* of the bronze appliqué are the only ones to hold birds.

The hats of the four Middle Archaic winged bronzes (representing Turan, an unnamed divinity, or possibly a protective genius) and of the Volterran Late Archaic figure are more elaborate than that of 77.AO.84. Their head coverings are either wrappings of long ribbons of cloth or wrapped hats.[24] The hats worn by the Middle Archaic bronze korai are the best parallels for the pitch of the hat, the form of the crown, the thick, rolled rim, and the lack of visible hair on 77.AO.84. In contrast, the cart attachments wear their hair long and unbound, and the Volterran figure wears hers tucked up inside the hat (it is just visible beneath the rim in front and back).

Although there are numerous illustrations in Etruscan art of women wearing tunics, heavy cloaks, or boots, there does not seem to be any other figure wearing this combination. The nearest relative is the uniquely dressed Etruscan votive bronze of a woman from Brolio in Florence (Museo Archeologico Nazionale 561), dating to the late sixth to early fifth centuries, already mentioned above for its Chiusine style.[25] Instead of a tunic, Florence 561 wears an old-fashioned long Ionian chiton, along with boots, a pointed hat, and a cloak (which is pulled up over the hat). Florence 561 must represent a divinity.

Not only is the dress of 77.AO.84 unusual, so is its subject. There are only three other *kourotrophoi* of amber known to this author, and all are Etruscan. These include 77.AO.85, a *kourotrophos* in a London private collection that must date to the mid-seventh century,[26] and a tiny amber *kourotrophos* in the Metropolitan Museum of Art that is strongly Ionian in style, very like a number of Chiusine bronzes, and of fifth-century date.[27] Each of the four amber *kourotrophoi* holds the child on the left side; otherwise, they differ from one another in details of pose, dress, and style.

Kourotrophoi have an ancient history in and around the Mediterranean.[28] Without other specific information, such as an inscription or burial context, the images of women, possibly nurses, holding children cannot be associated with any one divinity or function. 77.AO.84 may represent “any of the multifarious lists of mythical mothers and nurses who were so popular, and often venerated, in early Italy.”[29] Although its hat brings the amber figure close to the Metapontine terracottas of Artemis with the pointed hat, there are no examples of this hatted variety as a *kourotrophos*.[30] The child, who can be interpreted as standing upright, may specifically allude to the Greek Artemis. Artemis could cause deformities in children, particularly of the foot or leg; conversely, she could be called upon to protect the child from such deformities. That the child is held up and represented as well formed could be read as emphasizing the divinity’s protective role.[31]

Alternatively, it is possible that the adult figure of 77.AO.84 represents a divinity similar to the Latin Mater Matuta. The solar aspects of amber would make it an attractive material for such an image. *Matronae* prayed to this goddess of light and childbirth and presented to her, not their own child, but their sisters’.[32] The exceptional figured ambers from (the priestess’s) Tomb VI at Satricum, where the Mater Matuta was worshipped, lend weight to this hypothesis.

Whatever the identity of 77.AO.84, her pose and the child are significant. Her right hand is placed on her breast, atop the cloak opening. The gesture of 77.AO.84 is the same one made by the two figures of 77.AO.81.1 (cat. no. 3) and perhaps by the adult figure of 77.AO.85. It is of great antiquity and is found on many contemporary images, including a considerable number of Early Etruscan sculpted works. It has been variously read, resulting in correspondingly varying interpretations of the figures making it—and vice versa. The gesture makers have been identified as divinities, priestesses, votives, supplicants, adorants, adherents, and mourners.[33] Although most scholars agree that the hand-on-breast gesture is Oriental in origin, there is less consensus about its meaning. It has been interpreted as a sign of thanksgiving (signifying gratitude to a deity for a favor conferred), as a sign of adoration or of offering, and as one of mourning. Some have seen it as one variant in a system of female gestures that call attention to the primary and/or secondary sex characteristics.[34] In his discussion of the marble *Dame d’Auxerre* (Musée du Louvre), who also makes this gesture, Jean-Luc Martinez is cautious in assigning a precise signification, in particular a funerary one, to the sculpture.[35]

For Etruscan sculpture in small and large scale, Bonfante regards the gesture as one of mourning.[36] If the figures are indeed mourners, the gesture would support the thesis of a funerary role for the pendant. However, if the figures are ancestors (including heroes) or divinities, identifying the gesture as funerary is a less sustainable conclusion. Almost every known Etruscan figure making the “hand on breast with thumb up” gesture has come from a tomb, and it could be argued that some represent divinities. Notable examples are the seventh-century stone Figure A from Casale Marittimo,[37] a number of early *bucchero* caryatid figures,[38] the early-sixth-century bronze female divinity from the Vulcian Polledrara cemetery “Isis Tomb,”[39] one of the limestone figures from the Pietrera tumulus,[40] the woman atop the much-restored Chiusine “Paolozzi urn,”[41] several of the stone female divinities from Chiusi, and a number of the Chiusine enthroned “canopus” figures of young men.

Bonfante notes the pose of a figure on a gold plaque from Rhodes.[42] A comparable right-hand gesture is made by some East Greek plastic vases in the form of a female bust; it is also made by some of the Artemis Metapontine terracotta figures from the San Biagio sanctuary, representations that Olbrich considers to be of the goddess herself, not votives.[43]

The pose of 77.AO.84—the position of the body, head, arms, and hands—gains from being read as one movement in a sequence "frozen" at its most characteristic point. This is the case with many Egyptian images, as R. H. Wilkinson explains: "Sequential gestures exist where a certain pose or gesture occurs within a sequence of continuous action.... The specific gesture usually illustrated was perhaps chosen because it represented the most important or recognizable part of [a complex] ceremony, but it must be remembered that gesture can only be understood in terms of the meaning of the larger ritual in which it was embedded."[44] If this is the case with 77.AO.84, then it may be that pose is one that incorporates a fertility gesture and is at the same time one of promise: the divinity avows the deceased the gift of rebirth, the activity of the left hand that of carrying the child and that of the right, avowal. That the object was ultimately funerary may have modified or even added to the pose's meaning.

This glittering, large ornament was a potent amulet, the subject of which could place the wearer under the divinity's protection. As such, it joins many of the earliest images of females in world art, which are small in scale and functioned magically, many as protection. Amulets in the form of a standing female figure, often suckling a child, were popular in Egypt as early as the New Kingdom.[45] In the ancient Near East, amulets of seated female figures occur as early as the eighth millennium.[46] A third-millennium fertility goddess from Cyprus wears a miniature of her own image.[47]

The figural groups are all carved fully in the round and well secured by the system of attachment for suspension. Were the pendants intended to swing freely? Were they an attachment to clothing, a larger ornament, or even a structure? Were they used in life or only for the rituals of death, and thereafter in the tomb? Who made the ambers, following which models or according to which recipes?

Who placed them with the deceased? Whoever was buried with the ambers, and whoever saw to it that these ambers accompanied the deceased, must have had the appropriate knowledge and status. If the original owner was the deceased, might the owner have been a ritual political-religious specialist? Were 77.AO.84 and the other ambers part of her ceremonial properties? On the other hand, might these fabulous figured ambers have been grave gifts, even insignia, of another ritual political-religious specialist? Whatever the answers, something similar must also be the case for the amber ensemble from Tomb VI at Satricum.[48]

NOTES

1. Richardson 1983, p. 29. See Waarsenburg 1995, pp. 438–40, nn. 1179–92, for a discussion of the side-by-side pose and twinned figures, with special attention to Orientalizing imagery in amber and in ivory.
2. British Museum 43: Strong 1966, pp. 66–67, no. 43, pl. XIX.
3. The fragmentary cloaked female figure amulet is University of Pennsylvania Museum of Archaeology and Anthropology MS 2536: see Warden 1994, pp. 134–43, no. 3, figs. 13.7–9; Turfa 2005, pp. 226–27, no. 242.
4. Richardson 1983, p. 29.
5. Ibid., pp. 28–29.
6. Ibid., pp. 45–47, with earlier bibl., including J.-C. Balty, "Un centre de production de bronzes figurés de l'Etrurie septentrionale (deuxième moitié du VIIe–première moitié du VIe siècle avant J.-C.): Volterra ou Arezzo?," *Bulletin de l'Institut historique Belge de Rome* 33 (1961): 5–68. Jürgeit 1999, pp. 26–27, provides a concise analysis of the arguments for the date and origin of related types of votive bronzes and their possible connection to Volterra, and dates the Karlsruhe example to about 570 B.C. Haynes 1985, pp. 251–52, considers the London bronze (1907,0311.1) to be "Northern Etruscan" and dates it circa 600–575 B.C.
7. Richardson 1983, pp. 45–47.
8. Ibid., pp. 46–47, figs. 44–45.
9. From a votive deposit at the Montecchio farm at Brolio (Val di Chiana): Florence, Museo Archeologico Nazionale 561. It has been dated to the mid- to the third quarter of the sixth century. See Torelli 2000, p. 622, no. 275; Zamarchi Grassi 1992, p. 205; Colonna 1985, p. 164; Richardson 1983, pp. 55–56; and A. Romualdi, *Il deposito di Brolio in Val di Chiana* (Rome, 1981), pp. 10–11, 26–29, no. 17.
10. Bonfante 2003, p. 46.
11. Richardson 1983, p. 31, notes that the closest parallels to the Etruscan bronzes are figures on reliefs from Nimrud and Kuyunjik, dating to the end of the eighth century B.C. Warden 1994, p. 140, provides other excellent North Syrian parallels from Zincirli, Carchemish, Maraş, and Sakçagözü, and to ivory carving attributed to the region. The exhibition and catalogue Bartoloni et al. 2000 is essential to the understanding of the Orient and Italy. E. Di Filippo Balestazzi, "L'orientalizzante adriatico," in *I*

Greci in Adriatico 2, Hesperià 18: Studi sulla grecità di Occidente, supplement del convegno internazionale, Urbino, 21–24 octobre 1999, ed. L. Braccesi and M. Luni (Rome, 2004), pp. 57–100, adds significantly to the evolving picture of interaction with North Syria. The Hittite parallels suggested by each student of this material point the way for further understanding.

12. Haynes 2000, p. 121.

13. Antiquarium di Poggio Civitate, unknown inventory no. Haynes 2000, pp. 120–25, summarizes the Near Eastern aspects of these friezes and their connection with the terracotta plaques from Metaponto. The terracotta frieze plaques from Serra di Vaglio (Basilicata) are directly related to these.

14. Such a rim not only would have stabilized the hat (and perhaps reduced stretching), but also would have increased its heat retention, a critical feature of cold-weather hats, one possibly fundamental for the origin of the hat (and wearer).

15. See Bonfante 2003, pp. 68, 71, 76–77, nn. 8–13, 48–49, 80–88. The conical hat and its typology, especially for early Italy, still deserve further study. See also M. Pipili, "Wearing an Other Hat: Workmen in Town and Country," in *Not the Classical Ideal: Athens and the Construction of the Other in Greek Art*, ed. B. Cohen (Amsterdam, 2000), pp. 150–79; Smithers 1988, pp. 214–15, with reference to M. Bonghi Jovino, *Capua preromana: Terrecotte votive*, vol. 1 (Florence, 1965), p. 14, and vol. 2 (Florence, 1971), pp. 70–71; and Olbrich 1979. For the related form in helmets, see A. Bottini, *Armi: Gli strumenti della guerra in Lucania* (Bari, 1994); and A. Bottini et al., *Antike Helme* (Mainz, 1988).

    The seventh-century B.C. *bucchero* lady from Falerii Veteres (Richardson 1983, p. 32, n. 41), is one of the oldest examples of a woman wearing the pointed hat, and some of Richardson's Early Etruscan Ladies (pp. 49–51) wear a small pointed hat under the veil. Some Late Archaic bronzes wear small conical hats stacked one atop another.

    The Assyrian conical hat is constructed from a soft material, so that it does not stand up, and its crown is creased, with a sagging tip. This is more than likely a felt hat, as are the hats worn by a number of (possibly hairless) elite male figures engaged in various ceremonial activities (perhaps including augury) on a number of bronze *situlas*, such as one found at Vaće, Slovenia (Narodni muzej Ljubljana P581), and another excavated at Magdalenska gora near Smarje, Slovenia (Narodni muzej Ljubljana P4281). These are soft, pointed hats with rolled rims articulated with diagonal lines.

    A pointed-hat type common to Cypriot Archaic figures is almost identical to the "bonnets" worn by the later-fourth-century B.C. *bambino in fasce* votives from the Capua region. For the Campanian material and its relationships, see Smithers 1988.

    In addition to the possible antecedents gathered by the authors listed above, relevant comparisons for 77.AO.84 include the unusual hat worn by Naramsin on a basalt stele of circa 2220–2184 B.C.: it has a raised edge and is decorated with both horizontal and vertical lines. D. P. Hansen in *First Cities* 2003, p. 204, no. 130, writes: "Although the shape of the cap is perplexing, it clearly is not the horned crown associated with divinities.... A conical cap is worn by certain heroes on Akkadian cylinder seals, and it has been noted that it resembles the military cap of Ebla."

16. This extraordinary example is a seated goddess (with hair showing in front and a twisted braid in back), represented on the silver rhyton terminating in the foreparts of a stag in New York, Metropolitan Museum of Art 1989.281.10 (Empire Period, circa fifteenth to thirteenth centuries, presented by Norbert Schimmel Trust, 1989). The goddess holds a falcon (possibly) in her left hand, a cup in her right. Her tall, seamed hat has a diagonally striated turnup; the horns (or perhaps uraeus) are represented in profile.

17. Olbrich 1979, chap. 4, distinguishes the hat found on many of the Metapontine terracottas of Artemis (always worn over long, flowing hair) from that of the Etruscan *tutulus* and points out its parallels at Samos, Cyprus, Rhodes, Assos (Troas), Sicily, Etruria, Lucania, and Apulia and in Magna Graecia (Taranto and Metaponto-Pisticci). She also charts the relationship between the hat of the San Biagio terracottas and the Phrygian hat of the Artemis Bendis type.

18. New York, Metropolitan Museum of Art 17.190.2067, Gift of J. Pierpont Morgan, 1917.

19. Paris, Petit Palais, Dutuit Collection. The flying figure carrying an amphora is Dut 1600 (5), and the bee-divinity is Dut 1600 (6). See "The Archaic and Afterward" in the introduction, n. 219, for bibl.

20. The painting on the back part of the Tarquinian Tomb of the Funeral Couch, in the view of Steingräber 2006, p. 139, presents the space as a festival tent or baldachin on posts, "dominated by a large empty bed reminiscent of a catafalque, with two light shrouds, two pillows, two wreaths, and two conical caps resembling the pilos caps of the Dioscuri." Represented is either a deceased aristocratic couple or a divine duo. If the latter is the case, Steingräber believes the representation to be a *theoxenia* or a *lectisternium*, and the hats then represent the divinities aniconically. However, compare Haynes 2000, p. 237: she suggests that the two hats may be funerary *cippi*. Is there a connection between the bulbous conical shape of Etruscan *cippi* and the similarly shaped protuberances of seventh-century B.C. Daunian steles? For the latter, see Nava 1988.

21. The Middle Archaic bronzes are distributed between Perugia and Munich. A recent proposal for the placement of the winged figures on the four corners of the box of the *carpentum* in the reconstruction by S. Bruni is convincing (see summary by him in Torelli 2000, pp. 580–85). See also Emiliozzi 1997, pp. 82–86. The bronzes are generally thought to date to around 570 B.C. and have been compared to the repertory of Etrusco-Corinthian pottery. For the wingless kore figures in Perugia, in addition to Bruni and Emiliozzi, see Richardson 1983, pp. 269–70; and Höckmann 1982.

22. For the bronze kore from Volterra, now in Munich (Antikensammlungen 3678), see Richardson 1983, pp. 268–69. Paraphrasing Richardson, the figure wears a properly understood Ionian chiton, a rarity among Etruscan korai, a dress that illustrates a drapery style of some competence, contrasting with her "thoroughly un-Greek" heavy, round head, ugly ears, and broad, smiling face. Hair peeks out from under the hat brim in front and back. The conical hat has a similar turned-up rim with diagonal markings and is wrapped (clockwise) with a long strip of cloth in a pattern distinct from that of the above-mentioned Middle Archaic korai. Richardson singles out the Munich kore as one of the finest of the Mannerist korai, as well as the biggest. Her unparalleled costume, a mixture of Ionian and Etruscan fashion, her pose, and her (perhaps) youthful proportions of large head and smaller body, characterized by a rather planklike modeling, also set the kore apart. Might the Munich bronze be an updated reflection of an early statue, or phenotype, of Artemis?

    The face of the adult figure is very like those of two sculptures in the British Museum, a bronze statue of a woman from the Polledrara cemetery "Isis Tomb" who holds a horned bird (GR 1850.2–27.15), and a gypsum statue of a woman said to be from the same tomb (GR 1850.2–27.1). As is the case for 77.AO.84 and the other five ambers, the gypsum statue shows the influence of prototypes from Greece, particularly Crete and the Peloponnese, as well as from Phoenicia and the Near East. For the "Isis Tomb" sculptures, see Roncalli 1998; and S. Haynes, "The Bronze Bust from the 'Isis Tomb' Reconsidered," StEtr 57 (1991): 39, where she proves that Roncalli's theoretical reconstruction of the bust is untenable.

23. Florence, Museo Archeologico Nazionale 74913. The standing (perhaps) figure on the apex of the kyathos handle wears a chiton, a conically shaped hat under a veil, and boots, and holds what looks like a small raptor on her right hand. G. C. Cianferoni in *World of the Etruscans* 2001, pp. 26, 91, no. 165, dates it to the last decades of the sixth century B.C.

24. More needs to be understood about the "wrapped" conical hat, the headdresses made from cloth bands, and the so-called twisted hat. Bonfante 2003, pp. 142–43, has unraveled much, including the occasions for wearing such headgear and the gender of the wearers.

25. For Florence 561 from Brolio, see n. 9, above.

26. The unpublished amber pendant in a London private collection is similar in physiognomy and style to two amber pendants of women from the Circolo dei Monili, Vetulonia (see, for example, Bissing 1931, pp. 49–52), and very like several of the bronzes in Richardson's Geometric Overlap Series C and Orientalized Geometric Series A, B, and C.

27. New York, Metropolitan Museum of Art 17.230.52, Rogers Fund, 1917: *Art of the Classical World* 2007, pp. 295, 473, no. 340; and Richter 1940, p. 32, figs. 104–5. Although there are no known amber parallels for the style and format of the fifth-century pendant B.C. in New York, it compares well with the Ionizing sixth-century B.C. sculpture of the Chiusine area, as is shown by comparison with a bronze kore in Paris, Bibliothèque nationale, Cabinet des Médailles 204: see Richardson 1983, pp. 265–66, figs. 605–6. That the child of the New York pendant is carried in a sitting position might indicate that the subject is one of presentation or abduction.

28. For the *kourotrophos*, see T. Hadzisteliou Price, *Kourotrophos: Cults and Representations of the Greek Nursing Deities* (Leiden, 1978); V. Tran Tam Tinh, *Isis Lactans* (Leiden, 1973), with a review by L. Bonfante, AJA 80 (1976): 104–15; L. Bonfante, "Dedicated Mothers," in *Visible Religion III: Popular Religion* (Leiden, 1984), p. 13; L. Bonfante, "Daily Life and Afterlife," in Bonfante 1986, p. 240; L. Bonfante, "Votive Terracotta Figures of Mothers and Children," in *Italian Iron Age Artefacts in the British Museum: Papers of the Sixth British Museum Classical Colloquium*, ed. J. Swaddling (London, 1986), pp. 195–201; I. E. M. Edlund, "Man, Nature, and the Gods: A Study of Rural Sanctuaries in Etruria and Magna Graecia from the Seventh to the Fourth Century B.C.," in *Papers in Italian Archaeology IV: The Cambridge Conference, Part IV, Classical and Medieval Archaeology*, BAR International Series 246, ed. C. Malone and S. Stoddart (Oxford, 1985), pp. 21–32; and Smithers 1988, esp. chap. 2.

29. On the ubiquity of the *kourotrophos*, Brendel 1995, p. 240, summarizes: "An unnamed *kourotrophos* occurs quite frequently among the artless statuettes which worshippers deposited as ex-votos, to please the sacred spirits of the place." He lists the *kourotrophoi* of Italy, among them Diana, Mater Matuta, Minerva, Persephone, Turan, and Uni, and in Greece, Artemis, Athena, Demeter, Eileithyia, Eirene, Ge, Hekate, Hera, Hestia, Ino/Leukothea, Leto, and Persephone. (Ino/Leukothea's role as the young Dionysos's nurse probably gave her the character of a protectress of small children.) On Leukothea, see I. Krauskopf, "Leukothea nach den antiken Quellen," in *Akten des Kolloquiums zum Thema "Die Göttin von Pyrgi," Tübingen, 16.–17.1.1979*, Bibliotheca di Studi Etruschi 12 (Florence, 1981), pp. 137–48.

30. A headless terracotta *kourotrophos* from the San Biagio sanctuary, with the image of a standing child scratched into its planklike body, is a *unicum*: Olbrich 1979, no. B14b.

31. This reference comes from Callimachus's *Hymn to Artemis*. At 3.128, Artemis is called out for inflicting her grievous anger when she causes wives "to give birth to children of whom none stands on upright ankle." The Getty *kourotrophos* pendants, thus interpreted, could be amulets of the "frightening-the-demons" type. Here, too, the apotropaic nature of amber reinforces the subject. Although she writes of objects of a later period, Stephanie Leitch's explanation is relevant: "Demons can see and the pagan prescriptions for avoiding evil, most notably, were prescriptions that were activated through sight and seeing.... Among the methods chosen for foiling an evil force was the use of a bright and dazzling object" to distract it from its intended victim. S. Leitch, "Seeing Objects in Private Devotion," in *Pious Journeys: Christian Devotional Art and Practice*

in the Later Middle Ages and Renaissance, ed. L. Seidel (Chicago, 2001), cited by R. Mellinkoff, *Averting Demons: The Protecting Power of Medieval Visual Motifs and Themes*, 2 vols., ed. C. Lanham (Los Angeles, 2004), p. 47.

32. See Waarsenburg 1995, pp. 438–40, 460–61, with key bibl. for the Mater Matuta.

33. Waarsenburg 1995.

34. The interpretation of the arm and hand positions has been used as integral evidence in the naming of figures and their role in the tomb or sanctuary. Waarsenburg 1995, p. 432, n. 1136, believes that "on the old discussion of whether female votive statuary represents goddesses, priestesses, or possibly adorants ... at least for the nude female statuary the goddess interpretation is the most feasible option." I. E. M. Edlund Berry, "Whether Goddess, Priestess or Worshipper: Considerations of Female Deities and Cults in Roman Religion," in *Opus Mixtum: Essays in Ancient Art and Society*, Skrifter Utgivna av Svenska Institutet i Rom, 8, vol. 21 (Stockholm, 1994), pp. 25–34, provides an excellent discussion of the topic, especially in reference to Rome.

    Bonfante 2003, p. 219, pries open the question again in discussion of stone sculptures from Casale Marittimo, noting that Figure A reaches up to the neck in a gesture characteristic of female mourners. Germane Etruscan votive bronzes include the Middle Archaic bronzes Florence 230–31 (Richardson 1983, pp. 261–64, figs. 579–80, 597–98) and the Late Archaic bronzes Ny Carlsberg Glyptotek H224 (ibid., pp. 295–96, fig. 700) and Arezzo 11603 (ibid., p. 282, figs. 654–55).

35. J.-L. Martinez, *La Dame d'Auxerre* (Paris, 2000), pp. 20–22.

36. Most recently, Bonfante 2003, p. 219.

37. Ibid., n. 36.

38. Ibid., p. 219.

39. London, British Museum GR 1850.2-27.15. See n. 22, above.

40. Bonfante 2003, p. 71, n. 456; and Richardson 1983, p. 39, n. 1056: Florence, Museo Archeologico Nazionale 85148854. However, the hands of another of the Pietrera tumulus figures (85148553) are flat on the breast, the right one over the left, in a gesture common in the ancient Near East to show reverence and submission: see J. K. Choksy, "In Reverence for Deities and Submission to Kings: A Few Gestures in Ancient Near East Societies," *Iranica Antiqua* 37 (2002): 7–29. Haynes 2000, p. 83, questions whether the Pietrera tumulus sculptures "are meant to represent mourners or ancestors of the buried aristocrats."

    See also H. Damgaard Andersen, "The Etruscan Ancestral Cult: Its Origin and Development and the Importance of Anthropomorphism," *Analecta Romana Instituti Danici* 21 (1993): 7–65; A. Minetti, "Le necropoli chiusine del periodo orientalizzante," in *Chiusi etrusca*, ed. A. Rastrelli (Chiusi, 2000); and Bartoloni et al. 2000, p. 306, no. 424.

41. Chiusi, Museo Civico 63092, circa 630–600 B.C. See Bartoloni et al. 2000, p. 306, no. 424 (where M. Iozzo summarizes the convincing explanation by Cristofani 1978, pp. 125–27; and Damgaard Andersen 1996) (see n. 40, above), p. 35, n. 26. See also the impasto in Florence: Sprenger and Bartoloni 1981, p. 90, no. 50; and Gempeler 1974, pp. 55ff., no. 44, pls. 12, 15.

42. Bonfante 2003, p. 139.

43. One example of a plastic vase in the form of a female figure who places her open hand on her breast is Berlin 30733: see U. Gehrig, A. Greifenhagen, and N. Kunisch, *Führer durch die Antikenabteilung* (Berlin, 1968), p. 43, pl. 35; and J. Ducat, *Les Vases plastiques rhodiens archaïques en terre cuite* (Paris, 1966), p. 35, no. C26, pl. 5.3. For the Metapontine examples, see Olbrich 1979, chap. 4, pp. 70–98.

44. On independent and sequential gestures, see Wilkinson 1994, p. 205: "Symbolic gestures may utilize the positioning or movement of the body, head, arms or hands, and are usually 'frozen' at their most characteristic point in representations. Functionally, two types of gestures may be differentiated—'independent' and 'sequential.'"

45. Andrews 1994, passim.

46. White 1992, passim.

47. The picrolite cruciform figurine from Yialia (Cyprus Museum 1934/1112/2) confirms that many Neolithic tiny figures were used as pendants and tomb offerings. The Yialia picrolite wears a nearly identical image around her neck. See L. Vagnetti, "Stone Sculpture in Chalcolithic Cyprus," *Chalcolithic Cyprus*, Bulletin of the American Schools of Oriental Research 282/83 May/August 1991: fig. 1. J. Mertens reminded me to look at the Yialia figurine.

48. Waarsenburg 1995; and N. Negroni Catacchio, "L'ambra nella protostoria italiana," in *Ambra, Oro del Nord*, exh. cat. (Venice, 1978), p. 199, although, as Waarsenburg notes, it lacks supporting arguments. On priestesses in early Italy, in addition to the bibl. assembled in Waarsenburg 1995, nn. 1310–19, see M. Beard, J. North, and S. Price, *Religions of Rome* (Cambridge, 1998); and M. Beard, "The Sexual Status of the Vestal Virgins," *Journal of Roman Studies* 70 (1980): 12–27.

2. Pendant: Female Holding a Child (*Kourotrophos*) with Bird

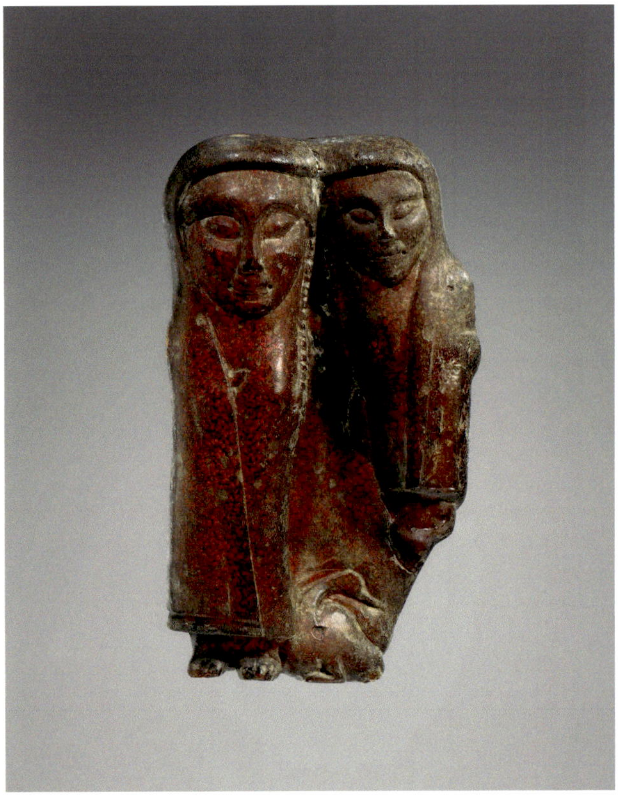

Accession Number	77.AO.85
Culture	Etruscan
Date	600–550 B.C.
Dimensions	Height: 83 mm; width: 50 mm; depth: 50 mm; Diameter of suspension holes: 2 mm; Weight: 48 g
Subjects	Bird; Etruscan culture; Fertility; Funerary use of amber (also Burial); Ionia, Greece (also Ionian, Greek); Kourotrophos; *Potnia Theron*

Provenance

–1977, Gordon McLendon (Dallas, TX), donated to the J. Paul Getty Museum, 1977.

Condition

The pendant is in good condition, with a firm, smooth, stable surface. Before its entry into the Getty Museum, the two broken sections of the pendant were reattached and a synthetic fill was added to the break that runs along the left contour below the feet of the smaller figure. There are additional small chips on the reverse along the break and on the boot toes of the larger figure. There are visible inclusions in the fissure at the center, between the two figures, and in the midsection of each figure. The back surface and much of the front are covered with a dusty, light-yellow-ocher layer of degraded amber. In ambient light, the piece is reddish brown with some translucent areas; in transmitted light, the object is translucent and ruby-red.

Description

The pendant is conceived fully in the round and is composed of two frontal figures in a side-by-side pose, with a long-necked waterbird at the lower right. The human figures are identified as a woman and child because of their proportions, morphological (facial) features, dress, hair, and forms of the upper torso. The woman extends the full length of the amber and fills approximately one half of the composition. The child is carved into the upper section of the other half; beneath its feet is a spur of amber, which might be read as a groundline. Below the child, at the bottom, is the bird. It stands on the same plane as the woman. Since the bird is represented only on the obverse and the triangular depression above its head is inside the garment, it should be read as standing within the shelter of the woman's outer garment.

Despite the difference in scale between the figures and in some of their details, both share the same head-to-body proportion, as well as dress and hairstyle. Their facial characteristics are comparably fashioned (even if they are not identical): the forehead, eyebrow ridge, temple region, and nose are conceived as a single modeled unit composed of a continuous curving form from the top edges of the head to the end of the nose. The slightly bulging, almond-shaped blank eyes are fitted neatly beneath the eyebrow ridge, the outer canthi higher than the inner ones, and the right eyes slanted higher at the outer canthi. Their noses are long and narrow (that of the woman is slightly wider), with delicate nostrils. The mouths are wider than the noses. The lips curve into slight smiles; the lower lips are slightly wider than the upper. The cheeks and lower faces are wide and rounded. The chins are short; in profile, they protrude to the level of the root of the nose. While the two faces are very similar, there are minor differences between them. The child's face is finer in structure, her features smaller, and her chin more pointed. There is a distinct nasolabial line on the woman's face; there is none on the child's.

Both figures wear a similar undergarment. There is no neck detail; the garment is indicated only by the hem and lower section of a long skirt. Both also wear close-fitting veils over their heads. The front of the hair is just visible at the brow. The woman's left frontal hair lock descends from her temples to just below her breasts; the child's (on her right side) ends at the shoulder. The same heavy outer garment covers both figures. The line parallel to the front edge of the mantles may be a turnback or fold of the cloak; alternatively, it may represent the seam closing the lower edge of the sleeve. With her left arm and hand, the woman encircles the child; she places her right hand on her own chest. The tear-shaped form emerging from the border of the cloak may be the top of her thumb, although it is very large. Alternatively, it might represent the tip of a lotus blossom. The child's arms are not visible. The woman is barefooted: four toes and the instep are delineated. There is no elaboration of the child's feet. The plump bird's body and legs are in a resting pose (the feet are visible), the neck is stretched back, with the head reverted, and the right wing is raised.

As is the case for all of the other amber objects in this group, the original form of the amber blank appears to have played a key role in the composition. The nodule's shape is suggested by the placement, size, and stance of the figures, and by the depressed area between the adult and child.

There are small drill holes in the corners of each figure's mouth. Abrasion marks are visible underneath the chins, along the left body contour, around the head and neck of the bird, and between the feet of the adult. There are engraved lines around the eyes, separating the lips, along the front edges of the mantles, and in the hair plaits. The single perforation has two holes, one exit between the two heads at the position of the ears, and the second exit in the indentation between the two heads. The pendant probably was suspended from a strand or strands knotted at the base of the lower hole so that the piece hung as if the figures were standing.

Discussion

There is no exact parallel for this pendant. In style, 77.AO.85 is very like others in the Getty group; it shares comparisons with them and is equally complex in its relationship to contemporary and earlier art, including Greek (Cretan, Ionian, and Peloponnesian), Cypriot, and Near Eastern objects. The subject, like that of 77.AO.84 (cat. no. 1), is a *kourotrophic* divinity. The iconography is underlined by the compositional format and the chosen material, amber. The dress and the hairstyles of the two figures identify them as female. The smaller figure likely represents a child rather than an infant, since it "stands on upright ankle."[1] The active pose of the goose contrasts with the stillness of the human figures. Because the adult and child are frontal, standing, and stationary (and the adult wingless), it is unlikely that an abduction scene is represented.[2]

The bird of 77.AO.85 is schematic but telling. The carver indicated some salient features that suggest that a particular species is represented: the round head, long, undercurved bill, and distinctive form of the tail feathers aid in identifying the fowl as a white-fronted goose.[3]

There are many similarities between this pendant and the *Female Holding a Child (*Kourotrophos*)* (77.AO.84) presented above. There are also differences in the pose, hairstyle, and manner of covering the heads. Only 77.AO.85 includes a bird. The woman of 77.AO.84 is hatted, no hair is showing, and the hood section of her mantle is down. No hair shows on the child, and it is wrapped in a mantle. Both figures are shod. In contrast, both figures of 77.AO.85 have their heads covered by a common mantle, show hair at the brow, and wear temple locks. The woman of 77.AO.85 is barefooted. No long chiton or other undergarment is delineated for the figures of 77.AO.84, yet in this pendant, the bottom hem of the undergarment is shown. If the protrusion emerging in the area of the chest from the front closure of the cloak of 77.AO.85 is a thumb, and the hand is thus flattened on the breast, the gesture is similar to that of 77.AO.84 and to those of both women of the *Addorsed Females* pendant (77.AO.81.1, cat. no. 3). As discussed above in the entry for 77.AO.84, this hand gesture has been variously interpreted. It is likely one with complex meanings, but it certainly had a fertility aspect and perhaps a funerary one. Avowal or promise also may be inherent. On the other hand, the droplike shape could represent the tip of an unopened lotus blossom, a subject of great antiquity in ancient art and a symbol of youth, fertility, and rebirth.

The lotus blossom may have been thought especially apt for an unfurled young life, as it was in Egypt.[4]

The type of undergarment worn by each figure of 77.AO.85 cannot be determined, since only the hem and lower edge of a skirt are indicated. There is no articulation at the neck. This might suggest that the carver neglected these aspects of dress or that the figures of 77.AO.85 wear only skirts beneath their mantles. If this is the case, one possible parallel is the bronze divinity from the Vulcian "Isis Tomb," whose only garment may be a skirt.[5]

Amber comparisons for 77.AO.85 include the other five in the Getty Orientalizing group and two others: an unprovenanced pendant in the form of two standing figures in the British Museum,[6] and a pendant in the form of a female figure, possibly from Ascoli Piceno, in Philadelphia (University of Pennsylvania, Museum of Archaeology and Anthropology).[7] An ivory carving of a seated female figure, part of a furnishing from Pianello di Castelbellino, has the distinctive short neck, short-set body, and facial profile of the ambers.[8]

For the figures of 77.AO.85, the best comparisons among Etruscan small bronzes are found in Emeline Richardson's Early Etruscan Ladies, Series B, Group 1,[9] the same group that helps to situate 77.AO.84 and 77.AO.81.1. Bronzes in Florence (Museo Archeologico Nazionale 225) and London (British Museum 1907.3–11.1) are particularly relevant for their body proportions, facial features, and overall combination of dress elements. Florence, Museo Archeologico Nazionale 27, and Leiden, Rijksmuseum van Oudheden H3 ZZZZ 1, provide the best parallels for the cloak of 77.AO.85. The "Etruscan-ness" of 77.AO.85 and the related ambers and bronzes is brought out further when they are compared to large-scale Etruscan figures. They share with the gypsum figure in London (from the Polledrara cemetery at Vulci), the pair of limestone figures from Casale Marittimo, and some early Chiusine limestone figures a solidity and retention of the permanent materials in the sculptures.[10]

Potnia Theron figures stamped on a number of *bucchero* kyathos handles—two excavated at Poggio Civitate and others likely from Chiusi[11]—are important comparisons, not only for composition and style, but also for specific features such as the birds and the figures' temple locks. As Larissa Bonfante has outlined, the Greek-influenced Etruscan fashioning of temple locks was popular from the end of the seventh through the first half of the sixth century.[12] Comparable temple locks are worn by some funerary female busts from Chiusi, by the bronze divinity from the Vulcian "Polledrara Tomb" or "Isis Tomb," and by the standing female bronze from the Brolio deposit.[13] The amber figures' locks are most like the latter's.

Not only might the hairstyle be Greek-derived, so too might aspects of the style and iconography. Both Peloponnesian and South Ionian stylistic aspects of 77.AO.85—and of the *Divinity Holding Hares* (77.AO.82, cat. no. 4), and the other *Kourotrophos* (77.AO.84)—are brought out by comparison to certain Arcadian and Sicyonian bronzes of Hermes Kriophoros and of other unnamed shepherds.[14] The ambers and bronzes have a related solidity of sculptural forms and similar modeling of the bodies beneath the dress and relative proportions (head-to-body and torso-to-leg length); they also all have thin arms and small hands and feet. (The small hands and small feet are also characteristic of the four largest figures from the Brolio find, the bronze statuettes of a female and three warrior males.) The backs of 77.AO.85 and 77.AO.82 are especially like those of the Peloponnesian bronzes. A comparison to the *Man in Cloak* in Providence, to cite one example, is telling.[15]

The South Ionian aspect is apparent when the ambers are compared to the most "Samian" of Etruscan bronzes. For instance, the "Kneeling Archer" in Providence is akin in facial details, general physical type, sculptural proportions, and smooth modeling.[16] The South Ionian aspects of the amber pendants are elicited by comparison to an ivory of a horse-tamer and to a wood sculpture of two figures, both thought to be Samian.[17] Alfonsina Russo suggests the existence of an Ionian, specifically Samian-influenced, amber-carving atelier in the Metaponto area, with two examples: the seated amber figure from a grave at Tolve and another from Tomb 122 at the Rutigliano-Purgatorio Necropolis.[18]

The common mantle and the goose may be iconographic details that help one to interpret the meaning and functions of 77.AO.85. The mantle shelters the figures beneath it and separates them from the outside: it can serve both literally and figuratively as a sign of protection.[19] The common mantle can be interpreted as an ancient fertility motif, a signifier of matrilineal descent, a symbol of marriage and procreation, and more simply a protective device.

How does the goose function in this pendant? Is it a symbol or attribute, or does it perform some temporal or narrative role? Long-necked birds are among the earliest sculpted objects: one of the earliest is the ducklike (perhaps) bird, seen in profile, from Uruk, of about 3000 B.C.[20] In Egypt the goose is one of the forms of the solar god Atum. Early in Etruscan art, in illustrations of both landscape and the built environment, waterfowl are in residence, and they are represented as resting, standing, or in action. Birds, especially waterfowl, feature prominently on the bronze objects from Iron Age Italy. Ducks, geese, and swans are among the most numerous subjects of figured ambers found at sites in Greece, Etruria, and Latium.[21] Long-necked and short-legged waterfowl may be the most frequent of all faunal decoration in earliest Etruscan imagery, embellishing countless objects found in tombs. Images of female divinities with waterfowl, usually in the schema of the goddess known as *Potnia Theron*, are found on bronzes, including vessels and ornaments, as early as the eighth century B.C.

Early Etruscan sculptural images of divinities, male or female, defined by attributes are relatively rare, and it is significant that among them are goddesses with birds, mainly waterfowl and raptors. Among the sculptured representations are the early-sixth-century bronze divinity with a horned bird from the Vulcian "Polledrara Tomb," or "Isis Tomb," and a slightly later freestanding bronze statuette in Cortona with a large bird of prey (perhaps an eagle) perched on her head.[22] The latter is comparable to the Laconian (or possibly Tarentine) divinity that forms the handle of a bronze *hydria* of about 570 B.C. found at Grächwil, Switzerland.[23] Female divinities with birds are to be found in Etruscan *bucchero*, painted vases, and gold objects of adornment (namely earrings, pendants, and plaques). Many are in the *Potnia Theron* schema, and some are represented in the bird-atop-the-head pose.[24] Divinities with birds (again both waterbirds and raptors) on contemporary Greek vases (primarily Corinthian and Laconian) and on a series of ivory plaques from the Spartan sanctuary of Artemis Orthia include depictions of both schemata.[25] An Etruscan mirror support of fifth-century date is a later relevant example: it represents an old-fashioned kore figure wearing what appears to be a pointed hat with an upturned rim.[26] The join to the mirror is in the form of addorsed, upside-down swans.

Above are listed the images of female divinities with birds. With the possible exception of the lion- or hare-wielding Mistresses of the Animals, no other divinities as such are represented with birds or other animals. The only other example of a *kourotrophos* with a bird known to me is a much later type of Etrusco-Latial terracotta votive statue from Satricum, of fourth-to-second-century date. In these terracottas the woman is seated, the child is in her lap, and a bird is standing in front. B. M. Fridh-Haneson posits that this and related multifigured, single-mantled terracotta votives are Orphic-Dionysiac, and that they represent rebirth to eternal life by divine adoption, a hoped-for assimilation and identification with Dionysos.[27]

What roles are played by the bird of 77.AO.85? Might the fowl act as an attribute, signify the location of the figure's divine actions, or point to a specific activity? After all, the bird is in action, in contrast to the static pose of the figures. Might the goose signify transit and rebirth[28] or designate the figure as the Greek Artemis? It is perhaps not a native Italian divinity, such as the Etruscan Artumes (or Artames or Aritimi), who "never became mistress of the wild animals or even goddess of the hunt, as she had been in Greece."[29]

The elaborate perforation system of the pendant, which when strung would have maintained the upright posture of the figures, strongly suggests that 77.AO.85 was suspended or worn or was attached to something before its ultimate burial. As a shining ornament, 77.AO.85 was a large, glittering jewel figured with potent imagery. As a permanent amulet, it could have been considered as theomorphic, one that would have offered its wearer, on earth, in the tomb, or in the afterworld, the protection of the deity represented. Both material and subjects were the province of persons of elevated social rank, members of the religious and political elite. In life, its owner could have shown herself to be a votive of the divinity represented: the combination of material and subject would have played a powerful danger-averting and protective role. In the tomb, 77.AO.85 might offer special protection and even guidance to the deceased in the fraught voyage to the afterworld.

NOTES

1. See cat. no. 1, n. 31.
2. For a recent discussion of images of pursuit and abduction in Etruscan art and the possible ambiguities of meaning, see A. Carpino, *Discs of Splendor: The Relief Mirrors of the Etruscans* (Madison, WI, 2003), pp. 14–16, 19–21.
3. For the identification of the bird, see Houlihan 1986, pp. 57–59. On the conventions of representing birds, see Ruuskanen 1992. Douglas Causey (pers. comm.) corroborates the identification of the image as that of a white-fronted goose. The bird "breeds in parts of northern Europe and Asia and winters in parts of Europe, Asia, the Middle East, and North Africa" (Houlihan 1986, p. 57). In ancient Egypt, the standard hieroglyphic sign for a goose is generally taken to represent this species. In captivity, they are sociable and peaceful birds and thus would have been excellent geese to domesticate, unlike some other species. Then as now in Egypt, the white-fronted goose is a delicacy, and it appears that ancient Egyptians looked upon it as one of the

more desirable table geese (ibid., p. 59). The species is still found in Italy today.

4. If this is the tip of a lotus blossom, the amber might be compared to an unusual type of Egyptian New Kingdom statuette and to certain Greek terracottas and plastic vases of youthful figures holding a single lotus bud, or to the lotus-bud jewelry depicted in Greek sculpture and vases. As E. Russman in *Hatshepsut* 2005, p. 42, proposes for the Egyptian Eighteenth Dynasty images, the symbolism may have been thought especially apt for untimely deaths. On the symbolism of the lotus-blossom jewelry worn by Phrasikleia, a funereal archaic marble kore, see M. Stieber, *The Poetics of Appearance in the Attic Korai* (Austin, TX, 2004); and R. Higgins, *The Aegina Treasure: An Archaeological Mystery* (London, 1979).

5. British Museum GR 1850.2-27.15. On the dress, see Bonfante 2003, p. 223, n. 31; Haynes 2000, p. 154; Haynes 1985, pp. 252–53, no. 21; and Roncalli 1998. Bonfante asks a question (p. 223, n. 31): "The bust is wearing a necklace and tight belt: is it naked or dressed in a 'transparent' linen chiton?"

6. Strong 1966, pp. 66–67, no. 43, pl. XIX.

7. Philadelphia, University of Pennsylvania, Museum of Archaeology and Anthropology MS 2536: Turfa 2005, pp. 226–27, no. 242; and Warden 1994, pp. 134–43, no. 3, figs. 13.7–9.

8. Seated female figure, Ancona, Museo Archeologico Nazionale 4417 (from Pianello di Castelbellino): Rocco 1999, pp. 50–51, cat. no. 36, pls. XXVIII–XXIX.

9. Richardson 1983, pp. 44–47.

10. The gypsum statue of a woman in the British Museum is GR 1850.2-27.1 (Sculpture D1) (see cat. no.1, n. 22). For a survey of the Chiusine sculptures, see Hus 1961. For the Casale Marittimo sculptures in Volterra, see, for example, *Principi Guerrieri: La necropoli etrusca di Casale Marittimo*, exh. cat., ed. A. M. Esposito (Milan, 1999).

11. For the related *bucchero*, see Berkin 2003, pp. 38–40, nos. 22–23, figs. 13–14, pl. 67 (his Type 1). For the related *bucchero* at the J. Paul Getty Museum, see CVA, *United States of America*, fasc. 31, *The J. Paul Getty Museum, Malibu*, fasc. 6 (Malibu, 1996).

12. For the temple locks, see Bonfante 2003, pp. 70–71, nn. 40–42. As she points out, the numerous spiral hair holders excavated from Etruscan graves indicate the long popularity of the fashion. Among them are gold rings with amber disks.

13. For the bronze from Brolio, see cat. no. 1, n. 9.

14. For the Hermes Kriophoros bronzes in Boston, Museum of Fine Arts (the larger is 99.489, H. L. Pierce Fund, and the smaller, 1904.6), see M. True in Kozloff and Mitten 1988, pp. 77–85, with references to the related bronzes the Hermes Kriophoros from the Stathathos collection (unnumbered), the Hermes from Ithome (Athens, National Archaeological Museum 7539), the Hermes from Andritsaina (Athens, National Archaeological Museum 12347), the two Arcadians (Berlin, Staatliche Museen 30552 and 10781), and the Hermes Kriophoros in New York (Metropolitan Museum of Art 1972.118.67, Bequest of Walter C. Baker).

15. Providence, Rhode Island School of Design 20.056: D. G. Mitten, *Classical Bronzes. Catalogue of the Classical Collection, Museum of Art, Rhode Island School of Design* (Providence, 1975), pp. 41–45, no. 12.

16. Providence, Rhode Island School of Design 47.792: ibid., pp. 102–5, no. 29.

17. The ivory horse-tamer in Samos (Vathy Museum), early sixth century B.C.: B. Freyer-Schauenburg, *Elfenbeine aus dem samischen Heraion: Figurliches, Gefässe und Siegel* (Hamburg, 1996), pp. 26–28, pl. 3b; and Marangou 1969, p. 196.

18. A. Russo in *Magie d'ambra* 2005, p. 116. The seated figure from Tolve is illustrated on p. 114.

19. For a discussion of multiple goddesses under a single mantle, see G. Koch Harnack, *Erotische Symbole: Lotosblüte und gemeinsamer Mantel auf antiken Vasen* (Berlin, 1989); H. G. Buchholz, "Das Symbol des gemeinsamen Mantels," Jdi 102 (1987): 155; B. M. Fridh-Haneson, *Le manteau symbolique: Étude sur les couples votifs en terre cuite assis sous un même manteau* (Stockholm, 1983); E. Simon, *Die griechischen Vasen* (Munich, 1976), p. 53; K. Schauenburg, "Iliupersis auf einer Hydria des Priamosmalers," RM 71 (1964): 68–70; and M. Guarducci, "Due o più donne sotto un solo manto in una serie di vasi greci arcaici," AM 53–54 (1928–29): 52–65. On the role of the common covering cloth and protection, see the far-reaching study of M. S. Gittinger, "Selected Batak Textiles: Technique and Function," *Textile Museum Journal* 4, no. 2 (1975): 13–19.

20. See the example published by E. Heinrich, *Kleinfunde in den archäischen Tempelschichten in Uruk* (Berlin, 1936), pl. 13c, referenced in Bonner 1954, p. 140.

21. See the introduction for the subject of birds in amber; for a listing of amber waterfowl found in Italian and Greek sanctuaries and graves, see Mastrocinque 1991, pp. 65–88.

22. For the bronze statuette in Cortona, Museo dell'Academia Etrusca 1571, see Richardson 1983, p. 339, figs. 800–802. The bronze divinity in London from the Polledraran "Isis Tomb" at Vulci is British Museum GR 1850.2-27.15 (Bronze 434). See n. 5, above. Once a full statue, the fragment likely represents a native Italic deity, perhaps one of fertility, as the hand-on-the-breast gesture and the other hand holding a bird may indicate. "The horned bird was often depicted by the early peoples of Italy and north of the Alps, and may have had some significance in local cult worship": http://www.britishmuseum.org/explore/highlights/highlight_objects/gr/b/bronze_bust_of_a_woman.aspx. Haynes 2000, p. 155, notes the supernatural bird's Villanovan antecedents, the possible religious significance of bronze cup handles with images of a human figure flanked by birds and quadrupeds in the tombs at

Bisenzio and elsewhere, and that birds associated with priestesses or goddesses (Atargatis, Artemis Ephesia, and Artemis Orthia) are known from Syria, Ephesus, Dodona, and Sparta.

On the identity of ancient images of female deities with birds of prey, especially Hittite, see J. Vorys Canby, "Falconry in Hittite Lands," *Journal of Near Eastern Studies* 61 (2002): 161–202; and Ridgway 1977, p. 112, n. 35: "Several Phrygian statues of the Goddess Kubaba are characterized by the attribute of a raptor pressed against the chest, perhaps significantly with the left hand." It has been pointed out that in hieroglyphic Hittite, a raptor is the second sign in Kubaba's name (*Expedition* 6 [1964]: 28–32). Tanaquil understood the actions of raptors. She, "like most Etruscans, was expert in interpreting celestial prodigies and delighted at the omen" of an eagle snatching and replacing her husband's cap as they entered Rome, presaging his own, the Tarquins', and Etruria's future, as related by Livy (1.34ff).

23. For the *hydria* from Grächwil (Bern, Historisches Museum 11620), see *Die Hydria von Grächwil: Zur Funktion und Rezeption mediterraner Importe in Mitteleuropa im 6. und 5. Jahrhundert v. Chr.; Akten Internationales Kolloquium anlässlich des 150. Jahrestages der Entdeckung der Hydria von Grächwil, 12.–13. Oktober 2001*, ed. M. A. Guggisberg (Bern, 2004); Stibbe 2000; C. Stibbe, "Exceptional Shapes and Decorations in Laconian Pottery," in *Sparta in Laconia: The Archaeology of a City and Its Countryside*, ed. W. G. Cavanagh and S. E. C. Walker (London, 1998), pp. 72–73, with reference to his previous discussion of the work; and H. Jucker, "Altes und Neues zur Grächwiler Hydria," in *Zur griechischen Kunst: Hansjörg Bloesch zum 70. Geburtstage am 5. Juli 1972*. = AntK Beiheft 9 (1973): 57–78.

24. Among the earliest images of a bird atop the head (perhaps a bird hat) is the tiny eighth-century B.C. (possibly Vetulonian) amber in the Metropolitan Museum of Art (1992.11.14a, Purchase, Renée and Robert A. Belfer Philanthropic Fund Foundation, Patti Cadby Birch, and The Joseph Rosen Foundation Inc. Gifts, and Harris Brisbane Dick Fund, 1992). A significant parallel is the late-sixth-century B.C. necklace from Vulci (possibly); its nine pendants, made from sheet gold, are in the form of a crowned and necklaced bust of a female divinity with a resting duck on her head (Rome, Museo Nazionale Etrusco di Villa Giulia 53486): A. M. Moretti Sgubini, ed., *La Collezione Augusto Castellani* (Rome, 2000), pp. 180–81, no. 134, with references to comparanda in Hamburg and Munich. The *bucchero* oinochoe in Florence (Museo Archeologico Nazionale 3179), from Chiusi, may be another representation of the same divinity. The neck of the vase is made into the lower part of her head, the lid into a bird-topped hat: see *World of the Etruscans* 2001, pp. 28, 29, 95, no. 180. The handle of an unparalleled large bronze kyathos from Bisenzio (Florence, Museo Archeologico Nazionale 74913), of the later sixth century B.C., has three figures represented on it, two walking and greeting figures on each side of the handle, and a third at the apex, whose position is difficult to understand: she may be interpreted as supported by one figure on the inside of the handle (whose eyes are cast downward) or as seated at the apex of the handle. The divinity (for so must she be) wears a chiton, a conically shaped hat under a veil, and boots, and holds what looks like a small raptor on her right hand: see *World of the Etruscans* 2001, pp. 26, 91, no. 165.

25. Canby 2002 (in n. 22, above); and Marangou 1969.

26. Etruscan, first quarter of the fifth century B.C.: H. B. Walters, *Catalogue of Bronzes, Greek, Roman and Etruscan, in the Department of Greek and Roman Antiquities, British Museum* (London, 1899), no. 551; and Richter 1968, p. 108, no. 203, fig. 644.

27. B. M. Fridh-Haneson, "Votive Terracottas from Italy: Types and Problems," in *Gifts for the Gods*, ed. T. Linders and G. Nordquist (Uppsala, 1985) = *Boreas* 15 (1985): 67–75; see also Fridh-Haneson's 1983 study (in n. 19, above), pp. 27ff.

28. On the symbolism of the waterbird, see "Early Iron Age and the Orientalizing Period."

29. Jannot 2005, p. 147.

3. Pendant: Addorsed Females

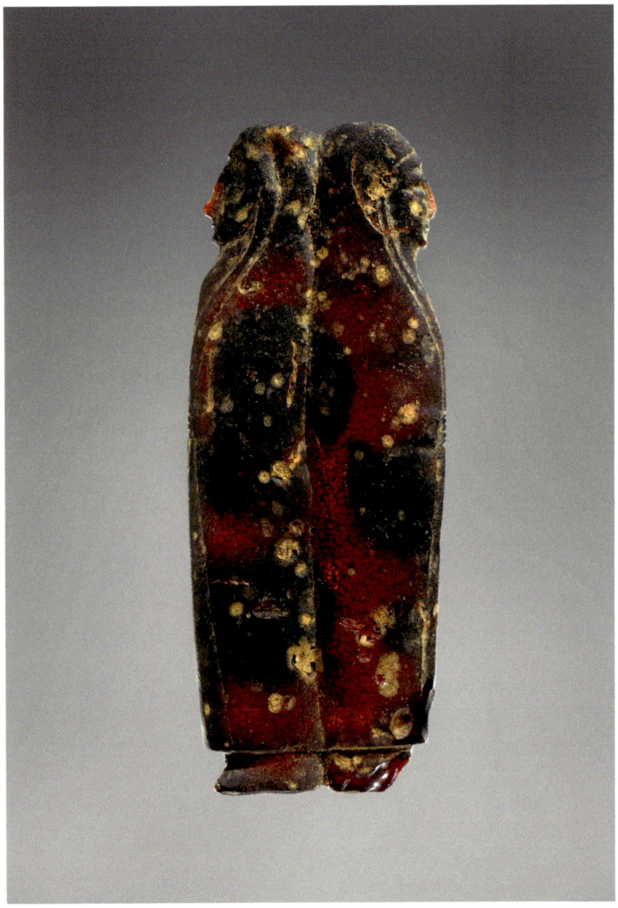

Accession Number	77.AO.81.1
Culture	Etruscan
Date	600–550 B.C.
Dimensions	Height: 100.4 mm; width: 39.9 mm; depth: 13 mm; Diameter of suspension holes: 2.5 mm; Weight: 39.3 g
Subjects	Egypt; Etruscan culture; Magic

Provenance

–1977, Gordon McLendon (Dallas, TX), donated to the J. Paul Getty Museum, 1977.

Condition

The pendant is intact and in good condition. Although its two sides mirror each other, the obverse is distinguished by a more plastic rendering of the figures and superior carving, especially of the left-hand figure (hereafter referred to as Figure A). Figure A has minor chips on the reverse shoulder and small breaks on the feet at the hem area. Figure B has a chip on the reverse shoulder. The piece is a dark reddish brown color in ambient light and is dark ruby-red in transmitted light. There are many large inclusions, including those at the heads, torsos, knees, and feet of both figures.

Description

The pendant is carved from a relatively flat piece of amber and represents two figures standing back to back. An indentation and an engraved line separate the rigidly posed, profiled figures. The two figures are identified as females because of their facial characteristics and dress. Although similar, they are not identical: the faces are especially idiosyncratic.

On both sides of Figure A, the forehead-to-nose line is smooth, with only the slightest bulge at the brow ridge and a faint indentation for the root of the nose. The eyes are amygdaloidal, with the outer corners higher than the inner. The angle of the nose is close to the facial plane; the nose itself points slightly downward, and overhangs the level of the short chin. The upper lip area is short; the upper lip protrudes over the lower. The face of B is different: the forehead slopes more acutely; the root of the nose is more deeply indented; the nose is attached lower on the face; and the angle of the nose is farther from the facial plane. The eyes of Figure B are also almond-shaped but are narrower. The figure's lips have an even profile. Her chin is short and small, and there is a hint of a double chin. The mouth furrow, cheek modeling, and area under the chin are features that make Figure B appear older than A.

The two wear the same type of dress, a long chiton that descends to the ankle; a tightly fitting veil, which covers the hair and falls to the hem; and a heavy, ankle-length cloak. There is no indication of hair on either woman.

Swellings and indentations on the surface of the piece describe the shape of the arms and upper torsos. Each figure places one hand flush on the chest and with the other holds the left edge of her garment at the waist area. Figure A has her right hand up and the left below; Figure B has her left hand above and the right below. The small feet of Figure B are pointed, without any toes delineated, suggesting that both figures were meant to be represented as shod. (The toe area of Figure A is broken off.)

Tool marks include visible abrasion traces in the sections closest to the engraved line that separates the two figures, at the necks, and at the ankles. The perforation for suspension passes from front to back at the center of the pendant, at the nape of each figure's neck.

Discussion

This amber pendant is unique in composition and iconography. The combination of dress elements worn by the figures of 77.AO.81.1—the long chiton, veil, and heavy, enveloping overgarment, Emeline Richardson's Heavy Cloak—is otherwise unattested. Generally, the best parallels for the figures of 77.AO.81.1 are the two other pendants from this group with standing female figures—77.AO.84 (cat. no. 1) and 77.AO.85 (cat. no. 2)—and their comparanda, including the group of four ambers likely from Ascoli Piceno in Philadelphia (University of Pennsylvania, Museum of Archaeology and Anthropology)[1] and the pendant of two standing figures in London.[2] The female figure in the latter, also presented in profile, is especially close in physical type and dress, although she is of a huskier physical type. 77.AO.81.1 also has correspondences with three other kinds of archaeological material: Etruscan bronzes, ivory and bone carvings, and *bucchero* ware. In each case, the traditions informing the imagery are similar.

The votive bronzes most like 77.AO.81.1 are in Richardson's Early Etruscan Ladies Series,[3] the same group that provides good parallels for 77.AO.82 (cat. no. 4), 77.AO.84, and 77.AO.85. Three, now in London, Leiden, and Florence, respectively, are the most similar.[4] The salient parallel for 77.AO.81.1 is the last, Florence 27, which is the latest in Richardson's B Series. The sculptural contours and facial profiles are nearly identical, and, in both cases, the shawl curves over the forehead in the same way. The small feet are also comparable, although the Florence bronze wears *calceoli repandi* (curled-toe boots)[5] rather than the simple boots of the amber women. British Museum 1907.3–11.1, among the earliest in the Ladies Series, is most like the figures of 77.AO.81.1 in facial type, in the small, articulated hands, and in the shod feet. The Leiden bronze is most similar in dress. Although more simply rendered than on 77.AO.81.1, it is the same garment.[6]

The ivory and bone carvings related to 77.AO.81.1 include an identical pair of bone plaques from a chair or other furnishing from a tomb at Belmonte Piceno, representing a large winged goddess flanked by two small female figures;[7] and a number of bone pendants, one from the Large Building excavations at Poggio Civitate,[8] and others, all votives, from the Stipe di Sant'Omobono, Rome.[9] The small figures from the Picene bone plaques wear clothing different from that of 77.AO.81.1 but are similar in their arm and hand positions and their addorsed poses. The unique figure-seal from the Sant'Omobono deposit is analogous in style and figural type but differs in the position of the arms. The arms of all of the Sant'Omobono votives (and probably that of the Poggio Civitate pendant) hang at their sides.

The images of *Potnia Theron* decorating the handles of some Orientalizing *bucchero* kyathoi are significantly similar in format and style to 77.AO.81.1. The figural type, proportions, and modeling are comparable to a series of kyathoi handles from Poggio Civitate,[10] and to another type found at several sites, but which, too, may have originated at Chiusi. This type is adorned with images of a full-figured *Potnia Theron* who is winged and holds long-necked birds.[11]

There are few carved ambers formed into doubled subjects, but among them is another work from the same Getty group, *Paired Lions* (77.AO.81.3, cat. no. 6). Two other approximately contemporary examples, both Picene, are the fragmentary pendant with addorsed figures from Castelbellino[12] and the Philadelphia pendant (MS 2538) in the form of a draped standing female backed with a nude figure.[13] Earlier doubled-figure compositions in amber include the Getty *Addorsed Sphinxes* (78.AO.286.2, cat. no. 28) and a number of the pendants from Tomb VI at Satricum (circa 640–630 B.C.): two pairs of twinned nude females, a pair of twinned, possibly masked figures, a seated monkey with a small, fetuslike, possibly human figure on his head, and two pendants of conjoined foreparts, one a lion-centaur and the other a lion and (perhaps) dog.[14] Two parallels much later than 77.AO.81.1, dating to the fourth century B.C., are pendants of addorsed human heads. One is the central pendant of a votive necklace from the sanctuary of the goddess Mefite in the Valle d'Ansanto (a rare example of a figured amber excavated from a sanctuary),[15] and the second a nearly identical pendant, also from a string of very similar-looking amber human-head pendants.[16]

Addorsed compositions and doubling generally were already age-old by the sixth century B.C. Doubled figures, some addorsed, are not uncommon among ivory and bone carvings from Egypt, the Near East, and the Phoenician world. This compositional motif was often used for objects such as handles[17] and pommels, as well as in weaving or embroidery. Doubled images may represent the same figure twice or two different figures. Doubling and twinning may always have been inherently magical. The doubling of a phrase in a magical spell was understood to increase its efficacy, and the same may have been true for imagery. In amulet making, doubling and addorsed compositions were well-established conventions.[18]

Many kinds of objects in the ancient world include two figures represented side by side, in both seated and standing poses, but addorsed compositions are rare. Among the exceptions are terracotta vessels, bone and ivory carvings, and the above-noted ambers. In some cases, the figures are identical; in others, they are distinct.[19] The significance of the addorsed pose is not clear.[20]

In Egypt, as C. Desroches Noblecourt points out, addorsed figures are a convention used to represent figures that are actually side by side.[21] Stelae of the New Kingdom and later periods frequently show two deities back to back.[22] The conscious pattern seen in the countless examples of paired deities, other figures, or symbols is unmistakable. As Richard Wilkinson observes, "Often in fact, a pair of deities—especially goddesses—are depicted identically in dress and appearance and differ only in name, as though their very duality gave them significance enough."[23] One Etruscan example is a funerary object from Orvieto dating to the second half of the sixth century, a tufa *cippus* in the form of back-to-back busts of female figures, likely sphinxes, and a suitably Egyptian subject for an addorsed composition.[24]

Objects with identical addorsed female subjects in Egyptian, Eastern Mediterranean, and Orientalizing Greek art may point the way for further study of the meaning of the subject type in Italy. Because 77.AO.81.1 is made of amber, the figures must be divinities, heroines, or demons. The hand gestures of the pair may emphasize the fertility and funerary aspects of the amber. If a divine subject, the pendant may represent a "duplicate divinity." That is, as T. Hadzisteliou Price has summed up, a doubled divine image may represent one deity with two names or two aspects, natures, or cults; a deity that appears in the plural, such as the Eileithyiai; or a duplication of one goddess for the purpose of strengthening the deity's quality.[25] However, because of the subtle differences between the two figures of 77.AO.81.1, it may be that the pendant represents a dyad: that is, two discrete divine individuals.

NOTES

1. Warden 1994.
2. Strong 1966, pp. 66–67, no. 43, pl. XIX.
3. Richardson 1983, p. 45.
4. British Museum 1907.3–11.1; Leiden, Rijksmuseum van Oudheden H3 ZZZZ 1 (from Montalcino in Chiusine territory); Florence, Museo Archeologico Nazionale 27.
5. The famous Etruscan curled-toed footwear does not appear in Etruria earlier than the second quarter of the sixth century B.C. See Bonfante 2003, pp. 60–62; and Richardson 1983, p. 47 (see also introduction to part 2, p. 33, no. 11).
6. For a discussion of the garment, see Bonfante 2003, pp. 45ff.; and Richardson 1983, p. 46.
7. Ancona, Museo Archeologico Nazionale 1154 (from Belmonte Piceno Tomb 83): Rocco 1999, pp. 135–36.
8. Poggio Civitate, Antiquarium 71–100: Phillips 1993, p. 78, fig. 121.
9. Rome, Antiquario Comunale (bone figures from the Stipe di Sant'Omobono) 27876 (single figure) and 27877 (figure-seal): *Civiltà degli Etruschi* 1985, pp. 276–77, 10.18.c 7,6. The seal has a flat, square base engraved on its underside with a lion in whose maw is a human leg. Attached to the head of the figure-seal is a large, disklike flange, perforated for suspension. The flange may derive from the sun disk worn by some Egyptian deities; it recalls the sun disk–like appendages on some bronzes from Satricum. See Richardson 1983, p. 267, for references to three bronzes with head appendages from Satricum in the Villa Giulia (Museo Nazionale Etrusco di Villa Giulia 10513–18, 110922). She presumes all these images to be of the Mater Matuta herself. The disk and the lion of the Sant'Omobono figure-seal may be interpreted as links to Hathor.
10. Berkin 2003, pp. 47, 97, no. 35, fig. 17, pl. 11 (his Winged Goddess Type 1, with Murlo nos. 73268–69).
11. The full-figured type was excavated at Poggio Civitate: see Berkin 2003, pp. 46–47, no. 34, fig. 16, pl. 10 (his Winged Goddess Type 2, with parallels at Pescia Romana, Ischia di Castro, Vulci, and Poggio Buco). Owls perch above the latter type. Another *bucchero* comparison is to be found in the mirrored, addorsed female heads (of a long-haired Daidalic phenotype) decorating the handle crests of another type of kyathos. One example of this addorsed-heads phenotype was excavated at Poggio Civitate: see Berkin 2003, p. 45, no. 31, fig. 15, pl. 9 (with parallels cited from Cortona, Chiusi, and Vulci).

12. Marconi 1933, cols. 413–14, fig. 46.

13. Philadelphia, University of Pennsylvania, Museum of Archaeology and Anthropology MS 2538: Warden 1994, no. 2, figs. 13.4–6; Turfa 2005, p. 226, no. 241.

14. Museo Nazionale Etrusco di Villa Giulia 12024–26, 12028, 12033. The definitive study is Waarsenburg 1995. For twinned figures, see his pp. 438–41. For the Picene amber, see Warden 1994. Waarsenburg allows the possibility that the figures are masked and compares them to vessels isolated by R. De Puma, "Nude Dancers: A Group of Bucchero Pesante Oinochoai from Tarquinia," in *Proceedings of the 3rd Symposium on Ancient Greek and Related Pottery, Copenhagen, August 31–September 4, 1987*, ed. J. Christiansen and T. Melander (Copenhagen, 1988), pp. 130–43.

15. The votive amber necklace with multiple profile female heads (at least six) from the sanctuary of the goddess Mefite (Valle d'Ansanto) is in the Museo Provinciale Irpino, Avellino. See Losi et al. 1993, p. 210, n. 20; NSc 30 (1976): 503–4, no. 1309g; and G. Colucci Pescatori, *Il Museo Irpino* (Cava dei Tirreni, 1975), p. 33, pl. IX.

16. This group of ambers was on the Zurich art market in 1981.

17. Perhaps related in meaning are eighth-century B.C. ivory handles in the form of addorsed, partially clothed females, including the closely related fan handle from Nimrud in the British Museum (WA 118102: Bartoloni et al. 2000, pp. 102–3, no. 1). Another ivory analogue is a janiform "Astarte" figurine excavated from a well at Kameiros (British Museum GR 1864.107.671), which L. Schofield suggests is a local Rhodian product made under the influence of Oriental ivory carving: see L. Schofield, "The Influence of the Eastern Religions on the Iconography of Ivory and Bone Objects in the Kameiros Well," in Fitton 1992, p. 174, pl. 1a. The tufa *cippus* with addorsed busts (possibly of nude female figures or sphinxes) from Orvieto is an important architectural manifestation of the schema (Florence, Museo Archeologico Nazionale 73138): *World of the Etruscans* 2001, p. 87, no. 149.

18. Among the most common of Egyptian addorsed-figure amulets are those figured with the *pataikos*, joined with Bes or with a falcon-headed dwarf: see Andrews 1994, p. 39.

19. Among the most notable examples of nonidentical figures are the East Greek terracotta alabastra in the form of two addorsed female figures.

20. Does the addorsed pose have a special antidemonic potency? Johnston 1995, p. 364, notes that in European tales of witches' sabbaths the participants are portrayed as dancing back to back instead of face to face.

21. Desroches-Noblecourt 2006, p. 189.

22. Wilkinson 1994, p. 130.

23. Ibid.

24. The tufa *cippus* with addorsed busts (perhaps sphinxes) is Florence, Museo Archeologico Nazionale 73138: *World of the Etruscans* 2001, p. 87, no. 149.

25. T. Hadzisteliou Price, "Double and Multiple Representations in Greek Art and Religious Thought," JHS 91 (1971): 48–69. See also V. von Graeve, "Neue archaische Skulpturenfunde aus Milet," in *Archaische und klassische griechische Plastik: Akten des internationalen Kolloquiums vom 22.–25. April 1985 in Athen*, vol. 1 (Mainz, 1986), pp. 23–30; and C. Sourvinou-Inwood, *"Reading" Greek Death to the End of the Classical Period* (Oxford, 1995), p. 244.

4. Pendant: Divinity Holding Hares

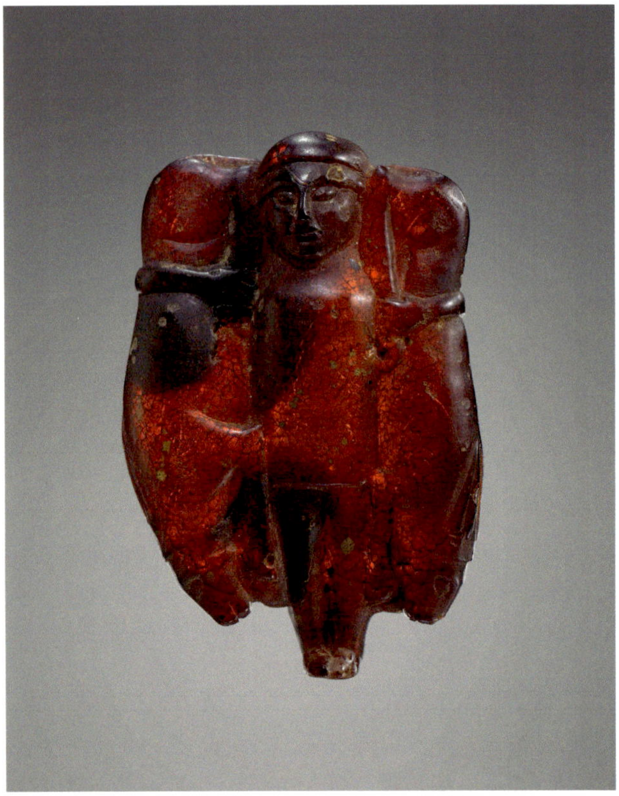

Accession Number	77.AO.82
Culture	Etruscan
Date	600–550 B.C.
Dimensions	Height: 97 mm; width: 64 mm; depth: 24 mm; Diameter of suspension holes: 2.5 mm; Weight: 76 g
Subjects	Artemis; Childbirth; Egypt; Etruscan culture; Hare; Ionia, Greece (also Ionian, Greek); Magic; *Potnia Theron*

Provenance

–1977, Gordon McLendon (Dallas, TX), donated to the J. Paul Getty Museum, 1977.

Condition

The piece is in excellent condition, and its surface is uniformly hard, smooth, and shiny. The amber is crazed overall and has numerous minute cracks and fissures, which are especially noticeable on the figure's feet and in the head and body of the hare to her left. The broken tip of the left hare's nose is the only significant loss to the pendant. In ambient light, the piece is dark red-orange and translucent; in transmitted light, the amber is transparent and bright red-orange. There is a large, cloudy inclusion in the upper body of the hare on the figure's right side, and other inclusions are scattered throughout the piece.

Description

The pendant is worked from a large piece of amber, rounded on both the obverse and reverse sides. The figure is identified as youthful because of the lack of beard hair and the relative body proportions. Its sex is not evident.

The figure wears a short, simple, schematically rendered short chiton, or *chitoniskos*. There is no indication of the front neckline or the sleeve hems, and no belt. The top edge of the footwear is not indicated, but the smooth, close-fitting shape and pointed toe box suggest that they are boots. The figure stands in a rigid frontal pose. In each arm, the figure grasps the hind feet of a large crouching hare held head downward. The space between the figure's neck and the hares' bodies is undefined.

The figure's head is large in proportion to the body; its narrow shoulders slope slightly. The broad chest protrudes in the breast area, but there are no individualized breasts. The hips are narrow. The back is full through the upper area and concave in the lumbar region, and the rounded buttock area protrudes. The long, thin arms end in small, rounded hands. Each hand shows four fingers and the roots of the thumbs. The legs are thick, short, and sturdy, with full thighs and calves, and are contiguous for their full length. No knees are indicated. The feet are small.

Geometry and pattern infuse the face: it is round, broad, and flat; the curve of the hairline in front is echoed by that of the eyebrow ridge, and the line of the lower eyelids mirrors that of the chin. The eyes are blank, elongated ovals that are plastically rendered and outlined with raised eyelid rims. The outer canthi are higher than the inner ones, and the left eye is narrower and slanted farther upward than the right. Both the frontal and parietal eminences of the cranium are wide. The otherwise flat forehead bulges slightly at the brow ridge, but there is no indentation for the root of the long nose, which lies at an angle close to the face. The upper lip area is short, the mouth small, and the mentolabial sulcus shallow. The mouth is wider than the nose, the lower lip wider than the upper. Grooves indicating the mouth furrow run from the top of the nose wings to the corners of the mouth. The lips, formed by two nearly parallel bars separated by a groove, have their corners punctuated by drilled indentations. The chin is also wide and projects slightly, the under-chin area full and angled downward to the neck. In front, the smooth cap of hair is raised above the plane of the brow; in back, the hair follows the shape of the skull and is wide through the crown and indented below the occipital protuberance. A point of hair at the neck extends below the juncture with the spine (in the cervical region) and must overlap the neckline of the tunic in back.

The hares are large compared to the human figure and are powerfully built, with compact, muscular shoulders and legs. Their forelegs are stretched out under their chins, and the hind legs are tucked up close under the bodies. On each animal, an incision line separates the front of the face from the back of the head. The hares' ears are long and pointed and lie flat against their shoulders. Their longish tails hang down their backsides, with the tip of each tail touching the head of the human figure, creating a parallel line across the top of the pendant.

There are abrasion and scraping marks in many places on the pendant. Drill holes define the indentation between the toes and footpads of the hares' feet and punctuate the corners of the figure's mouth, the corners of the hares' mouths, and the roots of the hares' ears.

The pendant is drilled for suspension with two long perforations, both of which exit at the top, in the crown of the figure's head. The perforations' lower ends exit on the reverse: one hole is between the foot of one hare and just below the figure's left ear, and the other is in the crook of the hare's foot near the figure's right elbow. The least visible and perhaps the strongest method of suspending the pendant would have required two filaments, each tied in a knot at the lower end of the reverse exit holes, the two meeting at the top hole and knotted at this juncture.

Discussion

Although it is comparable in style, technique, and sculptural conception to the other five pendants in the Getty Orientalizing group—77.AO.84 (cat. no. 1), 77.AO.85 (cat. no. 2), 77.AO.81.1 (cat. no. 3), 77.AO.81.2 (cat. no. 5), and 77.AO.81.3 (cat. no. 6)—there are no close parallels for this pendant in any media. All six of these pendants demonstrate stylistic and iconographic connections with Near Eastern, Peloponnesian, South Ionian, Cypriot, and native Italian sculpted objects in bronze, ivory, terracotta, and amber. Like those of the other five, the composition of 77.AO.82 is of an ancient type and shows well the broad range of visual schemata, techniques, and styles current in northern inland Etruria.

No definite conclusion has been reached about the sex or age of the figure. As to identity, the composition, the age-old "Mistress" or "Master of the Animals" schema, announces that the figure is heroic or divine (or both). The proportions, head-to-body and legs-to-torso, the short neck, high chest, and small feet, and the relative scale of hares and figure seem to suggest that the figure is that of a child, but there exist many images of adults with these characteristics in Greek and Etruscan art of this and earlier periods.

The somatic parallels with some images of children in Hittite art are striking. A critical comparison is the rock-crystal figure of a child in the Walters Art Museum, which Jeanny Vorys Canby demonstrates to be a "Hittite expression of the miraculous child concept."[1] The short hairstyle, head-to-body proportions, and general sculptural form are remarkably alike.

Neither the hairstyle nor the dress of 77.AO.82 indicates with certainty the sex or age of the central figure. The hair is short and somewhat similar to what Emeline Richardson calls the Etruscan Ducktail, the masculine chin-length cut worn by Early Etruscan votive bronzes of elite males, swordsmen, spearmen, and kouroi.[2] It is not far from the fashion of the sacrificing youth of the Campana Panel, which Sybille Haynes interprets as hair shorn in mourning.[3] The amber figure's hair can be compared to certain images of Artemis, but it has no analogue. A bronze statuette of Artemis (or an adorant or juvenile dedicant) from Lusoi has hair that is cut short in back and with short bangs, a boxy hairdo. This bronze dates to the end of the sixth century B.C. and is thought to be a variant of a cult statue of Artemis Hemera, goddess of Lusoi. It conserves much older traits.[4] Overall, it is the object most similar to 77.AO.82 in hairstyle, dress, and blocklike body. Parallels for the hair may also be drawn with some ivories of Artemis Orthia from her Spartan sanctuary. In these, the divinity is young and has cropped hair; in some cases she wears a feather headdress.[5] These Spartan figures wear a long, smooth-fitting dress that contrasts with the shorter one worn by 77.AO.82 and the shorter chitons of the Lusoi bronzes. The Spartan ivories belong to what may be the oldest type of Artemis representations, that of the goddess in the form of a frontal draped standing figure.[6]

The amber figure's costume—the plain, short dress and boots (or bare feet)—is not unusual for male figures, such as hunters, archers, shepherds, charioteers, riders, athletes, and flying figures, but it is uncommon for females. An important comparison is one of a pair of seventh-century B.C. female stone figures from Casale Marittimo (Volterra). Figure A wears a long braid and a *perizoma* over a short tunic (and her disproportionately long arms and hands are held across her chest in the "mourning gesture").[7] Later Campanian and Etruscan bronze archers (of circa 530–500 B.C.) wear leggings under a very similar short costume, but many of the latter are not clearly sexed; they are often also called Scythians and Amazons.[8] Early Etruscan masculine parallels for the clothing of 77.AO.82 are the tunics worn by three *bucchero* athletes (who also wear the *perizoma*) and at least one of the hunters on the Bernardini sheath,[9] as well as by the riders on the terracotta relief plaques from the principal building at Poggio Civitate.[10] Later sixth-century B.C. examples are two of the male subjects of the antefixes from the Temple B cell-row building at Pyrgi, the winged and rayed sun-god Usil and a bird-headed figure, who is probably Lucifer, the morning star.[11] Important Greek masculine parallels for the amber's dress (and for its iconography) are the sixth-century Arcadian and Sicyonian bronzes representing Hermes Kriophoros and other unnamed shepherds,[12] certain Ionian bronzes of active figures,[13] and an ivory figure of a horse-tamer from Vathy, Samos.[14] All these works present the short garment as a close-fitting, smoothly rendered sheath of thick cloth.

Among the Greek female parallels for the short chiton are various active figures, athletes and hunters most notably. As Eva Parisinou outlines, the short hunting dress is worn not only by Artemis, but also by the hunters Atalante and Prokris, by the hunting female demons the Gorgons and the Erinyes, and by female athletes.[15] The short garment may denote the status of the figure and its activity; the same was true for girls who wore the *exomis* at the Olympian Heraia.[16] One of the earliest antecedents for 77.AO.82—if it is female—is a unique and early Greek bronze of an armed female from Thermon, variously dated between the twelfth and eighth centuries.[17] She wears a short chiton, necklace, and boots, and carries a bow in her right hand. Her hair is gathered up under a pointed hat. This may be the same subject as a bronze of a running "girl" from Samos of about 600 B.C.[18] Etruscan parallels for the chiton of 77.AO.82 include those of the figure (Thesan/Dawn) controlling winged horses who decorates another of the six astronomical antefix types from the south side of Temple B of Uni/Astarte at Pyrgi;[19] the bird-grasping divinity represented on a series of Etruscan terracotta antefixes from Capua;[20] and a bronze found at Pietrabbondante. The bird-tamer's chiton is longer than the hare-tamer's, but it is similarly rendered—simple in its line, it suggests easy movement. (The Capuan figures and the amber hare-tamer are also comparable in their general proportions, youthfulness, face shape, short neck, and long, thin arms.) The bronze from the Pietrabbondante sanctuary wears a shortened dress, and her short hair forms a point in back.

The Ionian aspects of the amber hare-tamer's style are brought out by comparison not only to the Samian bronze of the running girl and to the Ionian-looking Peleus from the Loeb Stand C,[21] but particularly to the ivory horse-tamer in Vathy. This last and 77.AO.82 also share iconography (both figures have short hair and wear unbelted chitons and smooth-fitting boots) and a similar technique (low relief, smooth modeling, and an analogous use of graving tools).

Seven other Etruscan sculptural comparisons, all strongly Ionian in style, are related to 77.AO.82, not only in artisanal terms, but also in iconography. They are a large terracotta votive statuette from Portonaccio, Veii,[22] and four related bronze works: the "Herakles" from Valle Fuino (Cascia) in the Vatican, a bronze in Geneva (Musée d'Art et d'Histoire MF 1017), one in the Louvre, and another in the Getty Museum (96.AC.124).[23] All these figures are distinguished by tall, pointed headdresses or hats that curve forward and animal skins, which in most cases are arranged so the head hangs in front of the pubic area, in sporranlike fashion. Massimo Pallottino suggested that the Portonaccio terracotta represented an unnamed divinity of the Etruscan pantheon, and Giovanni Colonna convincingly associated it with the cult of Aritimi in the sanctuary at Veii.[24]

In answer to the question of who is represented in 77.AO.82, two depictions of masculine figures with hares are important. The winged and snake-tailed male demon of an Etruscan repoussé relief, dated to 600–585 B.C., holds a hare and a lotus flower in his left hand and a wader in his right.[25] A sixth-century B.C. haruspex on a gold ring, who holds a hare in the left hand, is a valuable testimony to the sacrifice of the hare in Etruscan haruspicy. The figure on the ring, probably from Vulci, was once identified by Rodolfo Siviero as Artemis.[26]

The composition of 77.AO.82 is also eloquent in establishing the perimeters of the identity. The image of a contest between a human figure and animals is known in Mesopotamian art since the fourth millennium and is generally interpreted as a symbol of power or control over nature, and possibly also as a symbol of protection.[27] While both male and female protagonists are known in the art of the Minoan-Mycenaean world, the female is more common. Images of the female figure with animals, first named *Potnia* on Linear B tablets, seem to have incorporated the Mesopotamian ideology relating to this composition. In Greek Olympian religion, this Mistress of the Animals disappeared, and her role and divine attributes were incorporated by other goddesses, among them Athena, Rhea, Hera, and Artemis. In Italy, there appears to have been incorporation by some unnamed divinities. By the seventh century B.C. in Greece and in Italy, the "mastery of the animals" schema was adapted to fit current needs, with variants sometimes differing significantly from the Oriental prototypes. Nevertheless, some elements remain constant, chief among them the symmetrical scheme, the frontality of the female figure, and the animals. Although in Italy, images of a "mistress" are more frequent than those of a "master," both subjects are found in Etruscan art. The "master" is usually shown between horses, and the "mistress" is shown as overcoming lions, most commonly, and birds (raptors and waterbirds, especially geese and swans).

While some scholars believe that the mastery imagery came directly from the Orient to central Italy, others have proposed that the subject was transformed in the Mediterranean East before its adoption on the Italian peninsula. There was direct contact with Cretan material in some instances.

Homer calls Artemis *Potnia Theron* (mistress of the wild beasts) in the *Iliad* (21.470). I. Krauskopf has argued that the *Potnia Theron* known in Etruria from the seventh century onward is more like a demon than a divinity proper (such as Artemis) and that there was no early connection between Artemis and *Potnia Theron*, since in Etruria the latter is never represented with Artemis's favorite animals. Noting that there is no identifiable Artemis (Artumes) in Etruria before the second half of the sixth century, she suggests that if there is a blending of the two, it is perhaps under Greek influence.[28] H. Damgaard Andersen suggests that the *Potnia Theron* figure of central Italy must be considered a local goddess and her cult a nature-worshipping one.[29] She proposes that the goddess's iconography changed in the Orientalizing period from the primitive Iron Age figure into one more like Near Eastern forms under the influence of luxury objects brought to Italy by Phoenician merchants and traders, "who in this way might have initiated an important development of the Italic religious beliefs or at least of the iconography of these deities."[30] Further, she argues that the importance of *Potnia Theron* diminished by the second half of the sixth century, since images of the goddess became scarce except in *bucchero*, and proposes that throughout the seventh and sixth centuries, as the Etruscan pantheon was slowly established, strongly influenced by the Greek, the old *Potnia Theron* seems to have been subsumed, possibly into aspects of more than one deity but including Ishtar/Astarte or Uni, Artemis/Artumes, or perhaps Vei/Ceres.

What Sybille Haynes sagely comments about Etruscan Orientalizing nude females of ivory and gold is relevant for female divinities of the period more generally: "It is possible that these images corresponded to preexisting concepts of nature divinities flanked by birds and quadrupeds that figured in earlier representations on open-worked bronze handles found as far apart as Bisenzio, Bologna, and Spadarolo."[31]

The hares, too, are critical to the identity of 77.AO.82. They bear a strong stylistic similarity to the lions of the group—77.AO.81.2 and 77.AO.81.3—especially in the detail of the line separating the faces from the upper part of the head, and to the comparisons presented for them in the catalogue entries below. Both in style and in subject, they relate to the pairs of lions held by *Potnia Theron* figures molded on the handles of a series of *bucchero* kyathoi from Chiusi and Poggio Civitate.[32] (See 77.AO.81.2 and 77.AO.81.3 for further discussion.)

Hares are an uncommon subject of jewelry and amulets in the art of Greece and ancient Italy, but they are much more commonly represented in vase painting of the seventh and sixth centuries. The few early examples of the hare as a subject of an Etruscan object of adornment (amulet, pendant, or ring) stand out. One is a tiny ivory pendant-amulet from Murlo, which this author identifies as a curled-up, couchant hare. Another is a carved carnelian with a crouching hare.[33]

The hares of 77.AO.82 are not newborns but large, mature animals. Wild hares, such as the indigenous Italian *Lepus corsicanus* and the *Lepus europaeus*, may grow to 75 cm in length.[34] (It should be noted that the females of the Italian hare are larger than the males of the species.) "When going it springs," says Xenophon of the hare. "No one has ever seen or will ever see a hare walking."[35] Hares possess excellent senses of sight, smell, and hearing; they were the fastest of the wild animals of ancient Italy; and they can dodge and change direction quickly or dive into streams if needed, as they are able swimmers.[36]

A successful chase would no doubt attest to the hunter's great fleetness of foot, command of the coursers, perhaps, and success with the throwing stick. Catching a hare might also be the result of a good snare. The hares of 77.AO.82 are in a crouching pose, alert yet carried head downward and encircled in the arms of the smaller figure, emphasizing the power of the hunter.

The wide-open eyes of the hares may specifically allude to the ancient belief that the animal slept with its eyes open and was thus always vigilant. Since the hare is also a burrower, a creature that moves between the earth and the subterranean realm, it had, like the snake, chthonic associations.

Hares rarely accompany figures in the mastery pose. Lions are the most common subject, long-necked waterfowl (especially waders) the second most common, and horses the third. There appears to be but one extant example of a hare-grasping female divinity: the main subject of the handle of a Laconian mid-sixth-century bronze *hydria* found at Grächwil, Switzerland.[37] The figure, called by some scholars Artemis and by others *Potnia Theron*, stands atop lions and bearded snakes and holds a pair of live hares firmly in her hands, one head upward and the other head downward. A raptor is perched atop her crown.

In some societies today, birds and hares are among the first wild prey children learn to hunt, chase, or trap.[38] Of course, hare hunting was not limited to the young in the ancient world; Xenophon's *Cynegiticus*, for instance, is especially concerned with chasing hares. As an animal of the wild, the hare belonged to the deities of the hunt. Xenophon recommends that a hound be loosed on a hare only after a vow has been registered to Apollo and Artemis the Hunter, and that hunters dedicate newborn hares to Artemis.[39] Not only did Artemis watch over the newborn, as noted earlier; she also protected the unborn.[40] Callimachus, writing in the third century B.C. and drawing on a wealth of ancient tradition, claims that Artemis's main pursuits are "the bow and the shooting of hares and the spacious dance and sport upon the mountains."[41] Callimachus also refers to Artemis as *Chitone*, and his Artemis asks Zeus to "give me to gird me a tunic with embroidered border reaching to the knee, that I may slay wild beasts."[42]

If the hare-tamer of 77.AO.82 does represent Artemis, or a related female divinity, the emphasis is on her aspect as young virgin goddess of the hunt, a double-edged role: as a fertility deity, she ensures the hunt and the well-being, safety, and reproduction of wild fauna, but as a hunter herself, she is lethal.[43] If the amber represents *Potnia Theron*, the hares may underline her chthonic aspect.[44] The Grächwil *hydria* handle may exemplify this. The fauna around the divinity represent her broad influence: she controls the earth dwellers, the lions; those who live both above and below the earth, the hares and snakes; and the hunter in the sky, the raptor—whose prey can be all creatures of the underground, the earth, and the sky. The hares and snakes reiterate the connection of the divinity with the realm below.

Hunting itself connects two realms, the outside, the wild, the nonlocal, and the unfamiliar with the inside, the domestic, the local, and the familiar, as Mary Helms suggests.[45] Hunters are often seen as shamans who mediate between these two worlds and have the uncommon ability to participate in unknown worlds.[46]

In Egypt, the hare was the sacred animal of Wenet, an anthropomorphic goddess who wore a standard on her head with a recumbent hare. According to Plutarch, the Egyptians esteemed the hare as a symbol of divine qualities, because of its swiftness and acute senses. The relation of the hare to Osiris, which has been variously affirmed, is unexplained but may have to do with the animal's burrowing. In Greece, the hare is linked with other female and male divinities as well. One of the earliest marble korai, a headless figure from Samos (Berlin, Staatliche Museen 1750), holds a young hare.[47] Previously interpreted as an offerant to Aphrodite, it has now been associated with Hera. Another Samian statuette of a kore holding a hare, a recent find, has also been considered a votive gift to Hera.[48] A marble kore consecrated in the Milesian sanctuary of Artemis Kithone/Chitone holds a bird in her hand; Katerina Karakasi interprets this kore as venerating Artemis not only as a fertility and vegetation goddess, but also as the goddess of the hunt.[49]

If the amber tamer is male, the hares may indicate a hunter-tamer, and Artemis's brother may be called up. Dionysos, too, is associated with the hare: youths carry dead hares in the company of the god; on two vases of the Amasis Painter, women (maenads) bring live hares to Dionysos; and on a vase by Lydos, a small satyr leans down to pet a hare in a scene with Dionysos and his entourage. Is the hare held by a maenad in the presence of Dionysos on the black-figure neck-amphora in Paris (attributed to the Amasis Painter) a gift to the god or an emblem of the wild nature of the god's followers?[50] However, as T. H. Carpenter points out, there does not seem to be a "consistent pattern of use for the hare in early Dionysian scenes."[51]

The Etruscan aspects of 77.AO.82 suggest that the subject had particular relevance in Etruria, and more particularly in central, internal Etruria. Might it represent a native Italian divinity such as that represented by the Vatican "Herakles" or the Portonaccio terracotta? Or could it be a rare illustration of one of the indigenous male hero-divinities that became absorbed into the Etruscan Hercle of the fifth century B.C.?

On balance, given the ancient connection between amber and divinities of light, and the iconographic and stylistic connections of 77.AO.82 to the Artemis Orthia ivories, the Laconian Grächwil *hydria*, the Lusoi bronzes, and the Samian bronze girl, the evidence seems to support a female identity for the figure. The pendant may represent the Greek (Laconian?) Artemis accommodated to a native female hunting divinity, one traditionally represented in the *Potnia Theron* schema. Whichever hunting divinity is represented, the pendant would offer protection. If Artemis, she would offer special protection for women in childbirth, not just because of her skill in midwifery, but also because she offers death to women for whom the pain of childbirth would be too great.[52] In the *Iliad*, Hera reminds Artemis, "It was against women that Zeus made you a lion, and granted that you might kill whichever ones you choose."[53] This more violent side of Artemis, from as early as the seventh century B.C., fueled apotropaic magic, and medicine.[54] In late antiquity, Artemis the bow-bearer was called upon in "aggressive" magic and medicine as quite literally a "pain-killer."[55] So, too, were Apollo and Herakles. In the rituals of death and in the tomb itself, this ornament-amulet would have offered protection of the most powerful sort. Still to be resolved is the possible relationship of this image with the iconography of the Hittite child. Might this figure, too, incorporate something of the miraculous-child concept?

Throughout antiquity, the hare's proverbial fertility made it a rejuvenating symbol, and it was used in direct magic to ensure regeneration.[56] As an ornament, this large, shining amber carving must have made a great impression with its potent imagery; as an amulet, it would have served its owner(s) well in life and death.[57] Among the few examples of amber hare pendants is another pendant in the Getty (79.AO.75.28, cat. no. 30), a "doubled" fertility amulet combining a hare and a cowrie.[58] In magical terms, the hares of 77.AO.82 may have mirrored, increased, or focused the fertility and healing aspects of the divinity they refer to, accelerating the speed with which the pendant could ward away danger (a hare's foot gives the owner the animal's fleetness of foot) or promote rapid healing.

NOTES

1. J. V. Canby, "The Child in Hittite Iconography," in *Ancient Anatolia: Aspects of Change and Cultural Development: Essays in Honor of Machteld J. Mellink*, ed. J. V. Canby, E. Porada, B. Ridgway, and T. Stech (Madison, WI, 1986), p. 68.

2. The Ducktail is described by Richardson 1983, p. 34, as "a smooth cap of hair that makes a low arch across the forehead, leaves the ears uncovered, and ends in a point at the top of the shoulders." She notes that it is worn by the later group of *perizoma*-clad spearmen and is "a stiff version of the characteristic haircut of the kouroi of the Middle Archaic period, [which] helps to date these figures in the middle or third quarter of the sixth century." Compare also the hair of the bronze kouroi in Siena (Museo Archeologico Nazionale di Siena 38720: ibid., p. 122, figs. 250–51) and in Volterra (Museo Etrusco Guarnacci 4: ibid., p. 121, figs. 252–53).

3. Painted wall panel of a youth before an altar from the Banditaccia Necropolis, Cerveteri (Louvre Cp 6626): Haynes 2000, p. 220.

4. Frankfurt, Liebieghaus 436: V. Mitsopoulos-Leon, "The Statue of Artemis at Lousoi: Some Thoughts," in *Sculpture from Arcadia and Laconia: Proceedings of an International Conference Held at the American School of Classical Studies at Athens, April 10–14, 1992*, ed. O. Palagia and W. Coulson (Oxford, 1993), pp. 33–39, with earlier bibl., including P. Bol, "Die 'Artemis von Lousoi': Eine klassische Wiedergabe eines frühgriechischen Kultbildes," in *Kanon: Festschrift Ernst Berger* = AntK Beiheft 15 (1988): 76–88; R. Tölle-Kastenbein, *Frühklassische Peplosfiguren: Originale* (Mainz, 1980), pp. 149–52; and LIMC 2 (1984), s.v. "Artemis" (L. Khalil), pp. 633, 738–40, no. 104, pl. 450.

5. For the ivory and bone plaques representing feather-crowned goddesses in various actions, see Marangou 1969, pp. 9–17, nos. 1–4, fig. 15. On feathered crowns in Etruscan art, see Bonfante 2003, pp. 69–70, 135–38, 139, 147.

6. LIMC 2 (1984), s.v. "Artemis" (L. Khalil), pp. 86–98, 631–32, 742–43, with reference to J. Boardman, "Artemis Orthia and Chronology," *Annual of the British School at Athens* 58 (1963): 1–17; and Dawkins 1929.

7. On Figure A as a female, see Bonfante 2003, pp. 219, 226, n. 36. The pair may be mourners but perhaps not ancestors: see F. R. S. Ridgway, "Near-Eastern Influences in Etruscan Art," in *Italy and Cyprus, 1500–450 B.C.*, ed. L. Bonfante and V. Karageorghis (Nicosia, 2001), p. 354. L. Bonfante refers to A. Maggiani, "Le

panther masks decorating the roofs? Haynes 2000, p. 120, notes the connection between the Murlo *sima* and the Tragliatella jug. For the Getty amphora, see R. De Puma in CVA, *United States of America*, fasc. 31, *The J. Paul Getty Museum, Malibu*, fasc. 6 (Malibu, 1996), pp. 13–14, no. 10, who identifies the animal as a "dog (hare?)." For the infundibulum in Copenhagen, see H. Sauer, "Ein etruskisches Infundibulum in Kopenhagen," AA (1937): cols. 286–308, figs. 1–3, 13–14; Riis 1938, p. 155, fig. 19; B. D'Agostino, "Il mondo periferico della Magna Grecia," in *Popoli e civiltà dell'Italia antica*, vol. 2 (Rome, 1974), p. 199; and B. D'Agostino, "Le genti della Campania antica," in *Italia, omnium terrarum alumna: La civiltà degli Enotri, Chone, Ausoni, Sanniti, Lucani, Brettii, Siculi, Elimi*, ed. C. Ampolo et al. (Milan, 1989), p. 572, fig. 555. For the Tragliatella jug, see, for example, Waarsenburg 1995, p. 449; and J. P. Small, "The Tragliatella Oinochoe," RM 93 (1986): 63–69. Haynes 2000, pp. 97–99, argues for a less mythic reading than most other recent interpreters do. Running hares were a favorite theme for the rims of Greek mirrors, most of which have been excavated from funerary contexts.

44. Christou 1968 (see n. 27, above) underlines the connection between the "Mistress of the Animals" and the underworld.

45. Helms 1993; Helms 1988; see also Y. Hamilakis, "The Sacred Geography of Hunting," in *Zooarchaeology in Greece: Recent Advances*, British School at Athens Studies 9, ed. E. Kotjabopoulou et al. (London, 2003), p. 240, with references to Helms's work.

46. Hamilakis 2003 (see n. 45, above), p. 240, refers to V. Turner, *The Forest of Symbols: Aspects of Ndembu Ritual* (Ithaca, NY, 1967); and W. E. A. van Beek and P. M. Banga, "The Dogon and Their Trees," in *Bush Base, Forest Farm: Culture, Environment and Development*, ed. E. Croll and D. Parkin (London, 1992), pp. 39–56.

47. Berlin 1750: Freyer-Schauenburg 1974, pp. 27–31, n. 87, with extensive bibl. She assumed the kore was consecrated to Aphrodite and notes that two votive gifts of marble hares were brought to the Heraion, as indicated in the sources. See also Karakasi 2003, p. 16, n. 59, pl. 11, with reference to Kyrieleis 1995.

48. Karakasi 2003, p. 17, nn. 64–68, explains, "Hera and Aphrodite at times fulfilled similar functions.... The hare was probably an appropriate votive offering for both deities, for both were seen as protectors of the female sphere and patronesses of conceptions and marriage. Whereas Aphrodite was more the embodiment of sensuality and erotic love, Hera was associated with the family, virginity, and marriage."

49. Ibid., p. 50, citing bibl., considers that the cult ceremonies associated with Artemis Kithone at Miletos can be considered initiation rites and that there was a "marriage market" aspect to the festivities. On Artemis Chitone, see also N. Strawczynski, "Artemis et Thesée sur le skyphos du peintre de Brygos Louvre G 195," *Revue Archéologique* 35 (2003): 3–24; Cole 1998, p. 43, n. 23; W. Günther, "'Vieux et inutilisable' dans un inventaire inédit de Milet," in *Comptes et inventaires dans la cité grecque*, ed. D. Knoepfler and N. Quellet (Geneva and Paris, 1998), pp. 215–37. Compare this to the high number of archaic terracottas of korai identified as Artemis holding hares.

50. Paris, Bibliothèque nationale, Cabinet des Médailles 222 (signed by Amasis as potter): ABV, 152.25, 687; CVA, *France* 7, III H e, pl. 36, 1–7, and pl. 7.

51. Carpenter 1986, p. 52.

52. See Faraone 1992, pp. 57–61, on the bow-bearing, death-dealing gods. Odysseus (*Odyssey* 11.171–73) asks his mother in the underworld, "Was it a lingering illness, or did the archer Artemis attack you with her gentle arrows and kill you?" On Artemis's role in childbirth, see, for example, N. Demand, *Birth, Death, and Motherhood in Classical Greece* (Baltimore and London, 1994). S. G. Cole, *Landscapes, Gender, and Ritual Space: The Ancient Greek Experience* (Berkeley, 2004), p. 212, n. 87, notes, "Artemis Eileithyia is more common epigraphically. Artemis Lochia is more common in literary sources."

53. *Iliad* 21.481.

54. See Faraone 1992, pp. 136–40 (appendix 4, "The Incarceration of Dread Goddesses"), with references; and Gager 1992.

55. On aggressive magic, see Bonner 1950, pp. 26–56. On weaponed divinities and talismanic magic, see Faraone 1992, pp. 136–40.

56. Hares have many other associations and meanings throughout ancient culture, especially in the circum-Mediterranean world, which may bear on this image. In Egypt, the Cape hare was a frequent subject of desert hunting scenes in tombs. By the New Kingdom, the scene of the desert hunt was already an age-old theme, one of many symbolizing regeneration. Amulets in the shape of a hare have a long history in Egypt. The earliest surviving example is dated to the Old Kingdom; they are occasional in the Middle Kingdom, and more common in the Late Dynastic and Ptolemaic periods. Faïence images of hares were deposited in tombs, probably because "the figures had some magical or amuletic significance" (Houlihan 1986, p. 70). Andrews 1994, p. 64, summarizes: "The hare was credited with powers of regeneration, but its swiftness of movement and the keenness of its senses were also well known: it was even believed to sleep with its eyes open. Its fecundity, of course, was proverbial. Thus a hare amulet could have worked in life to endow its wearer with fertility or rapidity of movement, or in death with hope of rebirth." Erwin R. Goodenough, *Jewish Symbols in the Greco-Roman Period*, vol. 8 (New York, 1958), p. 93, sees remarkable persistence in the type and use of hare imagery from the Egyptian, Hittite, and Greek past through Judaism and early Christianity until the early modern period, suspects it was a generally popular symbol of immortality, and believes "[it] represented Dionysus and all other fertility deities through whose destruction and love men came, usually in mysteries, to look for immortality."

57. The hare is a solar symbol in ancient Egypt. This is discussed in M. Caccamo Caltabiano, "Il simbolismo del 'lepre': Influenze ideologico-religiose dell'Egitto sull'area dello Stretto riflesse dai documenti monetali," in *L'Egitto in Italia dall'antichità al medioevo: Atti del III Congresso Internazionale Italo-Egiziano, Roma, CNR-Pompeii, 13–19 novembre, 1995*, ed. N. Bonacasa et al. (Rome, 1998), pp. 33–45.

58. See B. A. Kathman's entry for the Corinthian "Crouching Hare Toy(?)," in *Animals in Ancient Art from the Leo Mildenberg Collection*, ed. A. K. Kozloff (Cleveland, 1981), p. 113, no. 94.

5. Pendant: Lion with Swan

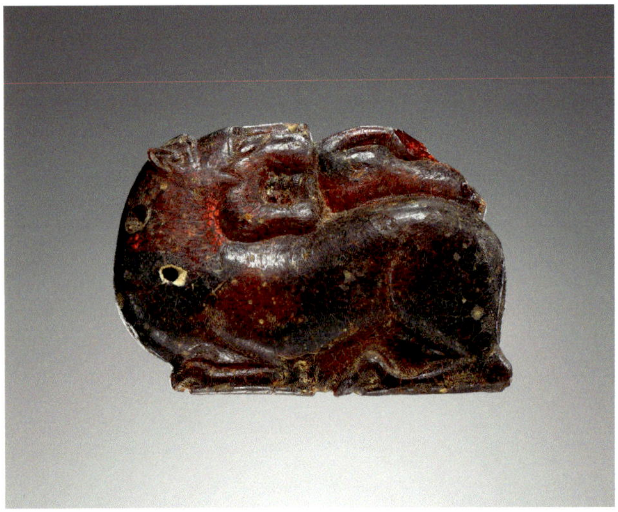

Accession Number	77.AO.81.2
Culture	Etruscan
Date	600–550 B.C.
Dimensions	Height: 42 mm; width: 60 mm; depth: 15 mm; Weight: 25.2 g
Subjects	Amulets; Bird; Etruscan culture; Lion

Provenance

–1977, Gordon McLendon (Dallas, TX), donated to the J. Paul Getty Museum, 1977.

Condition

The pendant is in a good state of preservation, intact with the exception of a chipped portion of the bird's head and bill. The surface is firm overall, with some abraded areas, minute cracks, overall crazing, and shallow chips. There is little surface deterioration or crust. Shallow chips are missing from the left side of the lion's jaw, the right front elbow, and the lower edge of the left rear leg. One sizable crack crosses obliquely across the lion's left shoulder. There are small internal fissures throughout the piece.

The amber is reddish brown. At the modern break on the bird's head, the interior red-orange color of the material is visible. It is translucent in ambient light, but when subjected to transmitted light, it is transparent and a rich red. Numerous inclusions (or, possibly, deterioration pits) are visible in the lion's chest, belly, and rump and in the base of the bird's neck. Before it entered the museum, the pendant is reported to have been lightly cleaned.

Description

This pendant is carved from an oblong piece of amber and is conceived fully in the round. The two main sides are nearly identical. Only the position of the lion's tail—which descends from the root downward, winds beneath the body and up around the right haunch, and terminates on the right hip—breaks the symmetry of the composition. The bottom surface of the pendant, including the legs and belly, is compressed more than are the two flanks. Between the body and legs on both the underside and the flanks are areas of the pendant without design (to be read as negative space). On the underside, the amber has been sharply cut-in to differentiate between the sunken undersurface and the body parts.

The lion looks back in a complete swivel. Its legs and feet are drawn compactly up under its body, the lower legs and feet stiffly parallel. The legs are angular, thin, and short, contrasting with the compact bulk and round curves of the lion's head and body. The pads and toes, especially of the back legs, are talon-shaped. The smooth blanket of the lion's mane is raised away from the face in front, standing up like a crewcut, and falls forward in a semicircular section between the ears. Around the face at the point of juncture of the jaws, the mane juts forward into a cusp of fur. A Z-shaped shallow groove comes to a V above the nose. The folded-back ears have a barely discernible pointed tuft of hair in front and lie well back on the head. The eyes are plastic and almond-shaped, with the outside corners sloping sharply upward. Outlined below by a continuous engraved line, the eyes are set off above by a line that also defines the length of the nose. An extremely fine ridge connects to the cusp. The lion's mouth is open, exposing the teeth (better preserved on the left side than on the right), including the sharp upper canines, extending down over the lower canines. The lower jaw is suspended squarely under the cheek, emphasizing the stylized, squared-off form of the head; the chin is rounded by the exaggeration of the semicircular flaps.

On the lion's back, and partly in its mouth, is a long-necked waterbird in the *reguardant* pose, with its head reverted. The neck is turned back so that the head and bill rest flat on its body. Although the breast of the bird is in the lion's mouth, the teeth do not indent the flesh. In proportion to the lion, the bird is large. A shallow indentation separates the bird's body from the large spread of the wing below it. The bird is more precisely rendered on its left side, where the triangular section below the wing (overlapping the lion's body) is more easily read.

In the area of the juncture of the lion's face and mane is a smoothed depression, well blended into the design. The perforation holes are the same diameter here as on 77.AO.82 (cat. no. 4) and the two *Kourotrophos* pendants, 77.AO.84 (cat. no. 1) and 77.AO.85 (cat. no. 2). Located on the right side of the neck is a stopped bore 4 mm in diameter. Engraved lines are visible throughout the piece. On the body, around the lower part of the lion's head, below the bird's body, circling the eye, and on the left flank are small hatchings and parallel series of scratch marks. The pendant was bored with a triangular set of perforations, one piercing the thorax from side to side, and one each from the exits of this through-bore to a hole on the front of the neck. When strung, the lion would have been suspended neck upward, with its legs perpendicular to the ground.

Discussion

This pendant has no exact parallel in subject or composition. For the style of the animals and bird, the closest comparisons are the lions of 77.AO.81.3 (cat. no. 6), the hares of 77.AO.82, the waterbird of 77.AO.85, and two amber lions' heads in the Louvre.[1] The squared-up compositional type of 77.AO.81.2 is not uncommon in Etruscan Orientalizing art. It is similar to the faunal compositions found in Etruscan goldwork, in bronze reliefs, and on *bucchero*. A nearly identical parallel for the amber is the pair of (possible) fibulae molded onto the shoulders of an enthroned canopic figure in Chiusi dated to the second half of the seventh century B.C.[2] A close comparison for 77.AO.81.2 is a double lion motif stamped on a number of impasto cup handles from Poggio Civitate (Murlo).[3]

In spite of its schematic rendering, the bird of 77.AO.81.2 can be classified as a swan and, more specifically, as a mute swan. The salient characteristics are its large size in relation to the lion, its extremely long neck, and the volume of its wing area, but perhaps most important is the diagnostic basal knob on the upper mandible.[4] The amber image corresponds to the modern bird as well as to Egyptian illustrations of the mute swan.[5] Did the artisan of the pendant know the lion and bird from firsthand knowledge or only from other representations? While there appear to have been no lions in ancient Italy, the mute swan is still a migrant visitor to the area.

The lions of 77.AO.81.2 and 77.AO.81.3 and the amber heads in the Louvre are related in style to a large group of archaic Etruscan lions, mainly bronze reliefs, first gathered together by W. L. Brown.[6] This large and varied group of lions is characterized by an amalgamation of early Etruscan lion types, which show strong links with the Near East, with Assyrian and Hittite types modified by East Greek stylizations and motif. The lions of the two lion-subject pendants in this group reveal their genealogies. The folded-back ear is a variant of the old Hittite heart-shaped form. The shape and slant of the eyes, the squared-off face, the curve of the lower jaw, and the form of the mane in front reveal their genetic ties with the ferocious lions of Assyria. The East Greek stylizations are brought out by comparison to lions from Ionia.

Two of Brown's lion groups are more closely related than others to the lions of 77.AO.81.2 and 77.AO.81.3. The first group includes two important parallels: the repoussé-formed lions of two reliefs from a chariot excavated from a "princely" tomb at Castel San Mariano di Corciano.[7] The lion's head of Thetis from the front of the chariot and the lion helmet of Herakles on the side panel are very close matches for the heads of these Getty amber lions. They have in common similar squared-off faces, jaws with deep mandibles, comparably formed eyes, folded-back ears, and smooth manes. The Thetis lion has the same distinctive cusp of the ruff on the cheek found on the lions of 77.AO.81.2 and 77.AO.81.3. The amber lions are also related to the lions of the "Loeb" cauldron from Marsciano in Munich.[8] They, too, have this distinctive cheek-ruff tuft. The Castel San Mariano chariot reliefs and the Loeb cauldron and stands reveal the impact of Ionian Greek art, especially that of Samos, as do the lions of 77.AO.81.2 and 77.AO.81.3. The second of Brown's groups of lion relatives are all lions' heads, each significantly larger in scale than the pendants. The comparanda include the Etruscan bronze chariot pole decorations and decorative bosses (tomb decorations of some kind) from Vulci and Tarquinia. The closest examples are the lion's

6. Pendant: Paired Lions

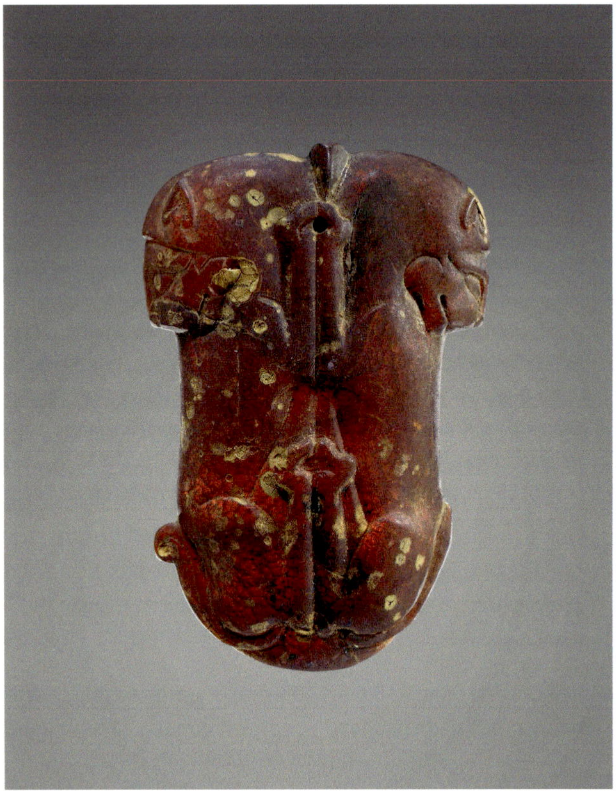

Accession Number	77.AO.81.3
Culture	Etruscan
Date	600–550 B.C.
Dimensions	Height: 56 mm; width: 82 mm; depth: 20 mm; Diameter of perforation holes: 2 mm; Weight: 54.4 g
Subjects	Egypt; Lion

Provenance

–1977, Gordon McLendon (Dallas, TX), donated to the J. Paul Getty Museum, 1977.

Condition

The pendant is in a good state of preservation. The surface is firm and stable, although it is laced with many cracks and has overall crazing and pitting. In ambient light, the piece is translucent and dark brownish black; in transmitted light, it is translucent and orange. There are some small fissures but no visible inclusions. Before entering the Getty Museum, the pendant is reported to have been mechanically cleaned.

Description

Although the two sides of the pendant are nearly identical, the side with fewer natural flaws is here designated the obverse. The pendant is composed of two compactly designed *reguardant* lions in repose that are book-matched along their ventral surfaces. At the top, functioning as a suspension device between the two felines, is a protrusion of amber. At the bottom is a section of amber that, together with the encircling tails, forms a calyxlike base.

On the reverse are two crevices, a deep, polished groove extending from the chest to the flank and a smaller indentation near the knee, both resulting from the removal of a fissure or inclusions before the pendant was figured. Other traces of manufacture—the engraved lines, scraped areas, and multidirectional abrasion marks—can be seen around the muzzles, haunches, and legs, and near the base of the tails. A drill was used for the interstices between the pads and toes of the lions' feet and at the corners of their mouths. The pendant has a triangular set of suspension perforations, each 2 mm in diameter. One extends laterally between the forepaws and intersects with two others, each of which goes from the top of the pendant (in the middle of the flange) to an exit of the lateral bore. When strung, the pendant would have hung with the lions' heads at the top.

Discussion

There is no exact parallel for this pendant. 77.AO.81.3 belongs to the same Getty group as 77.AO.84 (cat. no. 1), 77.AO.85 (cat. no. 2), 77.AO.81.1 (cat. no. 3), 77.AO.82 (cat. no. 4), and 77.AO.81.2 (cat. no. 5). The lions of 77.AO.81.3 are nearly identical with the lion of 77.AO.81.2 and to a pair of amber lions' heads in the Louvre.[1] They are also generally related in style to the hares of 77.AO.82. As is discussed in the entry for *Addorsed Females* (77.AO.81.1), there are also few other contemporary amber objects of doubled subjects with which to compare the two pendants. The only other amber of doubled lions is the (possibly) early-fifth-century B.C. fibula decoration from Belmonte Piceno, which represents the foreparts of two lions.[2] In 77.AO.81.3, the lions are ventrally positioned, or heraldically posed (rather than addorsed).

The pendant is drilled for suspension so that the lions' flanks would be presented. This is also the case with the other double-subject amber pendants, the twinned animal and human figures from Tomb VI at Satricum in the Villa Giulia,[3] the double lion with single face pendant from Pianello di Castelbellino in Ancona,[4] and the two-figure pendant in Philadelphia, perhaps from Ascoli Piceno.[5] In every case, as P. G. Warden pointed out, the pose "emphasizes the frontality of the figures and the possibly apotropaic nature of pendants of this sort."[6]

The two lions of 77.AO.81.3 appear to be identical, but they are not. The differences are minimal and may offer an opportunity to "see" something of the hand of the carver. The ease with which right profiles are worked (in contrast to the left ones) suggests that the carver was right-handed. The most notable difference between the two lions is the tips of their tails: one is curled back and terminates in a volute (obverse left), and the other is spear-shaped. On both obverse and reverse, the right-hand profiles of the lions are more carefully defined, with rounder mandibles and the heads slanted slightly upward; the left-hand profiles are squarer, with more angular mandibles and the heads carved parallel to the ventral line. Overall, the form of the pendant, especially with the calyxlike form of the tails at the bottom, is reminiscent of a pommel, and although no close analogue exists, it may indicate the experience of the carver or point to a specific model in another hard material: ivory, bone, wood, antler, or horn.

The pairing and style of the lions of 77.AO.81.3 is comparable to that of the pairs of lions held by a *Potnia Theron* on two types of *bucchero* kyathos handles found at Poggio Civitate and other northern internal Etruscan sites,[7] and on a *bucchero* infundibulum.[8] The schema is unusual because the divinity holds the confronted animals in front of her body. The similarity between the *bucchero* decorations and 77.AO.81.3 may be taken as additional evidence for locating the manufacture of the Getty Orientalizing ambers to northern internal Etruria.

Although the subject of two confronted lions is age-old, they are not usually presented in a crouching position or with the bodies joined heraldically. This may be a function of fitting the lions into the amber blank's original form, a challenge perhaps similar to that of squeezing them into a frontal composition for the handle of a *bucchero* kyathos. The antithetical compositional type, one especially popular for upright lions, has ancient roots in the Near East. The confronted-lion compositions employed in the Aegean Bronze Age and in Orientalizing Greece are related to Hittite types.[9] The composition bears remarkable similarity to the confronted lions of the Hittite Dagger God relief carving in chamber B at Yazilikaya.[10]

The composition of the doubled lions may have had a special set of functions, focusing or going beyond the meaning of the single-lion subject. The doubling may signify two lions, or it may represent repetition, a basic tool of magic. Repetition is an age-old formula for increasing the potency of any amulet, spell, or curse. Another way to read the imagery of the pendant is as an extract of the *Potnia Theron* schema, such as the pairs adorning the handles of the *bucchero* kyathoi. On these *bucchero* cups, the lions announce the presence of the divinity.[11] The divine and solar aspects of the amber may have automatically called up the presence of a divinity at the same time that the material underlined the meaning of the lions.

A lion image is one of the oldest of all signs. Considering the importance of the lion as a subject of Egyptian amulets, the most widely dispersed in the circum-Mediterranean world, 77.AO.81.3 likely embeds something of earlier amuletic symbolism, the Egyptian and Near Eastern particularly. In Egypt, the lion was a solar animal and a protective sentinel, a powerful symbol of defense.[12] As a desert dweller, the lion was believed to have regenerative powers. From early times there, a double-lion deity, Aker, the lions of yesterday and tomorrow, guarded the sun's passage through both ends of the day and protected the horizon.[13]

NOTES

1. Metzger 1991.
2. See cat. no. 5, n. 14.
3. For the Satricum Tomb VI material, see Waarsenburg 1995.
4. Ancona, Museo Archeologico Nazionale 4427 (from Pianello di Castelbellino): Rocco 1999, pp. 51–52, no. 37, pl. XX.
5. Philadelphia, University of Pennsylvania Museum of Archaeology and Anthropology MS 2538: Warden 1994, no. 2, figs. 13.4–6; and Turfa 2005, p. 226, no. 241.
6. Warden 1994, p. 139.

7. Berkin 2003, pp. 38–40, nos. 22–23, figs. 13–14.1, pls. 6–7 (his type 1), with reference to Valentini 1969. For the related *bucchero* at the J. Paul Getty Museum, see CVA, *United States of America*, fasc. 31, *The J. Paul Getty Museum, Malibu*, fasc. 6 (Malibu, 1996), pp. 17–20, pls. 305.4, 306.4, and 308.1.

8. Florence, Museo Archeologico Nazionale 72733: Valentini 1969, p. 428, pl. XI, no. 67.

9. See the excellent survey of the subject in Marangou 1969, pp. 63–65.

10. A. A. Yener, "Swords, Armor, and Figurines: A Metalliferous View from the Central Taurus," electronic version of article in *Biblical Archaeologist* 58, no. 2 (1995): http://oi.uchicago.edu/research/projects/gol/ba95.html. Yener refers to S. Kosak, *Hittite Inventory Texts (CTH 24150)* (Heidelberg, 1982).

11. For the concept, see Nagy 1994 (in n. 3 in the "Orientalizing Group" section) in relation to a group of terracottas from Cerveteri.

12. Wilkinson 1992, p. 69. For additional discussion of lion symbolism, see 77.AO.81.2, 76.AO.78 (cat. no. 31), and 77.AO.81.8 (cat. no. 32).

13. Wilkinson 1992, p. 159.

Ship with Figures

7. Pendant: Ship with Figures

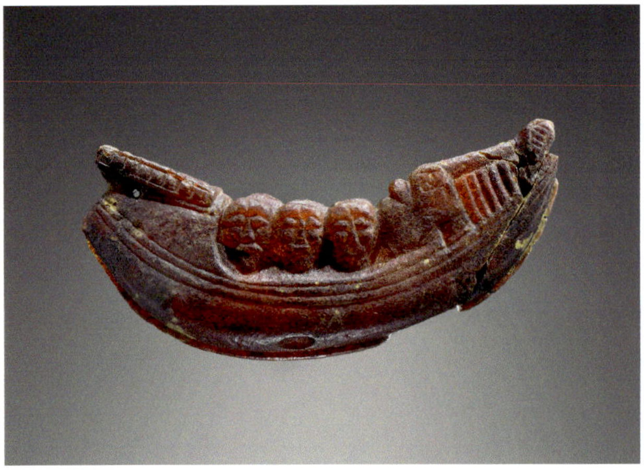

Accession Number	76.AO.76
Culture	Etruscan
Date	600–575 B.C.
Dimensions	Length: 120 mm; width: 35 mm; depth: 10 mm; Weight: 66 g
Subjects	Funerary use of amber (also Burial); Jewelry; Magic

Provenance

–1976, Gordon McLendon (Dallas, TX), donated to the J. Paul Getty Museum, 1976.

Condition

There is a break at the bow, a large chip in this break, a break at the bow set of suspension holes, and a number of small chips on the top side. A section of the keel is broken off, and there is an old fracture loss on the starboard side just above the keel. There are large fissures on the top, port, and starboard sides near the stern. The crazing is uniform overall.

The object is dark reddish brown in ambient light and translucent and dark reddish orange in transmitted light. There are a number of inclusions (or, possibly, deterioration pittings) throughout the piece, mostly on the port side.

While in the donor's collection, the piece was lightly cleaned and the two broken sections of the bow reattached.

Description

The pendant is worked fully in the round. The ship has a deeply rounded hull and a heavy keel. At one end, identified here as the stern, is a curved, knoblike protrusion that must be the *aphlaston*, or sternpost, which is shaped like a schematic bird-head device. In front of the *aphlaston* is a raised structure, probably the stern castle, articulated on its sides by five parallel vertical indentations, with a division down the center on the top, and uneven protrusions (the sheet and cordage?). At the bow, in front of the figures, is an undifferentiated section of amber and, above it, a rectangular form that protrudes over the bow. This forward structure, a bow screen, or possibly a flexible upper deck, is described by two narrow horizontal fillets within which are short vertical incisions; on its top, it is marked off with transverse parallel lines. Seven figures are aboard ship. On each side are three figures, represented by their frontal heads and necks. On each side is carved the profile of a long-haired, bearded man (head, neck, and a small section of the torso are represented); in front of him is a tied sack or other cargo. This figure sits in front of the stern castle and looks forward.

The six outward-facing figures are nearly identical. They have similarly shaped oval faces, centrally parted caplike hair, and cursorily modeled features. Their eyes are almond-shaped, blank, and bulging; their noses are triangular; and their mouths are formed as parallel bars. The bearded figure sits taller in the ship and is slightly larger in scale than the frontal figures. His almond-shaped eyes are set high in his face. He has a sharp, triangle-shaped nose, a prominent pointed beard, and long hair falling over his ears to the back hairline.

The pendant preserves evidence of its original form and its preparation before the figuration. Portside is a smoothed depression, and on the bottom is a smoothed lacuna in the keel. Three sets of through-bores perforate the pendant, at the bow, at the stern, and amidships between two of the passengers' necks. Each set forms a triangle: one hole perforates the body of the pendant from obverse to reverse, and the other two holes intersect this transverse bore at the exits. The latter two holes are drilled at an acute angle and meet at the topside of the piece, forming an apex. A filament or cord would have run from the apex of the triangle through the transverse bore and back up to the apex.

Discussion

There is no close parallel for 76.AO.76; only one other related amber carving is known, a ship with sailors from Padula (Lucania), which likely served as the bow of a fibula (see below). 76.AO.76 is generally related in style to the ambers discussed in cat. nos. 1–6 and was included in the same donation.

The shape of the pendant links it to a much older type of ornament, the crescent-shaped necklace, found in metal and amber, a fashion that flourished in the north and in Western Europe during the Beaker period and the early Bronze Age.[1] The form, materials, and symbolism of such pectorals were exclusive to high-status individuals. The pendant is also related to the engraved decoration of ships on two Daunian stone stelae of females.[2]

76.AO.76 may be "located" in Etruria (with debts to Greek art) and dated by the compact format of the carving and the style. The frontal figures (whether passengers, sailors, or marines) are related to the Getty *Kourotrophos* group—77.AO.84 (cat. no. 1) and 77.AO.85 (cat. no. 2)—and share many sculptural comparisons. The ship's frontal figures are akin to the Etruscan bronze votives in Emeline Richardson's Swordsman Series A, which she dates to the early sixth century B.C.[3] A work such as the male sphinx *acroterium* from Poggio Civitate (Murlo) of about 575 B.C. also helps to place it.[4] 76.AO.76 gives evidence of its connection to Etruscan art when compared to seventh-century ivories and caryatid figures from *bucchero* chalices.[5] The seated figure's profile is similar to that of Aristaios in the Spartan ivory plaque in London, and in its silhouette and harsh modeling is generally comparable to a variety of Etruscan works dating to between the late eleventh and mid-sixth centuries: some of the early male "canopic" heads from the Chiusi area, the limestone cinerary urn in the form of a seated man from Chiusi, and profile warrior heads in *bucchero*.[6] The ivory pyxis from Pania, Chuisi, of circa 620–580 B.C., likely carved in southern Etruria,[7] offers up important analogues for both frontal and profile heads and for the schematic illustration of the ship. However, there are differences: the unbearded figure of the Pania ivory is alone in the ship and handles double rudders (and his cargo consists of two amphorae). The ivory, bronze, impasto, and *bucchero* objects not only suggest an approximate date for 76.AO.76, but also suggest where in Etruria it may have been made and buried.

The ship of 76.AO.76 has many earlier and contemporary comparisons in the archaeological remains and in the art of the ancient Mediterranean, especially among small three-dimensional representations in terracotta, bronze, and other materials, carved stone reliefs, Etruscan wall painting, and most particularly Greek vase painting. The subject also figures on coins and gems. The Etruscan shipwrecks at Giglio, Antibes, Marseilles, and Pisa offer extraordinary information about the actual vessels. Taken together, the corpus of ancient material allows a very particular knowledge of ships and ancient seafaring, even though in the images "some elements of the ship's architecture [may be] telescoped, others expanded or otherwise exaggerated, while others [may be] disproportionately small or ignored entirely."[8]

The Getty ship is a different kind of vessel from that represented on the Pania pyxis. The Getty amber clearly represents a rounded hull but has no oars, oarlocks, or holes. It must be one in a class of merchantmen, or cargo-carrying ships, which the Greeks called a *holkadikon ploion* or *holkas* (towed ship)—that is, a vessel dependent on masts and sails to propel it. Unequipped with oars, the *holkas* would have been towed into and out of harbor.[9] Although the carver of 76.AO.76 has provided much detail about the ship (*aphlaston*, keel, stern castle, etc.), much is not indicated: there is no rudder, mast, sail, or cordage. Certainly, there were limits set by the form of the amber blank. As noted in the description, the lumpy material aft may be the stowed mast and sailing gear; alternatively, if the carriers extending from each of the three suspension points were joined up, they would form the schema of a mast and lines. The three sets of suspension holes would have enabled the pendant to hang from three separate points or to be attached to clothing or another, larger ornament, in three places. Then again, a carrier attached at each point might have been pulled up to a central point above the pendant, the lines of the carriers then forming a bisected triangle—a pattern like that of a mast and lines. With the pendant suspended this way, the rails would be parallel to the horizon, very like the position of a ship at sea.

Another important ship comparison is an ivory pectoral decoration from the sanctuary of Artemis Orthia at Sparta,[10] whose scene is perhaps a disembarkation. The subject has been interpreted as the carrying off of Helen to Troy.[11] Both ornaments are lunular in shape, with just a few figures.

There is only one other published amber in the form of a ship, from a burial at Padula, Lucania. First published by Amadeo Maiuri in a short, illustrated note in 1914, it was found along with numerous other objects during roadwork in the early twentieth century. Maiuri identified the disturbed find as the remains of a woman's grave.[12] While the Padula amber ship is generally similar to the Getty ship in subject, size, shape, suspension technique,[13] degree of relief carving, and the absence of a mast, they differ in artistic style, type of ship, characterization of the figures, and implied narrative.

The Padula ship is an oared galley with a dolphin-figured forefront or ram. Stowed nautical equipment—mast, sail, and cordage—is visible at the top on the starboard side and above and behind the figures on the port side. Aboard are four figures facing aft and a single figure facing forward. Only the upper parts of each figure are represented; all are in profile. The figure astern has a rudder (perhaps one of a pair). The two sides of the Padula amber are dissimilar. To port, only two of the sailors are fully indicated. To starboard at the prow is a large shield, the blazon of which is an eagle in flight to the left.

The Padula amber is generally similar to the Getty ship and calls to mind the same stylistic comparisons. The Padula pendant compares well with many early-to-mid-sixth-century B.C. objects, including Greek vases. This proposed early date is commensurate with that of the other amber objects illustrated by Maiuri, especially bulla-shaped amber pendants, which were popular throughout much of the Italian peninsula from the early seventh century. This ship amber, too, may have been carved in Etruria, perhaps in a southern Etruscan center such as Cerveteri.

The two ship pendants present many questions: why a pendant with a ship subject, for whom were they made, why were they buried in a grave (as is assumed for 76.AO.76), and once there, how did they work? The ship (with and without sailors) has a long history of funerary use, from early Egypt as well as Old Europe. The amber ships may have had a comparable function in tombs as that of ship models, wall paintings with ships, and vases decorated with ships (some with figures and some without).[14] The Tarquinian tombs with ships, the Tomb of Hunting and Fishing and the Tomb of the Ships, underline the high status of such imagery. The carving of ships with sailors on two Daunian stelae of women opens up many questions, some first posed by M. L. Nava.[15] Do these represent objects, such as pendants? Are they part of the dress? And what role does such prominent imagery play for the deceased individuals?

Although it is probable that the Getty pendant, like the Padula amber, was unearthed from a grave, both pendants may have had similar previous quotidian functions. If this is so, we might ask on what occasions or for what reasons the pendants were worn. The Getty pendant and the Padula ship offer no evidence of wear.

It is possible that the original forms of the amber lumps informed the imagery—that is, the form of the raw amber determined which kind of ship would be represented, and the narrative followed. If the ship imagery symbolizes the same thing, it was surely augmented by amber's magical aspects. It is conceivable that there was always a connection of amber with sailing, and with particular ships and voyages, real or legendary, which inspired the formation and embellishment of the raw material into these pendants.

Amber's natural buoyancy in saltwater may be one of the simplest explanations for the choice of a marine subject. The ship may represent, in a general sense, journey by sea, sea trade, or colonization. In the Archaic period, a successful sea journey held the promise of incredible wealth, but it was also a dangerous and risky undertaking. A ship-shaped amber pendant may have had a direct talismanic function. Was it a charm that brought good luck and good sailing?[16] A ship as an amuletic subject could also represent the "vehicle" on which danger could be sent away. A curse against a night-wandering demon commands, "Go away upon swift ships!"[17]

On the other hand, did this *portafortuna* once promise success in trading ventures or in the establishment of a colony? It must have been the elite who sent abroad trading ships, themselves symbols of status and prestige. A ship could also stand for some of the values of aristocratic society, such as the entrepreneurial spirit or the Odysseus-like cleverness and ingenuity that was required on dangerous journeys to unknown places. Successful marine enterprises, whether for business or for the establishment of colonies, contributed to the self-definition of the aristocrat. Figures such as Odysseus became *Leitbilder* for Greek and Etruscan aristocrats traveling far from home.[18] Owning, wearing, giving, or being given an object like the Getty ship pendant might offer or affirm an identity with Odysseus, might articulate a colonial Greek, Etruscan, or Italiote's ethnic identity, or might even affirm a genealogical connection to Odysseus.

At the least, because the pendants are made of amber, they can be seen as objects of high-status material embellished or reinforced with an inherently high-status subject.

The ship ambers are also open to more concrete epical and mythological readings. Some of the relevant Greek stories are Theseus's voyage to Crete, Jason and the Argonauts' search for the Golden Fleece, and Odysseus's adventure with the sirens. There is a close connection between the Argonauts and amber. In one of the amber-origin stories, the Argo sailed up the Eridanus River and the Argonauts came upon the body of water in which Phaethon had fallen. A Homeric subject might have had a particular relevance for an amber amulet. Homer's words were magical. "Quotations from his work could heal people when whispered in their ears or hung around their necks written on amulets, which should be preferably of gold."[19] Odysseus in particular was a magical figure, for he had survived transport within a mystically powerful, precivilized natural world; he went to locales where the heavens seemed to touch the land and sea, places between the celestial and earthly realms.[20]

Odysseus's challenge to the sirens had particular resonance in southern Italy: sirens ranged along the Tyrrhenian coast, and their home was an island offshore (perhaps real, perhaps mythical), or perhaps they were turned into rocks after their suicide.[21] The bearded figure of 76.AO.76 could depict Odysseus or perhaps the captain. The omission of Odysseus may have functioned as an intentional ellipsis; the wearer could symbolically replace the hero.

Another story known from the *Odyssey* that deserves our consideration—and one not, to my knowledge, previously associated with surviving images in Greek and Etruscan art—is an episode of Menelaus's Return. The bearded helmsmen of each pendant might, from this perspective, have been associated with Phrontis, the helmsman of Menelaus, "whom no man has yet surpassed in piloting a ship when storm winds blast." This is how, in the *Odyssey*, the horseman Nestor pays tribute to Phrontis, who was struck by Apollo "with his gentle shafts" as the ship bearing Menelaus and Nestor neared Sunium, the sacred headland, "while his hand held fast the steering rudder." "So Menelaus, although he was keen to journey on, stopped then at Sunium to bury and to honor his companion."[22] Menelaus's Return was one of the subjects painted by Polygnotos in a section of his murals in the Knidian Lesche at Delphi. Pausanias singled out the figure of Phrontis in his long ekphrasis of the Lesche program as "the only one with a beard," and, interpreting Homer's words, he added, "Menelaus was left behind to build Phrontis a tomb and to pay him the due rites of burial."[23] Certainly, a depiction that would recall Homer, Menelaus's Return, and Nestor would have carried strong associations with burial, honor, and magic and might even have been a compelling reason for the choice of subject and for the subsequent burial in the tomb of an elite personage—even if that personage were a woman.

In this regard, it is useful to recall the gold votive inscription from the coastal sanctuary at Pyrgi, which mentions "Uni/Astarte, Ino/Leukothea, and Eileithyia[,] all mother goddesses who were supposed to help mortal women in childbirth and watch over the growth of young children, as well as save sailors."[24]

Both the Getty and Padula ships may allude to the barque of the sun.[25] If there is a solar connection, the object may be linked to Bronze Age illustrations of the solar journey, the metaphor for rebirth,[26] and to a unique Etruscan depiction of the sun barque (with three passengers) sailing eastward, above the solar god driving his horses, on a mirror from Orbetello (Museo Archeologico Nazionale, inv. MA 73798). One part of the route to the resting place of the dead was over the sea, to a place "which evoked an occidental archipelago of the blessed, an Afterworld beyond the ocean, in the sector belonging to the netherworld's gods," as Jean-René Jannot writes, and it is "a sea monster or a ship of the high seas" that brings the dead to the other shore.[27] Such a vision of the afterworld may owe much to the Bronze Age in northern Europe, the source of the amber.

NOTES

1. L. Casson, *Ships and Seamanship in the Ancient World* (Princeton, 1971), p. 2. This was brought to my attention by J. Bouzek and E. Pleslová-Štiková. For an overview of the type, see E. Pleslová-Štiková, "A Crescent-Shaped Necklace from Velvary, Bohemia," in Beck and Bouzek 1993, pp. 147–52. A crescent necklace was the attribute of a high-ranking individual; the finds of amber and metal examples, and the representations of them on engraved anthropomorphic stelae, establish the early date.
2. Museo di Manfredonia 0806: M. L. Nava in *Ambre* 2007, p. 221, fig. 2, no. III.230.
3. Richardson 1983, pp. 64–70; compare, for example, Arezzo 11490, 11492, 11493, 11495 (ibid., pls. 24–31); and Volterra 18, 23, 24, 28 (ibid., pls. 29–32).
4. Poggio Civitate, Antiquarium 68–100.
5. Compare, for example, the pair from the Tomb of the Animals at Cerveteri (Rome, Museo Nazionale Etrusco di Villa Giulia S10V3:

Korai

8. Pendant: Standing Female Figure (Kore)

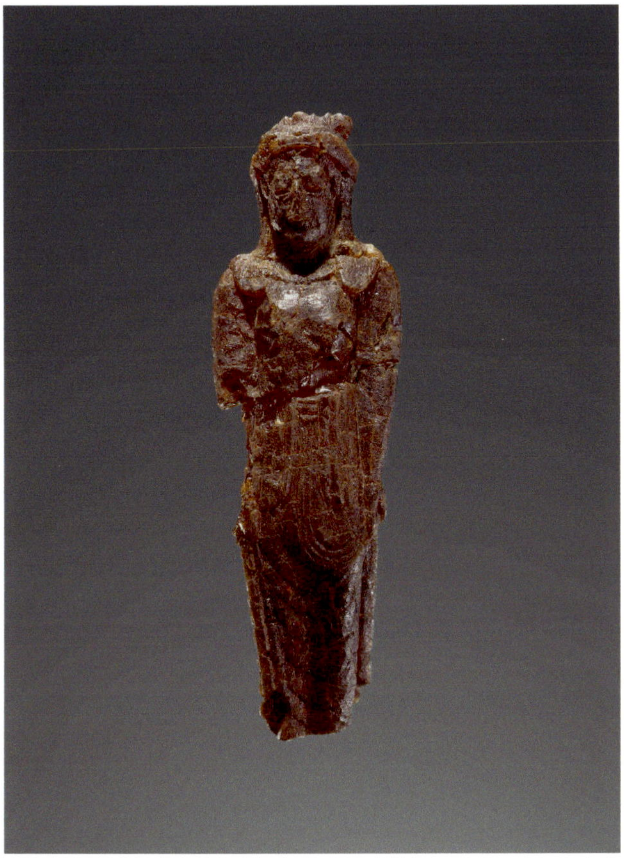

Accession Number	76.AO.77
Culture	Etruscan
Date	525–500 B.C.
Dimensions	Height: 67 mm; width: 20 mm; depth: 9 mm; Diameter of suspension holes: 2 mm; length: 9 mm; Weight: 6 g
Subjects	Artemis; Childbirth; Egypt; Fertility; Ionia, Greece (also Ionian, Greek); Magic

Provenance

–1976, Gordon McLendon (Dallas, TX), donated to the J. Paul Getty Museum, 1976.

Condition

Upon acquisition by the museum, the piece (which had been broken at the waist and reglued) was cleaned mechanically and treated with an amber-oil distillate. The treatment resulted in improved translucency and slight darkening. There are pinpoint losses over all the surface and small breaks along the proper right side, on the veil, the right breast, and the right arm from elbow to hand; and along the left side, on a section of the veil and the upper torso below the breast and below the left elbow. There are also breaks on the stephane, or crown, and at the proper right of the flange. The bottom of the figure from the lower hem area to the feet is missing. There is a hairline crack around the neck and shoulders. Two small spalls on the reverse reveal unweathered amber. In ambient light, this pendant has a uniform dark orange-red translucency, and in transmitted light, a bright orange-red color. There are no visible inclusions.

Description

The amber has not been chemically analyzed, but its appearance is consistent with Baltic amber. The object is flat on the reverse and concave on the obverse, suggesting the form of the piece of amber from which it was worked, but there are no depressions or grooves. The lack of visible inclusions and flow lines suggests that it was carved from amber formed inside a trunk. The flange at the top of the head is drilled from both lateral sides toward the center for the insertion of a carrier. Under strong light, the two borings are distinguishable. Some areas preserve the multidirectional scratches caused by the use of a fine abrasive: between the arm and the bodice, on the long folds of the garment, and at the juncture of face and hair.

The figure wears a chiton with belt, a veil, a crown, and bodice jewelry. She holds the chiton skirt in both hands and poses with her left leg slightly in front of her right. Her face is a full oval, the brow smooth. The smallish, narrow eye sockets are shallow and empty and likely held inlays. The nose is indented slightly at the root and is set at a low angle relative to the forehead and chin. The cheeks are flattish and full. The mouth, formed in a half smile, has a short upper lip (with the tubercle indicated) that protrudes over a full lower lip indented slightly at the center. The nodes at the corners of the mouth and the mouth angle furrows are indicated by short, nearly vertical indentations. The mentolabial sulcus is shallow; the chin is small and round.

The hair framing the forehead is parted in the middle, and the two sections are dressed to each side in a series of four rounded waves, each undulation plastically swelled. Each side section is pulled over the top of the ear and then behind it. The figure's hair is worn loose down the back in a fall that is curved at the bottom; it reaches the top of the thoracic vertebrae. Over her hair is a long veil of fine fabric. Atop the veil is a crown, worn at the position of the bregma. The pendant flange at the top of the head is carved with a bead and reel.

The torso section of the chiton falls into two vertical folds that overfall the waist in full sections as far as the position of her wrists. The skirt is drawn closely against the body. In her left hand, with her thumb over the cloth, she grasps the central portion of the chiton, which is delineated by three narrow pleats. In her right hand, the thumb also atop the cloth, she holds a section of the skirt, which is pulled horizontally. It forms a vertical section of several folds. The draping of the skirt forms six evenly spaced folds, patterned into almost parallel, horizontal sections. At the waist, three raised horizontal bands signify the belt. At the neckline is a curved raised area and, at about the clavicles, a pectoral ornament. It may be attached to the neck edge of the chiton or to the inside edges of the veil lappets. The joining up of the sides of the chiton to form "sleeves" is indicated on the tops of the sleeves by two adjoining parallel raised lines. The narrow hems at the elbow are clearly indicated. The veil falls forward to her shoulders, covers her ears, and forms deep lappets that reach to the level of her armpits. The veil leaves the tips of the shoulders free and drapes down her back in a straight fall to about the position of her ankles. In the back, the veil is patterned into a series of vertical pleats to a level just above the hem of the chiton. The center of the veil is flat, unpleated, and plain; the sides of the garment are turned back, folding onto the center portion. This results in the terminal edges patterning into swallowtails. Drop-shaped fabric weights (two) are discernible at the uppermost edges of the veil's zigzag folds.

Discussion

76.AO.77 is comparable in style and dress to a number of much larger objects (marbles, bronze statuettes, and terracottas) as well as to a number of other tiny figures in the form of either a kore[1] or a kouros made from other precious materials, including ivory and gold or silver sheet. The metal figures, in addition to serving as pendants, also adorned earrings, headwear, and dress. In addition to 76.AO.77, ten other miniature amber korai are extant. The ambers are all perforated, and seem all to have been designed to hang from carriers or pins. Nine are complete or nearly so, and two are fragmentary. The eleven ambers of female figures in the form of korai are (1) Malibu, J. Paul Getty Museum, 76.AO.77, figure in chiton and veil; (2) Perugia, Museo Archeologico Nazionale 101185, from Monteleone di Spoleto (Colle del Capitano), figure in chiton and himation;[2] (3) Berlin, Staatliche Museen (Antiken Museen, lost during the Second World War), figure in chiton and veil;[3] (4) Dresden, Albertinum 1384, larger figure in chiton, veil, and mantle;[4] (5) Dresden, Albertinum 1384 (*sic*), smaller figure in chiton with mantle;[5] (6) Belgrade, National Museum 689/1, from Novi Pazar, St. Peter's Church, kore in chiton, veil, and mantle;[6] (7) Belgrade, National Museum 688/1, from Novi Pazar, St. Peter's Church, kore in chiton, veil, and mantle;[7] (8) Belgrade, National Museum 692/1, from Novi Pazar, St. Peter's Church, kore in chiton and mantle;[8] (9) New York, Metropolitan Museum of Art, Philanthropic Fund, fragmentary figure in chiton and veil;[9] (10) Basel, Switzerland, private collection, kore in simple chiton;[10] and (11) Malibu, J. Paul Getty Museum, 82.AO.161.6 (cat. no. 9), fragmentary figure in chiton and veil. A related amber is the two-figure pendant in New York, a draped woman in chiton and mantle holding a child (the child's head is now missing), a *kourotrophos*.[11]

Only four of the above have secure documentation, the three from the large Novi Pazar "princely" burial and the single figure from a disturbed burial at Monteleone di Spoleto. Robert Heidenreich recounts that the Dresden ambers came from a grave near Rome. In the Getty Museum collection, 76.AO.77; three of the ram's-head pendants, 76.AO.82 (cat. no. 39), 76.AO.83 (cat. no. 40), and 77.AO.81.7 (cat. no. 41); and a pendant with the foreparts of a boar, 76.AO.84 (cat. no. 37), are related in style, technique, and state of conservation and are alike enough to consider the hypothesis that they may have been found in the same burial. The combination of subjects is plausible, given that the Novi Pazar burial included three korai and eighty rams' heads.

The pendants that retain their feet (all but 76.AO.77 and one of the kore from Novi Pazar, National Museum 688/1) appear to be wearing smooth, pointed boots. The Berlin and the larger Dresden korai stand on plate bases, but the lower part of the amber from Monteleone di Spoleto is difficult to discern. Apart from the three kore from Novi

Pazar, which were perforated for suspension in the feet area, all of the complete pendants were suspended from the head.[12] The suspension flange on 76.AO.77 closely matches that of an amber kouros pendant in London (British Museum 41). The only other kore pendant with a flange is the smaller Dresden draped female.

These objects and those in precious metal and ivory display a similar compact sculptural form, indicating that they were designed to be worn. The korai are typified by short, thick necks, long hair or veils, wide shoulders, long garments, legs and small feet close together, and arms and hands attached to the body. Some have plate bases. The nude kouroi stand with one leg slightly forward, a pose that helps the carver close the gap between the legs. These sculptural solutions provide extra support for potential weak spots such as the neck, arms, or feet.

76.AO.77 has features in common with each of the other pendants, but it is most like the similarly sized Berlin pendant. There is no extant photograph of the reverse of the Berlin figure, but the two korai are similar in contour, volume, stance, and hand position, as well as general body proportions: large head on diminutive neck (their heads make up roughly a fifth of their total height), wide shoulders and full upper arms, and a short stature. Both wear chitons with deep *kolpoi* and have similar long, capelike veils with lappets of cloth on the front and shoulders. Although the two pendants are close in style, and probably descend from a common artistic tradition, there are notable differences between them. Whereas 76.AO.77 has a bead-and-reel flange carved as part of the piece, the Berlin amber has a carrier hole bored through the head from temple to temple. The Berlin amber is more cursorily modeled than 76.AO.77. The expression of the Getty figure suggests the beginning of a smile; the face of the Berlin amber communicates greater gravity. (Is it a question of age?) The larger of the Dresden amber pendants is also veiled, but the veil does not form frontal lappets. 76.AO.77 and the Berlin, Dresden, and two of the Novi Pazar figures (688/1 and 689/1) are holding folds of drapery in both hands. Novi Pazar 689/1 displays the most marked instance of cloth grasping (as if the figure is starting a curtsy). The woman of the New York *kourotrophos* pendant balances the child with her right hand and holds the cloth of her garment in her left (a gesture mimicked by the child's left hand). 76.AO.77 and the other extant amber korai and kouroi (including the Novi Pazar ambers) were probably carved in Italy. The similarities of style and form among the amber korai suggest that there is a common invention behind them. The particularities of the dress and adornment of 76.AO.77 help to locate its origins.

The kore's chiton—belted, with long *kolpoi*, an emphasized verticality in the portrayal of the bodice folds, and a central bunching of folds—finds its best parallels among Ionian marbles, including (1) the figure on a votive relief from near Cyzicus (Berlin, Staatliche Museen 1851); (2, 3) two fragmentary marble korai figures from Miletos (Berlin, Staatliche Museen 1577 and 1744); (4) a marble from Didyma (Berlin, Staatliche Museen 1793); and (5) a Milesian-influenced marble from Theangela, Caria (London, British Museum B319).[13] The amber kore displays the South Ionian marble sculptors' feeling for the thickness of the chiton's fabric and a correspondingly rich plasticity in the modeling.

The kore's long veil is distinctly South Ionian, and perhaps characteristic of Miletos. The cloth surrounds the face and neck, covering the ears before falling down the back.[14] Two small lappets are formed at the shoulders, so that the presence of the garment is obvious from a frontal view.[15] In contrast with the tighter Samian arrangement, "around the temples the veil swells to suggest the hair underneath."[16] Characteristically South Ionian (and perhaps even distinctly Milesian rather than Samian) are the zigzag folds of the veil.[17] The cloth weights attached to the ends of the veil show how well the carver understood the light garment. 76.AO.77, which can be viewed in the round, may be one of the best available illustrations of the Ionian long veil, since all of the other extant examples from the area are partial figures or are represented frontally on reliefs.

The physical type might also be said to be Milesian. Analogues for the head include a group of terracottas attributed by F. Croissant to Miletos: two in London (British Museum 205 and 206), one in Athens (National Archaeological Museum 5669), and one in Paris (Louvre MNC 681),[18] which show a similar emphasis on horizontality in the design and placement of the features. The serene faces all communicate a sense of an immediate and living presence. Four marble heads, one from Miletos (Berlin, Staatliche Museen Sk 1631),[19] one from Samos (Berlin, Staatliche Museen Sk 1874),[20] one from Miletos(?) in Paris (Louvre Ma 4546),[21] and another in İzmir (Archaeology Museum 15136),[22] are similar in anatomical structure to 76.AO.77. The eyes of the tiny amber, though different in being hollow, are still distinctively alike in form.

The hairstyle appears to be unusual, but this may be owing to a paucity of comparisons. The curved lower contour is distinct from the common horizontal termination. Comparable coiffures on Greek kore figures include an early-sixth-century terracotta in London said to be from Tanagra, a marble kore from Andros in Copenhagen, a late-sixth-century bronze mirror stand in London (said to have been found in Rome), and a late-sixth-century (with later interventions) marble kore in New York, said to be from the neighborhood of Laurion.[23] The hair of the late-sixth-century B.C. marble Leto found in the Delian temple of Artemis (Athens, National Archaeological Museum 22)[24] is slightly longer than that of 76.AO.77, but the bottom-edge contour is the same. This "pageboy" style is seen more frequently on Middle Archaic votive bronzes (both korai and kouroi) in Etruria than among Greek Archaic korai. The most relevant examples include the Middle Archaic korai in Florence (Museo Archeologico Nazionale 266 and 231), and an unnumbered bronze in the Villa Giulia, Rome.[25] The Etruscan absorption of this style may help to locate where 76.AO.77 was carved.

Who is represented in 76.AO.77? It is hypothesized here that the amber represents a goddess. The material of the pendant and the crown determine its divine identity. If the figure represents an Olympian, Artemis, Leto, Aphrodite, and Eos would be possibilities. Each one is a brilliant goddess associated with light. However, details of the hairstyle and dress, and the South Ionian or even more specifically Milesian style, may preclude any identification other than Artemis or Leto. The veil and hairstyle, and perhaps the breast jewelry, are significant.

Although there is disagreement among scholars about the meaning of the veils worn by kore figures, they are generally agreed to be of considerable significance. Since, as Brunilde Ridgway summarizes, the kore type originated "probably as a divine image … heavily indebted to Oriental prototypes, both in rendering and in items of clothing[,] … [which was] further exploited to portray specific goddesses, usually by the addition of an attribute or extra garment,"[26] the context, the attributes, and the details of dress are all important. For Ridgway, the Leto from Delos raised questions about the interpretation of all East Greek and Samian veiled korai, and she came to the conclusion that the veil drawn over the head must imply more than regional fashions, that it

> *must at least stand for matronly status or even outdoors versus indoor attire…. Even strict dependence upon Oriental religious iconography would not ensure a similar divine identification for all of [them] … [but the] possibility however exists, and we should be open to it since we have too easily extended to all female figures that generic meaning of agalma which, to be sure, applies to many of them.*[27]

Lloyd Llewellyn-Jones summarizes the evidence for veiling in Greece and shows the multivalency of the veil's meanings, including a woman's place in the social order.[28] He reviews the complex range of the Greek terms for "veil" and the difficulty of matching them up with their artistic counterparts. Nevertheless, it is tempting to identify the veil of 76.AO.77 as the delicate and glistening *ampekhonē*.[29]

Besides the veil, the marble Leto from Delos wears another distinctive element, a double chain ornament (which once supported six pendants) that attaches the two veil lappets together like an old-fashioned sweater chain. 76.AO.77 wears a single, pendantless ornament across the breast. Is it attached to the garment, or does it secure the two lappets? In the final argument, Ridgway identified Athens 22 as Leto, partly because of the veil, and partly because of the marble female figure found with it, which she identified as Artemis. (There is no comparable jewelry on the smaller marble.)

The hair of Athens 22 is worn long and, like that of 76.AO.77, is curved at the bottom, which, as noted above, is not common in Greek sculpture, being more frequently seen in Archaic Etruscan small bronzes. This, and the formulaic rendering of the skirt of 76.AO.77, may aid in further situating the amber and even help to identify which divinity is represented.

The formula of representing a chiton skirt of fine fabric drawn into near-parallel folds is found on many marble representations of South Ionian female figures, standing and seated; examples are a votive relief and a fragmentary torso from Miletos.[30]

This is very similar to the patterning of horizontal folds found on Etruscan kore figures (even on figures with the feet placed close together). Two key parallels for the chiton skirt of 76.AO.77 are the Etruscan veil-wearers in Emeline Richardson's Tomba del Barone group of bronzes, and a superb "Ionian" bronze kore from Rimini in Copenhagen.[31] A very similar presentation of the chiton skirt is found on a number of sixth-century Etruscan (Caeretan?) gold plaquettes of korai, with three said to come from Cerveteri and one from Palestrina.[32] The divinities (since they are crowned and made from gold, they must be divinities)[33] fill the rectangular *aediculae* of the plaquettes, and in the London pair, the slope-sided hats overlap the edge. The gold figures wear diadems with colored glass or paste inserts, and the London pair each wears a necklace from which hangs a large inlaid pendant. The conical hats of the Etruscan examples are critical for the figures' identification: are they Artemis?[34] These Etruscan objects are further evidence of the impact of South Ionian artisans in Etruria, perhaps the same ateliers responsible for 76.AO.77 and the other extant amber korai and kouroi, including the Novi Pazar ambers.

The chiton grasping is a significant gesture of 76.AO.77 and all other korai figures and is a subject of considerable discussion. Helmut Kyrieleis has established the point that sculpted korai may represent movement in stasis (and not lose their essential meaning) in the way that a running or hunting Artemis painted on a Greek vase can communicate violent movement.[35] Read as if it were an Egyptian image, 76.AO.77 (and related images) represents a pose frozen at the most important or recognizable part of a ceremony, which "can only be understood in terms of the meaning of the larger ritual in which it was embedded."[36] For the tiny amber korai, is the pose frozen in running or dancing, specifically cultic and ritual dance? If a dance, lines from the Homeric *Hymn to Artemis* might be recalled:

> *This huntress who delights in arrows slackens her supple bow and goes to the great house of her dear brother Phoebus Apollo, to the rich land of Delphi, there to order the lovely dance of the Muses and Graces. There she hangs up her curved bow and her arrows, and heads and leads the dances, gracefully arrayed.*[37]

The active pose, the "attributes" of transparent clothing, veil, crown, chain, and long loose hair, and the material of 76.AO.77 suggest that the tiny pendant represents specifically Artemis or Leto. Given that Leto, like her daughter Artemis, is a birth goddess, and given the age-old association between amber and childbirth, it is likely that 76.AO.77 and her kin served as special ornaments, ones with fertility aspects, either in direct magic or as powerful danger-averting amulets. The clinging transparent clothing and the position of the figure's hand on the skirt both emphasize the pubic area, the location of the primary sex characteristic. This composition is comparable to the Egyptian manner of representing elite adult female fertility figures in the cases where nudity was inappropriate.[38] A fertility subject represented in a substance associated with fertility would enhance the meaning of the object, since subject and material are inextricably intertwined in all ancient amulets. Here the subject might well function as did its Egyptian counterparts, the fertility figures linking the process of giving birth in this world with that of being reborn into the next. If 76.AO.77 and the amber animals *were* originally buried together, the group might be compared to the more complex precious metal plaques from the eastern Mediterranean, such as the numerous multisubject Rhodian plaques (seventh century B.C.),[39] or a work like the semicircular gold plaque (Island Greek work) from Aydın (Tralles) of circa 600 B.C.[40] The most complex of the Rhodian ornaments include female figures (clothed or nude), human heads, animals and animal masks, pomegranate blossoms and other flowers, or solar symbols. A winged *Potnia Theron* with felines and a centaur grasping a fawn are also common subjects. The Aydın pendant is among the most elaborate compositions: the female figure is in the "mastery of the animals" pose and is joined by griffin's, bull's, and ram's heads and nonfigural solar imagery.

Much remains to be understood about the origin, meaning, and function of the large-scale kouroi and korai, but it appears that these sculptural types could serve both as votives and in funerary roles, to represent a divinity, a heroized dead person, or an idealized youth. Both kore and kouros could be a pleasing offering, an *agalma*.[41] Furnished with suitable attributes, the kore or the kouros could become a hero or divinity: one of the Dioskouroi, Apollo, or Athena, for example.[42] While much has been written about the life-size and larger korai and kouroi, there is a relative paucity of discussion about the miniatures.[43]

76.AO.77 and the other amber korai may have served for the living as well as for the dead, in body adornment, as amulet and ornament, roles that the large-scale objects were never called upon to play. The compact format of the pendants, the incorporation of suspension devices or plate bases into their designs, and the occasional evidence of use wear (rubbed surfaces, as on the Novi Pazar and Berlin korai), replacement of suspension holes (as might be argued for the Novi Pazar korai), or enlargement of the stringing holes (as on the Basel kore) suggest a significant use in life (or lives) before interment with the deceased. The ultimate "user" actually may have been the deceased. These tiny korai also may have been gifts and thus part of the funerary rituals. For the works that show no signs of wear, it is possible that they were made expressly, or acquired especially, for setting on the body or for placement in the grave at the time of burial. In death, the amber korai might have provided aid and protection during the fraught journey to the afterworld, if the beliefs attendant on such uses were current in Italy as they had been for millennia in Egypt.

NOTES

1. The term "kore," commonly used for the many kinds of images of Archaic standing draped females, is misleading but useful. See, for example, Ridgway 1977, pp. 85–86.

2. This kore pendant is one of two ambers (the other, 101184, is a bust of a female, possibly a siren) found in the disturbed sixth-century B.C. Tomb 21 from the group excavated at Monteleone di Spoleto, Colle del Capitano: see *Gens antiquissima Italiae: Antichità dall'Umbria a New York*, exh. cat. (Perugia, 1991), pp. 175–76, 356, fig. 3.4. M. C. De Angelis, "La necropoli di Colle del Capitano: Nuove acquizioni," in *Romagna tra VI e IV secolo a.C. nel quadro della protostoria dell'Italia centrale: Atti del Convegno* (Bologna, 1985), p. 288, fig. 9, considers the ambers to be local products influenced by the amber-working of the Basilicata.

3. Heidenreich 1968, p. 655, pl. 9.1 (Bernsteininventar Nr. 1).

4. Ibid., p. 655.

5. Ibid.

6. Palavestra and Krstić 2006, p. 136, no. 47, with earlier bibl.

7. Ibid., p. 137, n. 49, with earlier bibl.

8. Ibid., p. 137, n. 48, with earlier bibl.

9. Metropolitan Museum of Art 1992.11.21, Purchase, Renée and Robert A. Belfer, Philanthropic Foundation, Patti Cadby Birch, and The Joseph Rosen Foundation, Inc. Gifts and Harris Brisbane Dick Fund, 1992.

10. Unpublished.

11. Richter 1940, p. 32, figs. 104, 105.

12. The orientation of the figures may be significant. If the objects were perforated to hang feet upward and head downward, the orientation may have been purposely danger-averting, as it makes reference to the underworld (the opposite, or mirror, world of the dead). Alternatively, the orientation may allude to the nocturnal rather than just the diurnal phases of the sun or moon—that is, to their complete passage.

13. For Berlin, Staatliche Museen 1851, see Richter 1968, p. 93, no. 165; for Staatliche Museen 1577, Karakasi 2003, pls. 46 a–b, 47 c–d; and Richter 1968, p. 92, no. 161; for Staatliche Museen 1744, Karakasi 2003, pl. 22; for Staatliche Museen 1793, Richter 1968, p. 92, no. 162; for London, British Museum B319, Karakasi 2003, pl. 51 a–d; Tuchelt 1970, pp. 127 (L84), 150, 186; and Richter 1968, p. 93, no. 167.

14. Ridgway 1977, p. 97.

15. The lappets that form on the shoulders of a figure from a relief found in Caltidere (Myrina) in the İzmir museum (which E. Akurgal dates to about 570–560 B.C.) are very like those of the amber kore: see E. Akurgal, "Bemerkungen zur Frage der örtlichen und zeitlichen Einordnung der griechischen archaischen Grossplastik Kleinasiens," in *Studies in Classical Art and Archaeology: A Tribute to Peter Heinrich von Blanckenhagen*, ed. G. Kopke and M. B. Moore (Locust Valley, NY, 1979), p. 38, pl. 8.12.

16. Ridgway 1977, p. 97.

17. See, for example, Karakasi 2003, pp. 40–41.

18. Croissant 1983, chap. 2, group B.

19. Berlin, Staatliche Museen Sk 1631: Richter 1968, p. 59, no. 95, figs. 293–95; and Croissant 1983, passim, pl. 1.

20. Berlin, Staatliche Museen Sk 1874: Karakasi 2003, pl. 19; Richter 1968, p. 60, no. 98, figs. 30–45.

21. Louvre Ma 4546: Hamiaux 1992, p. 57, no. 49.

22. İzmir Archaeology Museum 15136: Karakasi 2003, pl. 50; and E. Akurgal, *Mitteilungen des Deutschen Archäologischen Instituts, Istanbuler Abteilung* 42 (1992): 67ff.

23. For the Tanagra terracotta (British Museum 75.39.20), see Richter 1968, p. 102, no. 61, 204–5; for the Copenhagen kore (Ny Carlsberg Glyptotek 1544), Johansen 1994, pp. 56–57, no. 13; for the mirror stand (British Museum 242), Richter 1968, p. 109, fig. 661; for the New York marble kore (Metropolitan Museum of Art 07.286.110), Richter 1968, p. 85, no. 138, figs. 441–44. Also compare the torso (from Paros?) in New York (Metropolitan Museum of Art 07.306): Richter 1968, p. 89, no. 151, figs. 483–86.

24. Ridgway 1977, p. 111. For relevant discussion of the pair from the Dodekatheon on Delos, see F. Zafeiropoulou, *Dīlos: Martyries apo ta mouseiaka ekthemata* (Athens, 1998), p. 225, no. 72; P. Jockey in *Sculptures déliennes*, ed. A. Hermary et al. (Paris, 1996), pp. 48ff., nos. 18–19; and Fuchs and Floren 1987, p. 167, n. 56.

25. For Florence, Museo Archeologico Nazionale 266 (votive deposit at Fonte Veneziana, Arezzo), see Richardson 1983, p. 262, figs. 588–89; for Florence, Museo Archeologico Nazionale 231, ibid., pp. 63–64, figs 597–98; for the Villa Giulia bronze, ibid., p. 264, figs. 601–2.

26. Ridgway 1977, p. 114.

27. Ibid., p. 112. Karakasi 2003 has made a strong case for the veil being worn during religious festivals.

28. L. Llewellyn-Jones, *Aphrodite's Tortoise: The Veiled Woman of Ancient Greece* (Swansea, 2003). See also D. L. Cairns, "The Meaning of the Veil in Ancient Greek Culture," in Llewellyn-Jones 2002, pp. 73–94; Ridgway 1977; U. Kron in *Athen* 1986, p. 56, n. 30; K. Tuchelt in ibid., pp. 32ff.; and Freyer-Schauenburg 1974, passim, on the meaning and interpretation of the veil as it relates to the kore. As Llewellyn-Jones 2003 outlines, the outer garment could denote stages in a woman's life cycle and appears to have played various social and symbolic roles throughout Greek culture. In Homeric epic, noblewomen (and in notable cases their serving women), the focus of the cycles, wear the veil, and in seventh-century B.C. painting, goddesses (Athena and Aphrodite) and well-born wives (Eriphyle and Helen) are depicted with various types of outergarments worn over the head. A girl's passage to womanhood was marked by her use of the veil; she offered her veil to Artemis on the eve of her wedding (when the bride comes under the protection of Aphrodite) before donning a special matrimonial veil; and as a

married woman she wore a veil in public, going without it only in the first stages of mourning.

29. Might the veil of the amber kore be the *ampekhonē* (or *ampekhonon*), the fine, expensive outer garment, most probably a veil, noted in antiquity for its delicacy and semitransparency? Llewellyn-Jones 2003 (in n. 28, above), p. 27, notes that although in texts *ampekhonē* is noted as associated with *hetairai* (prostitutes) and even an Arcadian shepherdess, it has divine associations. It is listed among the textile dedications to major goddesses, in several cases at sanctuaries of Artemis. "On the Athenian Akropolis, an *ampekhonē* is recorded as being draped over the statue of Artemis.... The word occurs three times in the clothing inscriptions at the Artemis Brauronia sanctuary: on two occasions the garment is draped around the statue.... One of [them] has woven into it, 'sacred to Artemis'": ibid., pp. 27–28.

30. Some examples are Berlin, Staatliche Museen 1792 and 1898 (Richter 1968, p. 51, no. 72), and an unnumbered marble (ibid., p. 51, nos. 70–71).

31. For Richardson's Tomba del Barone group (her Late Archaic Series A, Ionians, Group 2), see Richardson 1983, pp. 283–85. For Danish National Museum 4203 (from Richardson's Late Archaic Series A, Ionians, Group 1A, Traditional Sturdy series), see ibid., pp. 279–80, fig. 651. The Danish bronze has many older and unusual features. She wears two crescent-shaped diadems, one in front, one in back; two chitons and an Ionian diagonal himation; and scalloped waves around the face, two curls to each side, and a long fall in back. She holds her right hand on her breast, fist closed. Does the bronze represent an older type of statue, perhaps a cult figure, with an accumulation of related ornaments and dress?

32. Two of the plaquettes are in the British Museum (Jew. 1267–68, said to come from Cerveteri), and two are in Rome (Museo Nazionale Etrusco di Villa Giulia 40875, said to come from Cerveteri, and 53492, from Palestrina, Castellani Collection: Cristofani and Martelli 1983, pp. 300–301, nos. 192–94). Decorating some slightly earlier sixth-century gold earrings in Berlin (Staatliche Museen 30219, 442 a/b) are figures of *Potnia Theron*; she lifts high the left side of her skirt with her left hand, places her right hand on her breast, and stands between lions: ibid. pp. 291–92, no. 142.

33. L. Khalil (LIMC 2 [1984], s.v. "Artemis," pp. 738–40) surmises that the tiny gold images found in the sanctuary of Artemis at Ephesus probably represent the goddess herself, *because* of the precious material of which they are formed.

34. For discussion of the conical hat and Artemis, see 77.AO.84 (cat. no. 1).

35. H. Kyrieleis, "Der Tänzer vom Kap Phoneas," *Istanbuler Mitteilungen* 46 (1996), pp. 119–20, n. 43; also noted by Karakasi 2003, p. 50.

36. Wilkinson 1994, p. 205.

37. "To Artemis," XXVII, from *Hesiod: The Homeric Hymns and Homerica*, trans. H. G. Evelyn-White, Loeb Classical Library 57 (London, 1914).

38. On the representation of transparently robed elite female figures in Egypt, see G. Robins, "Dress, Undress, and the Representation of Fertility and Potency in New Kingdom Egyptian Art," in Kampen 1996, pp. 27–40.

39. Laffineur 1978.

40. For the ornament from Aydın in the Louvre, see Becatti 1955, no. 149; and Higgins 1980, p. 115 (as "almost certainly Island Greek"). See also Boardman 1980, p. 99, fig. 117; Laffineur 1978; and E. Coche de la Ferté, *Les bijoux antiques* (Paris, 1956), pp. 30, 44, 56, pl. 6.2. The ornament incorporates rosettes and zigzag patterns in granulation, as well as plastic figures of four heads in flat disks, two bulls and two rams. Higgins points to a Boeotian terracotta of the eighth century B.C. wearing a similar ornament and suggests that both show Syrian influence.

41. Ridgway 1977, pp. 49–59 (for the meaning of the kouros) and 108–13 (for the meaning of the kore). Convincing is C. M. Keesling's argument that Greek and Cypriot votive korai did not represent human votaries, since, without evidence to the contrary, the ancient viewer would expect to see a divinity in the image: C. M. Keesling, "Finding the Gods: Greek and Cypriot Votive Korai Revisited," in *Divine Images and Human Imaginations in Ancient Greece and Rome*, Religions in the Graeco-Roman World 170, ed. J. Mylonopoulos (Leiden and Boston, 2010), pp. 87–103.

42. Stewart 1997, p. 65. The extensive bibliography on the functions and meanings of kouroi includes the following, which were helpful in this study of amber kouroi and korai: B. S. Ridgway, *Second Chance: Greek Sculptural Styles Revisited* (London, 2004), p. 755; G. Ferrari, *Figures of Speech: Men and Maidens in Ancient Greece* (Chicago, 2002), chaps. 5–6; Kyrieleis 1995, pp. 119–20; G. Schäfer, "Gepickt und versteckt: Zur Bedeutung und Funktion aufgerauhter Oberflächen in der spätarchaischen und frühklassischen Plastik," Jdi 111 (1996): 25–74; Ridgway 1977; H. von Steuben, *Kopf eines Kuros*, Liebieghaus Monographie 7 (Frankfurt am Main, 1980); and Richter 1968. As Sourvinou-Inwood 1995 (see cat. no. 3, n. 25), p. 143, writes about the kouros as grave marker, "The metonymic sign of the deceased could represent metaphorically the now-perished beauty of the deceased, and it generally enhanced and colored positively the memory of that deceased." This might be true for a female or a child, as well as a male, deceased person. In this regard we might also ask, What was the role of the tiny bronze sheet kouros figures on the Etruscan chariot from Monteleone in New York (Metropolitan Museum 03.21.1, Rogers Fund, 1903)?

43. Perhaps something of the Egyptian approach to the minuscule was retained in the tiny korai and kouroi images, especially when they are made of high-status, magical materials. Miniature figures might provide magical, potent assistance for the deceased's afterlife. Wilkinson 1994, p. 42, in his section on

"the miniscule," registers the symbolic function of miniscule scale in commemorative ritual and mythical purposes, the Egyptian delight in intricately formed tiny objects as a display of skill, the suitability of the minute for objects of adornment, and the cases in which miniatures served as magical replacements for full-scale objects in funerary contexts.

9. Pendant: Head Fragment from a Standing Female Figure (Kore)

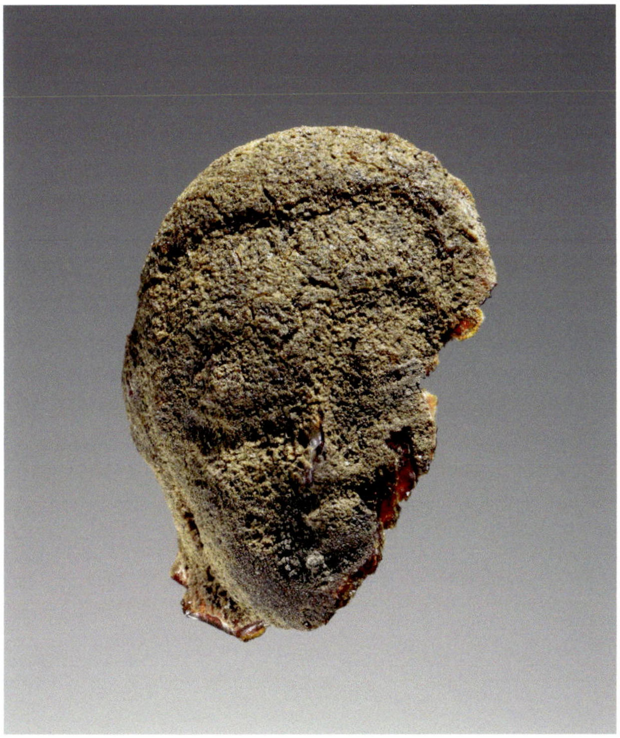

Accession Number	82.AO.161.6
Culture	Etruscan
Date	525–500 B.C.
Dimensions	Height: 31 mm; width: 20 mm; depth: 18 mm; Weight: 2 g
Subjects	Ionia, Greece (also Ionian, Greek)

Provenance

–1982, Jiří Frel, 1923–2006, and Faya Frel (Los Angeles, CA), donated to the J. Paul Getty Museum, 1982.

Condition

The amber is fragmentary: a large section of the left side of the head and the neck and a portion of the back of the pendant were broken off at some remote time, as is suggested by the weathering. Overall, the surface is worn and grainy but not friable. The surface is covered by a yellow-orange crust that obscures the underlying red-orange amber. There are new chips in several areas, especially along the edges, that have exposed glassy amber below. In natural light, the piece is opaque and yellowish brown. At the new breaks, which expose the interior, it is red-orange. It is not translucent under artificial light, and it is impossible to see if there are any inclusions. There is a stopped bore on the back of the head, and just below the right ear is a stopped bore with a plug of amber still inside it.

Description

This is a fragmentary and weathered object. What remains is the head and neck of a female figure. The brow is short, and the cheeks are modeled and full. The large, almond-shaped eyes have narrow rims. The nose is eroded, but its general form is still evident: it is triangular, short, and relatively narrow through the nares. The mouth is wider than the nose. The jaw area is wide. The hair at the brow is dressed in a wave pattern. In front of the right ear is a large coil of hair. The rest of the figure's hair is covered with a veil.

There is no surviving suspension device. However, the groove at the break on the left side of the head and the groove and break on the underside of the neck are likely two sides of a triangular suspension system.

Discussion

82.AO.161.6 was donated along with a large and varied group of amber beads and pendants about which little is known.

This fragment is unlike any other figured carved amber. Because of the odd break at the bottom of the piece and the remains of a triangular perforation system, it is proposed here that 82.AO.161.6 is the upper part of a standing figure and not a head-pendant. The three holes would have allowed the sculpture to hang upright. A number of other figured ambers from Italy, or said to be from Italy, and dating to between the seventh and fifth centuries were perforated for stringing in a similar manner.[1] The best parallel is another fragmentary amber pendant, in a London private collection, which represents a seated woman and child. It dates to the mid-seventh century B.C. and was probably carved at Vetulonia.

The figure of 82.AO.161.6 is similar in physiognomic type and dress to a group of sixth-century B.C. marbles from Ionia, from the area of Samos, Miletos, and Didyma. The telling comparisons are (1) a small marble head of a kore from Miletos(?) in Paris; (2) the statue of Dionysermos in the Louvre (from Samos? Miletos?); (3) two votive reliefs of standing women in *aediculae* from Miletos in Berlin; and (4) the fragmentary figures from the column bases found near the temple of Apollo at Didyma, also in Berlin.[2]

The amber, like the marbles, has a wide face with full, round cheeks, high cheekbones, slightly sunken areas at the lower edge of the eyes, and almond-shaped eyes. Those of the amber appear to be larger, but the shape and tilt are the same. The position of the veil on 82.AO.161.6, behind the brow-framing waves, is the same as that of the Louvre kore and the Berlin column base heads. The amber and the Didyma marbles have in common another detail of hairdressing, the thick curl of hair in front of the ears. The latter firmly places 82.AO.161.6 in the same sculptural tradition as that of the Louvre and Berlin marbles and aids in establishing a *terminus post quem*. The long veil of 82.AO.161.6 is also characteristic of South Ionian kore figures, as described in the previous catalogue entry. This is the same long veil intimated by the presentations of the Getty amber pendant of a female head, 77.AO.81.25 (cat. no. 26).

NOTES

1. 77.AO.81.12 (cat. no. 52) is a significant example of a pendant with multiple perforations.

2. For the Milesian kore head (Louvre Ma 4546), see Hamiaux 1992, p. 57, no. 49; for the statue of Dionysermos (Louvre Ma 3600 [MND 2283]), ibid., pp. 59–60, no. 51; for the two Milesian votive reliefs in Berlin (Staatliche Museen 1792 and 1647), Richter 1968, p. 51, nos. 70–71, figs. 228–29; and Karakasi 2003, pl. 42; for the column base figures in Berlin (Staatliche Museen F724–25), Richter 1968, p. 60, nos. 96–97, figs. 296–300.

Human Heads

The human head is the most common of all amber pendant subjects and enjoyed the longest duration, from the late eighth to the late fourth century B.C. From the beginning, they appear to follow a much older convention, whereby one part of a figure is sufficient to represent the whole. The symbolic meaning of the detached head as a pendant varies according to its type and use, but the head always represents a demon, hero, or divinity.

In this catalogue, I refer to this type as "head-pendants." These can be divided into four basic schemata: a frontal face that is plain on the reverse; a complete head (the neck or a part of the neck is included, and the pendants are made to hang frontally); a frontal head with the neck included, but a plain reverse side; and a profile head with the neck included, the reverse plain, and the pendant made to hang in profile. Female subjects (humans or anthromorphs), with dressed hair, ornaments, or head coverings, and satyrs are found in all four schemata. Heads of indeterminate sex (probably representing youths or sphinxes), Herakles in a lionskin helmet, and bearded male heads with human ears occur only in the form of a frontal face with no neck, like a mask, the strongest form of the facing head motif.[1] This is also the case with the rarely represented gorgoneion, the frontal face that precedes the appearance of the whole Gorgon in ancient art.[2]

All head-pendants of Greece and ancient Italy used in adornment have ancient small-scale antecedents in the art of Egypt, the greater Near East including the Syro-Phoenician area, and the Aegean, as well as in the image making of Ice Age Europe. The antecedents are of the utmost significance for the amber head-pendant: it should be seen as a late manifestation of a millennia-old tradition of wearing miniature decapitated heads, or heads *pars pro toto*, made in materials of high value, on the head, around the neck, or on the upper torso (and much more rarely elsewhere on the body).

The earliest surviving head-pendants in amber date to the late eighth century B.C., but they became more widespread during the second half of the sixth century, the same period in which the subject of a detached head was popular in other media, from architectural decoration to vessels and coinage. The format of the earliest documented en face amber head-pendant is very like that of the earliest Sumerian and Egyptian detached heads. The Sumerian profile heads used in inlays (various materials) date concurrently with the earliest Sumerian ornament-amulets in the form of frontal heads and faces. The frontal heads are in two basic types, those without horns (identified as goddesses) and those with bull horns (identified as gods). The earliest Egyptian amulets in the form of detached heads are flat-backed, front-facing heads and are exclusive to the late Old Kingdom and the First Intermediate period. Carol Andrews summarizes their appearance: short beard, prominent ears, and a suspension projection on the top of the head; made predominately of cornelian. Each is "intended to give its wearer the use of the senses in general."[3] Related to this type is another: human-head scaraboids, current from the New Kingdom onward. The enhancement of the scarab—the amulet par excellence of new life, regeneration, and resurrection—with a human face augmented its properties.[4] Mycenaean objects with detached heads include engraved gems and metal vessels. A signal Minoan work is the bead in the shape of a human head in the Jewel Fresco at Knossos.[5] The heads of female and male figures made of faïence (glazed composition) from Apadana, Mari, and other sites in Mesopotamia dating to the late second millennium B.C. must have had great influence on the development of the head-pendant as ornament and amulet.[6]

Antecedents for the amber head-pendants from Etruscan tombs are the protomes embossed in North Syrian metalwork, such as those on the bronze paterae and cups interred in the Barberini Tomb at Praeneste (those on the cups are winged);[7] the faces of possibly Syro-Palestinian

Tridacna squamosa shell figure-vessels, alabaster and stone cosmetic palettes, and ostrich-egg and other Oriental oinochoai;[8] the human heads on Etruscan *bucchero* vessels with relief decoration;[9] and the two types of ivory faces from the Barberini Tomb (the bearded faces and the single perhaps female face).[10]

The female heads on some late-eighth-century B.C. Cretan and Cycladic gold objects may be the direct ancestors of the earliest amber head-pendants. These include the frontal heads attached to a crescent-shaped gold pendant from the Khaniale Tekke Tomb from near Knossos,[11] and three gold "buttons'" from a tomb at Megara (and a related pendant from Naxos in Berlin).[12] Contemporary with these gold ornaments are the earliest surviving amber head-pendants: a pair of indeterminate sex from the sacred deposit at the temple of Artemis at Ephesus, and three undocumented faces from Italy (perhaps Vetulonia), that of a youthful female(?) figure whose hair is braided back from the brow, an unbearded, helmeted figure, and a female divinity (possibly) who wears an elaborate bird headdress.[13]

The next generation of frontal heads and faces includes the numerous seventh-century precious metal objects made in East Greece and Etruria. These include objects where the face is the single image, and others in which the face or detached head is one of many. Two objects exemplify the pervasiveness of this imagery: a pair of Etruscan armlets from Vetulonia, whose two terminal registers are made up of trios of faces (perhaps those of sphinxes), and an electrum temple pendant from Cameiros, a complex work that includes two frontal heads at the top.[14] Numerous bone and ivory protomes from East Greece and Sparta date to this period.[15] Although they are not independent works, the amber faces set into ivory and bone reliefs—a Laconian ivory sphinx and a pair of Laconian winged fertility divinities, flanked by smaller figures—should be mentioned.[16] The detached female heads protecting a number of sanctuary-dedicated Greek bronze vessels of the seventh century are related in both subject and function:[17] the female protective goddess represented on a vessel support from Olympia[18] is probably the same divinity as the one gazing outward from the Rhodian metalwork[19] and the early ambers. In addition, noteworthy seventh-century analogues for the gold, ivory, and amber heads are those represented on early electrum coinage.[20]

Although there are many instances of detached heads in seventh-century art, especially in East Greece, Crete, and Etruria, the subject was more widespread during the next century. The prominence of the detached head in Etruscan imagery, such as the so-called Canopic heads on the "metopes" of Caeretan *dolii*, or terracotta architectural decorations (on antefixes and raking *simas*), indicates the significance of the subject.[21]

The importance of the detached head in early-sixth-century Italy is indicated by the variety of materials in which it is found, the range of its use, and its geographic distribution.[22] The amber head-pendants of the sixth century gained currency at the same time that the subject was more frequently employed in other media: in Greek vase painting; in early Greek facing head coinage;[23] on contemporary gemstones; as precious metal adornment worn directly on the body (as on a necklace)[24] or attached to dress, or as ritual headgear;[25] as an attachment on bronze vessels; in architecture, as antefixes and on the raking *sima*; as the subject of terracotta votives; and for small votive(?) bronzes.[26] A key amber of this period is the early-sixth-century head of a female figure (a siren or sphinx) wearing a slope-sided hat, excavated from an amber-rich grave, Tomb 96 at Chiaromonte–Sotto la Croce.[27]

Further study of the female heads on Etruscan Orientalizing metal reliefs and Etruscan *bucchero* and on Greek and Etruscan vessels, *thymiateria*, and lamps,[28] as well as of the bearded male heads and demonic figures of "Phoenician" glass pendants—and their relationship to the early amber heads from Italy—should reveal a related purpose and iconography. They all must represent danger-averting divinities, heroes, and fantastic beings.

Amber head-pendants of the second half of the sixth century were made during a period of considerable amber availability in Italy, but most are small, from 10 to 50 mm. Notable examples are the six female faces from a tomb at Eretum: a frontal head-pendant of a woman from Certosa, two frontal female head-pendants from Tomb 102 at Braida di Vaglio (Basilicata), the Getty *Head of a Female Divinity or Sphinx* (76.AO.85.1 and 76.AO.86, cat. no. 10), and two heads (one possibly male) in the Louvre.[29] The earliest of the *profile* amber head-pendants date to the second half of the sixth century. From that point onward, until their virtual disappearance in the last quarter of the fourth century B.C., female subject head-pendants in both profile and frontal format coexist.

Just after the mid-sixth century, the first male head-pendants appear, first satyrs (in both facing and profile formats) and then, a few decades later, faces of the Cypriote-type Herakles and a type of unbearded males with human ears. The latter two types are uncommon. There are no male heads with bull's horns, a not uncommon subject in other media.[30] The earliest examples of satyr-subject head-pendants are the Getty *Satyr Head* (82.AO.161.1, cat. no. 13) and a trio (found together) in a New York private collection.[31] The satyr-subject head-pendants precede ambers representing full-bodied Bacchic revelers (perhaps all satyrs) by at least a decade, if not more, in the early fifth century, and continue to be a popular subject until the early fourth century B.C., when figured ambers ceased to be made. From their earliest appearance in Greek art, and enduring in popularity throughout late antiquity, satyr heads were considered especially efficacious in averting danger, evil, and particularly the evil eye.[32] In Italy and in Etruria especially, they were frequently employed as antefixes, often coupled with female heads, usually identified as maenads. Satyrs, in Etruscan art of the sixth century onward, were present in sacrifice scenes, and, as Jean-René Jannot points out, it is not clear whether this alludes to a cult of Dionysos or to one of nature more generally.[33]

The other identifiable male subjects of head-pendants include the unbearded Herakles in a lionskin helmet. Other male types are images of bearded males with naturalistically shaped ears and a number of unbearded faces of indeterminate sex that seem to represent youths. The latter, though nameless, probably represent heroes or divinities appropriate to the material of amber. Apollo is a good candidate; the god was worshipped as the Averter of Evil (Greek: *Alexikakos, Apotropaios*), the Protector (*Epikourios*), and the Purifier (*Katharsios*).[34]

Eye size varies among the earliest head-pendants of the seventh and sixth centuries: in some the eyes are naturalistically scaled; in others they are huge, staring, and rimmed with heavy lids. From the fifth century onward, large, "old-fashioned" staring eyes are the norm for both profile and frontal head-pendants, with the exception of a few classes, notably one attributed to Canosa and another to Campania.

From their earliest appearance, female head-pendants wear their hair elaborately dressed. In addition, they wear one or more kinds of head decoration or covering, sometimes in complex combinations: bands, crowns, caps, hats (various styles of *poloi* and various cone-shaped styles), *kekryphaloi* (kerchiefs), wrappings made from strips of cloth (sometimes over hats), and veils or other drapery-type head coverings. Not only are the facial types and artistic styles diverse, but the grooming and adornments also differ. Headgear, hairstyles, and jewelry vary, and the ambers in the Getty collection include most of these variants. A number of the female head-pendants, in various kinds of dress, are carved with a wing or wings. Among the earliest is a large profile head, 76.AO.85.2 (cat. no. 15). The winged head-pendants vary in type. Some, such as 76.AO.85.2, represent a youthful female. Others, such as 77.AO.81.5 (cat. no. 23), represent a heavier-faced, mature, and unsmiling type.

Most amber female head-pendants wear headgear in the Archaic Ionian Greek or Etruscan fashion, as most modern students of the material have commented. The figures are dressed in the manner of the elite—sometimes they are clearly divine or heroic figures, sometimes actual persons; in other cases, they represent persons in ritual roles as votives, celebrants, and offerants. Parallels are found in Greek, Campanian, and Etruscan art, for both style and dress, such as the women painted in Etruscan tombs and on vases, molded in relief on Etruscan *bucchero* or terracotta antefixes, engraved on Etruscan mirrors, or made into small bronzes, many of which are votives.

The findspot and context of only a small percentage of the hundreds of facing and profile amber heads that have come to light since the nineteenth century are documented, and of these, even fewer were excavated under controlled conditions. Seventh-century examples are few but are recorded as coming from sanctuaries where they were dedicated to the Greek divinities Aphrodite, Artemis, and Apollo Daphnephoros. The recorded examples of sixth-to-fourth-century date almost all came from female grave contexts (those of women or children), and in only one case from an Italic sanctuary, that of the goddess Mefite (a goddess with *kourotrophic* and chthonic powers), who is identified with Aphrodite/Venus.[35]

The identities of the various female figures represented in the amber head-pendants are still an open question. A number of proposals have been put forward, all with the apparent assumption that in every case, similar-looking heads represent the same being. A female goddess, a protective genius, or a maenad—these are some of the hypothesized identifications.[36]

In the view of this author, amber head-pendants represent a variety of beings, including divinities, supernaturals, and demons, and the identities are not fixed. The material of the head-pendants, the amber, seems to preclude identifying them as images of their owners, or even of mortal mourners. Although it is unlikely that the head-pendants represent votives, supplicants, or offerants—identifications often made for related figures in other media, such as Etruscan bronzes—it is still possible that they may do so in light of ethnographic analogies. Amber female head-pendants were made over centuries and descend from different iconographic types. The same image type may have been used to represent different beings, depending on the circumstances. In the case of buried heirlooms or prestige gifts, it is possible that a treasured piece may have had several owners and even different identities over time.

Roger Moorey's recent words about the identity of terracotta idols are pertinent to this vexing issue. In his Schweich Lectures, he proposed of anthropological research that

> in the first place *it highlights the fact that figures of similar appearance may have represented different beings, natural or supernatural; that the same type of figurine might have multiple functions; and that in one assemblage the same type might have had more than one function.* In the second place, *it indicates that terracotta anthropomorphic figurines do not have to conform to the tendency to regard them as necessarily representative of supernatural beings.... They may have embodied aspects of prevailing ideologies, whilst also reflecting contemporary society by encoding a variety of ritually significant knowledge relevant to the world of man and nature.... In light of ethnographic analogies ... clay figurines do not have to conform to our expectations for them to be representations of supernatural beings or forces rather than of living human beings acting as votaries or worshippers or perhaps of dead human beings as ancestors or ghosts.*[37]

Attributes may have modified the basic types into more individualized representations. The addition of a wing or wings, a diadem, or a necklace might signal an aspect of a divinity or demon and act as a determinant. Just as is the case with votive terracotta heads, the types might be modified by additions or adjustments. Identical terracotta votives may in one circumstance represent the offerant, in another, the divinity, and in still others, both at the same time—since the gods were represented in human form.

That the amber head-pendants are made from a potent material traditionally employed for divine, demonic, and heroic subjects, a material valued for its protective, apotropaic, and regenerative aspects, seems to limit the range of possible identities to certain female divinities, nymphs, protective geniuses, guides or psychopomps, demonic anthromorphs such as a sphinx or siren, spirits or souls, and, possibly, magical subjects like Medea or Circe.

What is critical is how these head-pendants might "work." The bodiless heads and faces were *pars pro toto* of the full body, and thus held special power. A head alone could convey the hieratic implications of the complete body. The facing head and the frontal eye—an excerpt of the facing head—were highly potent foci and functioned apotropaically.[38] The glaring eye, guarding against danger and averting evil with its "terrifying gaze" (*phobon blepon*), underscores the protective role of the image.[39] The number of head-pendants with large (disfiguring) holes in the faces may be critical evidence for one kind of use and identity. It is possible that they were disfigured by drilling in order to nullify the possible negative powers of the image.[40]

The identifiable pendants of Herakles are of the Cypriot type and underline the renowned potency of the immortal hero-divinity: Herakles is a savior and a healer, a protector of springs, and a danger-averting deity (in Etruria, his close connection with Uni clearly establishes him thus). In each of the amber Herakles heads, he is frontal and wears the skin of a lion, a symbol of his strength, prowess, and deadly force. His glaring eyes emphasize his affinity with wild and dangerous fauna: he could be *deinon paptainein* (terrible to behold). In the one instance where the context of the Herakles head-pendant is known, its figured amber counterpart is a satyr, servant of Dionysos.[41] This juxtaposition of ambers parallels the subjects of the main gate at Thasos, Herakles on one side with drawn bow and Dionysos on the other with the thyrsus.[42] The two dangerous gods acted as guardians of the city.

An *electron* amulet of Apollo might well connect his fiery missiles with amber's solar origin. Apollo, Herakles, and Artemis, Apollo's twin—the three bow-bearing gods—were, for good and ill, death dealing. The divine twins were sharp shooters in their murder of Niobe's children; the twin gods could also deliver peaceful deaths to the elderly.[43] What better all-around amulet than one that could protect, ward off death and danger, promise rebirth, and heal? (Apollo's son is Asclepius; fiery amber could function sympathetically to ward off fever; Apollo is one of the deities powerful for both the living and the dead.)

The dual natures of deities like Herakles and Dionysos or Artemis and Apollo are often played up in their iconography: their power to protect is directly related to their powers of destruction. In Italy, and specifically in Etruscan religion, the deities powerful for this life, the transition to the afterworld, and the afterlife itself were critical to funerary imagery. A special role was reserved for those deities with light and rebirth aspects. Erika Simon illuminates this eloquently: "Thus also the

Etruscans wished their dead to have light. They gave them amulets with astral symbols and painted the holy laurel grove of Apulu/Usil on the walls of their tombs."[44]

NOTES

1. The origins of the facing head must lie in the efficacy of the severed heads of real animals positioned as trophies in public or cultic places. In P. Erhart Mottahedeh's view (Mottahedeh 1979, pp. 274–75),

   > *there exists no better vehicle for conveying hieratic and demonic associations. Deity and demon alike were most powerful and impressive if encountered face to face, and in pictorial art only the facing head allowed direct confrontation between image and viewer. The facing head was, in effect, the pictorial equivalent of the cult image in the round. It could express more ably than the profile head the presence or actuality of deity or demon, for immediacy was potency. From earliest times the facing head motif reveals a fundamental duality and polarity in art; it exhibits, so to speak, two faces—one hieratic and godly, the other demonic and monstrous. Opposed to the profile head, it can assume either a positive or a negative role, signifying good or evil, sacred or profane, and similar polarities. This duality and polarity of the facing head motif was engendered by ancient religious beliefs in an elemental godhead which combined polarities—one both threatening and protective, beautiful and monstrous, fertile and barren.*

   She continues, "Its strongest form is the Mask, which is able to capture essence alone, undiluted by activity.... The motif was initially reserved for highly symbolic figures.... For human figures it served to strengthen divine nature, while for certain animals and hybrid creatures it served to underscore demonic force." For other discussions of the mask form and its apotropaic role, see the sources listed in Mottahedeh 1979, n. 198; Steiner 2001, pp. 196–97; Faraone 1992, pp. 37–38; and, most important, Frontisi-Ducroux 1991. Carpenter 1986, p. 97, discusses the eyes and masks of late-sixth-century B.C. Greek cups and amphorae. While he admits that they "may have had their original stimulus in magic or cult," he believes that their "meaning can be found in the realm of humour." However, he does not address the importance of humor as a time-honored aversion technique.

2. On Gorgo, Medusa, and the gorgoneion, see W. A. P. Childs and D. Tsiafakis in *Centaur's Smile* 2003; R. Mack, "Facing Down Medusa (An Aetiology of the Gaze)," *Art History* 25, no. 5 (2002): 570–604; S. R. Wilk, *Medusa: Solving the Mystery of the Gorgon* (Oxford and New York, 2000); Rocco 1999, n. 199 (with other relevant bibl. not listed here); J.-P. Vernant, "Death in the Eyes: Gorgo, Figure of the Other," pp. 114–15, and "In the Mirror of Medusa," pp. 141–51, in *Mortals and Immortals: Collected Essays by Jean-Pierre Vernant*, ed. F. I. Zeitlin (Princeton, 1991); LIMC 4 (1988), s.vv. "Gorgones" and "Gorgones in Etruria" (I. Krauskopf); and Mottahedeh 1979.

3. Andrews 1994, p. 69.

4. The scarab is the Egyptian solar subject par excellence. It represents the morning manifestation of the sun-god. Andrews 1994, p. 51, outlines: "Because of the underlying ideas inherent in its shape, the scarab form of itself offered the hope of new life and resurrection, but these magical properties could be enhanced even further by the inscription motifs or pictorial representations added to the flat underside." Amuletic scarab seals can also have human elements, and a human head with hair can replace the whole back, as on scaraboids produced in the Naucratis faïence factory at the end of the seventh and in the sixth century B.C.: V. Dasen, "Squatting Comasts and Scarab-Beetles," in Tsetskhladze et al. 2000, p. 91. Dasen establishes that the image of scarab beetles was associated with Dionysos (p. 95, with essential bibl., including her *Dwarfs in Ancient Egypt and Greece* [Oxford, 1991] and Hölbl 1979). Two other Egyptian amulet types relevant for the amber head-pendants are the faces of Bes and of Hathor. The relationship of the heads of Meskhenet to Greek and Etruscan female head representations may also prove significant. The head of Meskhenet was "frequently attached to a type of brick that Egyptian women crouched upon when giving birth. Meskhenet was also believed to appear at the time of an individual's death—perhaps to preside over 'birth' into the afterlife—and the goddess is sometimes depicted in this way in vignettes from the Book of the Dead": Wilkinson 1992, p. 41 (with reference to E. Russman, *Egyptian Sculpture: Cairo and Luxor* [Austin, TX, 1989], pp. 19–21, 214, n. 5).

5. Higgins 1980, p. 63, fig. 9.

6. For the faïence head-pendant from Susa (Apadana) in the Louvre (Sb 03588), see *Faïences de l'Antiquité: De l'Égypte à l'Iran*, exh. cat. (Paris, 2005), p. 65, no. 159. The following are all in the Louvre and illustrated in the above-cited catalogue: example from Ougarit (Minet el-Beida, Tomb VI, AO.15731), no. 154; two from Mari (Tomb 236, AO.19078, and Tomb 255, AO.19488), nos. 155–56; the other two from Mesopotamia (AO.07089, AO.06685), no. 157.

7. Large bowl mounted on three legs, Rome, Museo Nazionale Etrusco di Villa Giulia 13131, from the Barberini Tomb, Praeneste: C. Densmore Curtis, "The Barberini Tomb," *Memoirs of the American Academy at Rome* 5 (1925): 42–44, no. 79, pls. 26–27.

8. For the engraved *Tridacna squamosa* shell figures, sirens(?), falcons, and other creatures, see the recent discussion in B. Brandl, "Two Engraved Tridacna Shells from Tel Miqne-Ekron," *Bulletin of the American Schools of Oriental Research* 323 (2001): 49–62 (with earlier bibl., including R. A. Stucky, *The Engraved Tridacna Shells* [Sao Paulo, 1974]; A. Rathje, "A Tridacna Squamosa Shell," in *Italian Iron Age Artefacts in the British Museum: Papers of the Sixth British Museum Classical Colloquium*, ed. J. Swaddling [London, 1986], pp. 393–96; and D. Reese and C. Sease, "Some Previously Unpublished Engraved Tridacna Shells," *Journal of Near Eastern Studies* 52 [1993]: 109–28). For the other imported vessels, see A. Rathje, "Oriental Imports in Etruria in the Eighth and Seventh Centuries: Their Origins and

Implications," in Ridgway and Ridgway 1979; and A. Rathje, "Some Unusual Vessels with Plastic Heads on Their Necks," in *Studia Romana in Honorem Petri Krarup septuagenarii*, ed. K. Ascani (Odense, 1976), pp. 10–19. The head on the umbo of the British Museum shell (GR 1852.112.3) is identified as that of a woman on the museum's website.

9. For example, the bowl with twelve human heads from the Tomb of the Painted Lions, Cerveteri, third quarter of the seventh century B.C. (Rome, Museo Nazionale Etrusco di Villa Giulia 13234): Haynes 2000, fig. 39.

10. Rome, Museo Nazionale Etrusco di Villa Giulia 13396 (bearded males) and 13421 (female; possible sphinx) from the Barberini Tomb, Praeneste: M. L. Uberti, "Gli avori e gli ossi," in *I Fenici* 1988, pp. 404–21, 743, nos. 935–36; M. E. Aubet, *Los Marfiles orientalizantes de Praeneste* (Barcelona, 1971), pp. 165–68, pl. 25; and Curtis 1925 (in n. 7, above), pp. 31, 101–2, nn. 47–49, pl. 10.

11. Heraklion, Archaeological Museum: Musti et al. 1992, pp. 243–44, no. 53.2.

12. Paris, Louvre 13638: ibid., p. 246, no. 60 (with references).

13. The youthful face is in a private collection in Basel. In the Metropolitan Museum of Art are the helmeted male pendant, 1992.11.14b, and the female bird-hatted divinity(?), 1992.11.14a (both are Purchase, Renée E. and Robert A. Belfer, Patti Cadby Birch and The Joseph Rosen Foundation Inc. Gifts, and Harris Brisbane Dick Fund, 1992).

14. For the armlets in Florence (Museo Archeologico Nazionale 92600–92601), see Cristofani and Martelli 1983, pp. 137, 280, no. 96; for the complex electrum temple ornament, with rosettes, poppy flowers, a nude, necklaced female, a lion's face, and two frontal heads, in Paris (Louvre S 1208), Musti et al. 1992, p. 121, no. 80, with bibl.

15. See Laffineur 1978. For the plaquettes from Cameiros, Rhodes, in the British Museum (Jew. 1103-6, 1108), see Marshall 1911, pp. 85–87, pl. XI.

16. For the sphinx appliqués with amber faces from a couch in the Iron Age Celtic tomb at Grafenbühl, Asperg (Stuttgart, Würtembergisches Landesmuseum), the key first publication is H. Zürn and H. V. Hermann, "Der 'Grafenbühl' auf der Markung Asperg, Kr. Ludwigsburg, ein Fürstengrabhügel der späten Hallstattzeit," *Germania* 44 (1966): 83, 100–102, pl. 12. Later discussion includes Rocco 1999, pp. 83–85, (with a critical assessment); J. Fischer, "Zu einer griechischen Kline und weiteren Südimporten aus dem Fürstengrabhügel Grafenbühl, Asperg, Kr. Ludwigsburg," *Germania* 68 (1990): 120–21; J.P. Mohen *Trésors des princes celtes*, exh. cat. (Paris, 1987), pp. 24–26; B. Shefton, "Zum Import und Einfluss mediterraner Güter in Alteuropa," *Kölner Jahrbuch für Vor und Frühgeschichte* 22 (1989): 214, n. 34; Mastrocinque 1991, pp. 84–85; A. Mastrocinque, "Avori intarsiati in ambra da Quinto Fiorentino," BdA 10 (1991): 3ff. (where he assigns the Belmonte Piceno appliqués to the same atelier as the Asperg sphinx, and points to a Tarentine manufacture); and Rolley 1996, p. 389 (which identifies the place of manufacture as Picenum and the style as Laconian, but without ruling out Greek, possibly Tarantine, workmanship). For the Picene plaques with standing winged divinities between acolytes, see Rocco 1999, pp. 82–85, nos. 135–36, pls. XLIV–XLV.

17. Marangou 1969, pp. 154–58, nos. 88–101.

18. For Olympia Br 10881 (Athens, National Archaeological Museum 6201), see W. Gauer, "Gerät- und Gefässefüsse mit Löwenpranken und figürlichem Schmuck aus Olympia," AM 99 (1984): 38–40.

19. Among the plaquette-ornaments (garment decorations; closures?) is a group of eight silver-gilt objects from Bologna (Aureli necropolis, Tomb 11) dating to the last quarter of the seventh century B.C. (Bologna, Museo Civico Archeologico 25681–88): Bartoloni et al. 2000, pp. 362–63, no. 517. Is the severe and unsmiling figure *Potnia Theron*?

20. For early electrum coinage with faces and heads, see, for example, Kraay 1976; and R. W. Wallace, "The Origin of Electrum Coinage," AJA 91 (1987): 385–97.

21. Among the earliest heads used as antefixes in Etruria are the *Potnia Therons* from the Orientalizing "workshop" building at Poggio Civitate: see E. Nielsen, "Interpreting the Lateral Sima at Poggio Civitate," in De Puma and Small 1994, pp. 64–71. W. A. P. Childs in *Centaur's Smile* 2003, p. 64, proposes Artemis for the female heads interspaced with gorgoneia at the early temple of Hera at Corfu. N. A. Winter, *Greek Architectural Terracottas from the Prehistoric to the End of the Archaic Period* (Oxford, 1993), pp. 62–63, reiterates, albeit with skepticism, the proposal by M. Torelli ("Terrecotte architettoniche arcaiche da Gravisca e una nota a Plinio, *NH* XXXV, 151–52," *Nuovi quaderni dell'Istituto di archeologia dell'Università di Perugia* 1 [1979]: 307–8) that two Cretan immigrants to Italy invented antefixes decorated with heads. F. Kenfield, in his review of Winter (*Bryn Mawr Classical Review* 94.11.05), finds the Cretan-origin hypothesis attractive. On the subject of identity and meaning, Winter proposes that female-headed antefixes represent sphinxes. Kenfield calls attention to M. Mertens-Horn, "Una nuova antefissa a testa femminile da Akrai ed alcune considerazioni sulle Ninfe di Sicilia," BdA 66 (1991): 9–28, where she proposes an alternative interpretation specific to Sicilian female head antefixes. Kenfield suggests, "Female head antefixes could have different meanings in different places." The same is possible for the female amber head-pendants, and for comparable reasons. The metope-stamped *dolii* include a right-facing bearded male head, but no female heads or other female subjects, as L. Pieraccini makes clear in "A Storage Vase for Life: The Caeretan Dolio and Its Decorative Elements," in *Etruscan Italy: Etruscan Influences on the Civilizations of Italy from Antiquity to the Modern Era*, ed. J. F. Hall (Provo, UT, 1996), pp. 92–113.

22. For the origin and religious significance of the terracotta protome, see Croissant 1983, p. 18. On the origin of the Etruscan terracotta votives, see Smithers 1988.

23. For early Greek coinage, see, for example, Mottahedeh 1979; and Kraay 1976, pp. 20–29. Relevant examples are the "Apollo" staters from Colophon, the silver Aegean Dionysos type, and two electrum "satyr" types, from Phocea and Cyzicus, and the Cyzicus Athena head. The following comparable gemstones are illustrated in Boardman 2001: the chalcedony lentoid from Melos in Boston (Museum of Fine Arts 27.678) with a facing satyr (p. 137, pl. 274); the pale green steatite pseudo-scarab with a crowned double head (one bearded, one unbearded) in London (British Museum 480) from Cyprus (p. 180, pl. 281); also in London (British Museum 492), the Island pseudo-scarab of green steatite with a back in the form of a frontal satyr face, signed by Syries (p. 184, pl. 350); and the Greek pseudo-scarab of carnelian made in Etruria with Dionysos on the back in Boston (Museum of Fine Arts LHG 35 ter), by the Master of the Boston Dionysos (p. 186, pl. 408).

24. Some examples are the sixth-century B.C. gold head-pendants (female?) on a necklace found at Ruvo di Puglia in Taranto (Florence, Museo Archeologico Nazionale 6429): Guzzo 1993, p. 52, 191, CII A 1 (necklace), and pp. 71–73, 228, PV2 (pendants).

25. The silver-gilt frontal heads (and the kore, acorns, and lion's head) from a tomb in Taranto (Florence, Museo Archeologico Nazionale 12024–31) may have decorated a *polos*; there are no traces of holes for attachment, and they were presumably "glued on." For the group, see Guzzo 1993, pp. 106–8, 266, 332, L IV A 1. Comparable heads (female heads, gorgoneion, Herakles' face) are in the Getty Museum (96.AM.110–415: *J. Paul Getty Museum* 2002, pp. 122, 126–27), and a group of four (one male and three females) allegedly from Policoro are in the Ortiz collection, Geneva: *In Pursuit of the Absolute: Art of the Ancient World—from the George Ortiz Collection*, exh. cat., rev. ed. (Bern, 1996), no. 123. Several other heads were said to be found with this group.

26. Two small bronzes, an Archaic female head in Toronto and a Severe-style male head in New York (which also has a suspension loop in the top of the head), are apparently both from Italy. See S. Haynes, "A Bronze Head from South Italy," in *Miscellanea Etrusca e Italica in onore di Massimo Pallottino* ArchCl 43 (1991): 96–99.

27. For the head (inv. 210442), see *Magie d'ambra* 2005, p. 45; and *Ornamenti e lusso* 2000, p. 15, fig. 7, no. 152.

28. Two comparable ritual objects of seventh-century B.C. date are the terracotta lamp from Gela with rams' heads and *polos*-wearing female heads (Museo Archeologico Regionale di Gela 7711: P. Orlandini, "Gela: La stipe votiva del Predio Sola," MonAnt 46 [1963]: 33–41, no. 1, figs. 14–16, pls. 8a–c, 9a–b) and the marble lamp from the Selinuntine Malophoros sanctuary (Museo Archeologico Regionale "A. Salinas" di Palermo 3892: *Magna Graecia* 2002, pp. 292–93, no. 75).

29. For the amber heads from Eretum, see Losi et al. 1993; for the Tomb 102 Braida di Vaglio heads (Potenza, Museo Archeologico Nazionale "Dinu Adamesteanu" 95201, 95204), Bottini and Setari 2003, p. 40, nos. 136–37, pl. XLVI. The amber heads in the Louvre (originally pendants?) are modern additions to a gold bracelet (Louvre Bj 2347). The heads are separately inventoried as Louvre Bj 23471 a and b. Bj 23471 a is probably a male figure, as comparison with the bronze votive kouroi in Richardson's Middle Archaic Series A (Florence, Museo Archeologico Nazionale 62, 63, 68, 84 [from Arezzo]) suggests. The hairstyle of Bj 23471 b is comparable to that of contemporary sphinxes, such as the sphinx on an Etruscan gold fibula from Vulci in Munich (Antikensammlungen 2338): Cristofani and Martelli 1983, pp. 183, 296–97, no. 171.

30. It is significant that there are no extant amber head-pendants of male figures with horns, considering how popular the subject of a horned, bearded male figure is in pre-Roman art, in goldwork and ivory, and as a device on early coinage. This lack is even more surprising given the existence of four complete amber pendants in the form of recumbent bull-bodied, bull-horned anthromorphs, one in London (British Museum 68 ["Achelous?"]), another in Paris (Louvre Bj 2123), a third in New York (Metropolitan Museum of Art 1992.11.57, Purchase, Renée and Robert A. Belfer Philanthropic Foundation, Patti Cadby Birch, and The Joseph Rosen Foundation Inc. Gifts, and Harris Brisbane Dick Fund, 1992), and a fourth in a Geneva private collection. The style of the four taurine ambers is Ionian influenced; they have few comparanda apart from East Greek plastic vessels in the form of similar creatures. The full-body anthromorph types derive from the ancient Near Eastern bull-man and are often identified as Achelous (whose realm, as first stated in Hesiod's *Theogony*, is streams and water), but could be identified as Eridanus. Their antecedent is the human-headed bison or bull-creature that was associated with the sun-god Shamash; it is linked with the eastern sunrise. Since amber has both solar and aqueous associations (ocean, river, and stream), these amber pendants, which follow a millennia-old compositional format, may incorporate both the ancient Near Eastern solar associations and the watery ones. The four amber bull-men pendants revert their heads and were perforated to hang head downward. Does this pose connect them to funerary use? On the early human-headed bison or bull-man, see P. Hansen in *First Cities* 2003, pp. 230–31, nos. 157a–b. See also W. A. P. Childs, "The Human Animal: The Near East and Greece," pp. 49–70 (with key references, including LIMC 1 [1981], s.v. "Acheloos" [P. Isler], p. 13, no. 1), and S. Gavel, "Human-Headed Bull," pp. 108–10, no. 1, in *Centaur's Smile* 2003. Both Gavel and Childs emphasize the iconography of the bull-man as protector of flocks and of the Tree of Life. Mottahedeh 1979, pp. 104–6, addresses issues relevant for the ambers in her analysis of Achelous images on early Greek coinage.

31. Unpublished.

32. In Roman times, the satyr mask was thought to be effective in warding off the evil eye: see M. Henig, "Roman Sealstones," in

Collon 1997, p. 99. Faraone 1992, p. 557, discusses the use of heads in apotropaic imagery and gives the examples of the images of Hephaistos on the furnaces of bronze workers; the "laughable images" (*geloia*, called *baskania*), grotesque faces, and satyrs' masks on other ovens; and bird and animal protomes on other structures, all used in protection against ill will (*phthonos*, sometimes translated as "evil eye"). Faraone also cites the fragment of a lost Aeschylean satyr play in which effigies (*eidola* and *mimemata*, exact portraits) of satyrs are fastened to the exterior of a temple, from which vantage point they will frighten off wayfarers. In the tenth century A.D., Al-Beruni (*The Book Most Comprehensive in Knowledge on Precious Stones*, trans. H. M. Said [Islamabad, 1989], p. 181) stated, "The only reason for liking [amber] is said to be that it averts the evil eye." For the evil eye, see "Amber Medicine, Amber Amulets," n. 152.

33. Jannot 2005, p. 38.

34. Mottahedeh 1979, p. 277 (with reference to L. R. Farnell, *The Cults of the Greek States*, vol. 1 [Oxford, 1896]).

35. For the amber necklace with at least six heads from the Mefite sanctuary in Valle d'Ansanto in the Museo Irpino, Avellino, see Losi et al. 1993, p. 210, n. 20; NSc 30 (1976): 503–4, no. 1309g; and G. Colucci Pescatori, *Il Museo Irpino* (Cava dei Tirreni, 1975), p. 33, pl. IX.

36. D'Ercole 1995 suggests that the Lavello-Casino head-pendant may represent a protective deity (see cat. no. 55, n. 3). Losi et al. 1993, p. 203, incline toward identifying the head-pendants as representing "a female goddess or protective genius." Mastrocinque 1991, p. 151, wonders whether they might be maenads or nymphs, even perhaps the Heliades.

37. Moorey 2003, pp. 7, 49.

38. On the facing eye, see Winter 2000; and Faraone 1992, p. 379.

39. Faraone 1992, pp. 45, 58–59.

40. On the destruction of images to abort power, see Gager 1992; and Faraone 1992.

41. See "Amber Medicine, Amber Amulets," n. 171.

42. Faraone 1992, pp. 58–59.

43. Faraone 1992.

44. E. Simon, "Gods in Harmony: The Etruscan Pantheon," in De Grummond and Simon 2006, p. 48.

10. Pendant: Head of a Female Divinity or Sphinx

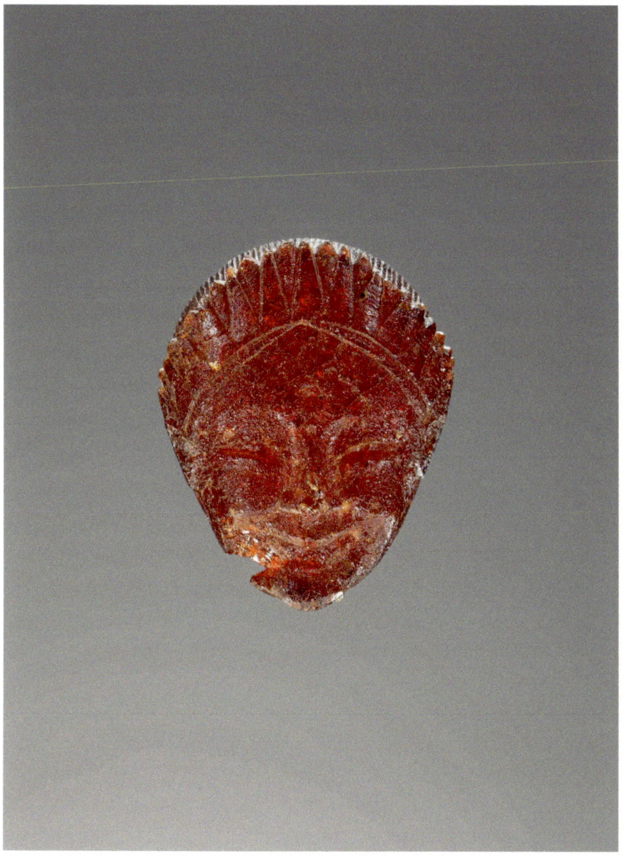

Accession Number	76.AO.85.1 and 76.AO.86
Culture	Etruscan
Date	550–525 B.C.
Dimensions	Height: 32 mm; width: 26 mm; depth: (face) 12 mm, (back) 5 mm, (joined) 17 mm; Weight: 9 g
Subjects	Jewelry; Samos, Greece (also Samian, Greek); Sphinx

Provenance

–1976, Gordon McLendon (Dallas, TX), donated to the J. Paul Getty Museum, 1976.

Condition

The two sections, 76.AO.85.1 and 76.AO.86, were acquired as separate objects by the donor and accessioned as such into the museum. After their entry into the collection, it was discovered that the two joined and composed one object. Before the donor purchased the pieces, they were cleaned to remove dirt and some encrustation. At the museum, the surfaces were treated with an amber-oil distillate, which made both pieces relatively more translucent but also darker. There are no visible inclusions in either section. The front of the face section, 76.AO.85.1, is in fair condition; it is covered with minute cracks and some chips, and the tip of the nose and a section of the right cheek are broken off. The back of the rear section, 76.AO.86, is in good condition. It retains a high polish on the exterior surface but is marked by opaque spots and tiny fissures, and there is a small loss on the left side. The interior surfaces of both sections are in good condition, with the exception of a small chip at the edge of the inside of the back section. Degraded amber is found in the abrasion scratches of both insides. In ambient light, the amber of 76.AO.85.1 is a deep reddish orange; in transmitted light, it is more transparent and a brighter orange. 76.AO.86 is dark red under strong light and almost opaque in ambient light.

Description

The two parts of the head were joined after being accessioned into the collection. When they are joined, the frontal aspect is an exaggerated egg-shaped oval, widest across the forehead, curved at the headdress, and almost pointed at the chin. In side view, the amber is a flattened oval. The wide forehead is arched at the top, with the brow line mirroring the jawline; the chin is small and pointed, protruding forward to the level of the zygomatic arches. The under-chin area is cut inward. Positioned so that the plane of the two joined sections is perpendicular to the ground, the face tilts slightly forward, the forehead is in front, and the chin is regressed toward the neck. In this orientation, the eyes appear to be downcast. There is a sharp cessation of the design at the back of the front section. Above the forehead is an ornament, a crown or ornamental band. It is decorated with a pattern of pinnate-shaped sections or rays in two levels. It emerges above a narrow fringe of hair, which is divided in the center and rendered by closely spaced vertical striations. The volute-shaped ears are high on the head, nearly parallel to the back plane, but so close to the head ornament that they seem to be attached to it. The mass of the hair is dressed in beaded tresses, like cornrows, which run in a rostrocaudal direction, the rows separated by closely engraved lines intersected by even more finely engraved ones.

The eye cavities are empty and flat, and likely once held inlays. The eye sockets are carved with an elongated sinuous opening, the outer canthus of each eye higher than the inner canthus. The eyes extend from the frontal plane around to the sides, so that in side view the eyes look hooded. The nose is indented at the bridge, with the line of the nose inclined at a modest angle away from the face. The corners of the mouth abut the inside curves of the cheeks, and both lips are pulled tightly upward (the upper overhanging the lower): this makes the mouth into a full smile. Since the mouth is recessed from the main facial plane, the effect of a prominent chin is increased, which also emphasizes the smile.

Tool marks or polishing abrasions are evident on the inner surfaces of both halves and just behind the ears on the face. The two sections fit together perfectly. There is no evidence of an adhesive.

Discussion

The lack of a suspension perforation, the high polish of front and back sides, and the sharp cessation of the hair at the circumference suggest that the two sections, 76.AO.85.1 and 76.AO.86, were fitted into a metal bezel, ring mount, or similar kind of setting. Light would have shone through it, and the large "drop" of amber would have glittered, marvelous to behold.

This amber has no exact parallel in style or form. It was likely unearthed in Italy and may have been made there, but the style is that of East Greece. 76.AO.86 is remarkably close to a group of terracottas put together by François Croissant, which he named H Group, Knidos(?).[1] Croissant linked the group to two of his Samian subgroups, Type A1 and Type A5, hypothesizing that a foreigner (a Knidian?) produced these terracottas in a Samian atelier. In comparison to the protomes of Croissant's H Group, 76.AO.86 has a similar facial structure—the same triangular form, eyebrow placement, wide and high arc of the brow edge, and sharp angle of the jawline, and a similarly shaped nose, recessed mouth, tucked-in lower lip, and prominent chin. 76.AO.86 has many features in common with Samian and other South Ionian works. Among them are (1) the ivory "Artemis" from the Halos deposit at Delphi;[2] (2) a marble head from Miletos in Berlin (Staatliche Museen Sk 1631);[3] (3) the over-life-size marble kouros from Samos now in Istanbul;[4] (4) a bronze horseman from the Heraion at Samos;[5] (5) a bronze statuette of a woman from Samos;[6] and (6) three Samian black-figure female head kantharoi, one from Vulci and a pair from Chiusi.[7] The potters of these kantharoi have exploited the open shape of the vessels, resulting in wide, heart-shaped faces, an exaggeration of the facial type of the amber, the comparanda above, and another Getty head (83.AO.202.12, cat. no. 21).

Who or what is represented in this amber? The physiognomy suggests that the subject is female. Because of the material, the type of head ornament, and the bright smile, it is hypothesized here that this *Head* represents a divinity or a sphinx, as must the amber head-pendant of a crowned female subject in London (British Museum 57).[8] The crown of 76.AO.86 is similar to the headdresses of some of the ivory and bone images of Artemis Orthia from the Spartan sanctuary,[9] to the feathered crown worn by the female divinity of the Laconian Grächwil krater (Artemis? *Potnia Theron*?),[10] and to some of the headdresses worn by many of the female heads on the handles of Laconian *hydriae*.[11] (The relationship of these headdresses to those of Hathor and Bes in Egyptian and Phoenician art may be more than a visual similarity.)[12]

The crowns and smiles of the tiny female divinities (*Potnia Theron*, Artemis, or her Etruscan counterpart?) embellishing a number of sixth-century Etruscan (Caeretan?) gold ornaments are important comparisons for 76.AO.86, not only for the identity of the amber, but also because of the East Greek connection of the gold working.[13] An earlier Greek parallel for this *Head*'s crown is the headdress worn by a seventh-century ivory sphinx from Perachora.[14] In turn, the Perachoran ivory's crown might be seen as a latter-day, schematic version of a Mycenaean fashion—such as those sported by the intensely smiling pair of sphinxes on an ivory lid from Mycenae, or the head ornament worn by the sphinx on an ivory from the Athenian Acropolis.[15]

If 76.AO.86 were mounted in a bezel, the mount may have resembled the earlier oval ring mounts (of gold, silver, or gilt silver) of a special class of seventh-century pendants, some of them scaraboids. Excavated examples include silver-mounted ambers from mid-seventh-century graves at Cumae and Veii, and a gold one from Vulci.[16] Three other hypothetical options for the setting of 76.AO.86, two Etruscan and one Egyptian, are the embossed gold *aedicula* cradle mounts bent around three Egyptian faïence *pataikos*/dwarf amulets for a necklace buried in a tomb at Vulci,[17] the embossed strip mount bent around a broken amber (for use as a pendant),[18] and the Egyptian gold settings for royal heart amulets and heart scarabs of precious nephrite.[19] However, while the silhouette of the Egyptian pendants is almost identical to that of 76.AO.86, the stones are flat on the reverse, not rounded like the amber.

NOTES

1. Croissant 1983, pp. 18–34.
2. Delphi Museum 10414: Lapatin 2001, no. 33. This ivory head is probably Samian, as first suggested by P. Amandry, "Rapport préliminaire sur les statues chryséléphantines de Delphes," BCH 63 (1939): 86–119, seconded by Croissant 1983, p. 38.
3. Karakasi 2003, pl. 44 ac; and Freyer-Schauenburg 1974. See also Croissant 1983, pp. 35–37.
4. Istanbul Museum 1645 (which fits on the body of the draped kouros Samos 5235): Freyer-Schauenburg 1974, pls. 41–46 (the views of the Istanbul head on pl. 42, upper right, correspond most closely to 76.AO.86).
5. E. Buschor, *Altsamische Standbilder I–V* (Berlin, 1934–61), figs. 198–99; and Croissant 1983, pp. 40–43, pl. 5. Compare the lower eyelids, which curve slightly upward in the center.
6. Samos, Vathy Museum B1441: Croissant 1983, pp. 129–40, pls. 38–39. The resemblance of the statuette to the Getty *Head* is enhanced by the similarly empty eye sockets. Comparable also are the structure of the head, the arch of the forehead, the head width (from ear to ear), the smile and placement of the lips, and the upward tilting and general shape of the eyes, although there are differences—the nose is more prominent, and the chin more recessed—reasons why Croissant placed this bronze in his chap. 5 Phocaean(?) group.
7. The head kantharoi, of both male and female subjects, provide revealing parallels. For the head kantharos from Vulci (Munich, Antikensammlungen 2014), see Walter-Karydi 1973, no. 484, pl. 56. For the Chiusi pair (Berlin, Staatliche Museen F4012–13), see ibid., nos. 482–83, pl. 57. Compare the short nose, wide smile, and elongated eyes.
8. Strong 1966, p. 73, pl. XXIII, describes the head ornament as "a stephane like the slats of a Venetian blind."
9. See cat. no. 2. The ivory plaque of the *Potnia Theron* / Artemis from a dress pin, of mid-seventh-century B.C. date, and a bone head of the goddess, of about 600 B.C., are illustrated in Hampe and Simon 1981, pls. 354–55.
10. For the Grächwil *hydria*, see cat. no. 2, n. 23.
11. For pertinent Laconian bronzes, see Stibbe 2000, passim.
12. See Bulté 1991, passim.
13. The smile is very like that of the winged figures in the *Potnia Theron* schema (she stands between lions and atop confronted recumbent waterbirds) on the pair of earrings in Berlin (Antikenmuseum 30219) and the single figures from Praeneste and Cerveteri in London (British Museum, Jew. 1267–68) and in the Villa Giulia (Museo Nazionale Etrusco di Villa Giulia 40875, 53492). For the goldwork, see Cristofani and Martelli 1983, pp. 58, 292, 300–301, nos. 144, 192–94. The plaquettes from Cerveteri are *aedicula*-like and are reminiscent of the Milesian votive reliefs with a standing figure in Berlin (see cat. no. 9, n. 2). For the East Greek aspects of the gold ornaments, see Laffineur 1978, especially pp. 56–66 ("Visages humains vus de face"), with critical references.
14. Athens, National Archaeological Museum 16519 (from the sanctuary of Hera Limenia): T. Dunbabin, ed., *Perachora*, vol. 2 (Oxford, 1962), p. 403 A I, pl. 171. Illustrated in Hampe and Simon 1981, pl. 411.
15. Athens, National Archaeological Museum 7634. The box lid is from the House of the Sphinxes, Mycenae: see Poursat 1977, no. 138, pl. 12; and Hampe and Simon 1981, pl. 332. For the crown of the Mycenaean ivory sphinx from the Athenian Acropolis, see Poursat 1977, no. 493, pl. 53; and Hampe and Simon 1981, p. 229, pl. 341.
16. Strong 1966, pp. 48–49. In addition to the examples in London (British Museum 12–15: ibid., pl. III), there are many others in collections both old (New York, Metropolitan Museum of Art) and new (the Getty Museum). There is the remote possibility that the mounting may have been of amber, on the basis of one known parallel: one of the possibly male heads (686/1) from the Novi Pazar, St. Peter's church, find was originally set into an oval amber frame, which is now lost but known from photographs. There is a notch cut above the right ear, but no perforations. As an argument against this idea, the back of the Getty pendant is rounded and polished, while the Novi Pazar example has a flat back and appears to be more roughly finished: see Palavestra and Krstić 2006, pp. 129, 131, fig. 57, no. 41 (with earlier bibl.).
17. M. Cristofani in Cristofani and Martelli 1983, pp. 134, 279, no. 93 (dated to the second quarter of the seventh century B.C.). One of the faïence figures is missing; one was given a gold *perizoma*. The mounts are stamped with lions, sphinxes, and monkeys.
18. The gold mount of the broken amber pendant are embossed with a meander and a rosette (D.C.A. collection, Geneva, Switzerland): *Art of the Italic Peoples*, exh. cat. (Geneva and Naples, 1993), p. 178, no. 80.
19. Compare, for example, the Eighteenth Dynasty heart amulet, "Royal Wife, Manhata," in New York (Metropolitan Museum of Art 26.8.144, Fletcher Fund, 1926): *Hatshepsut* 2005, p. 215, no. 137.

11. Pendant: Head of a Female Divinity or Sphinx

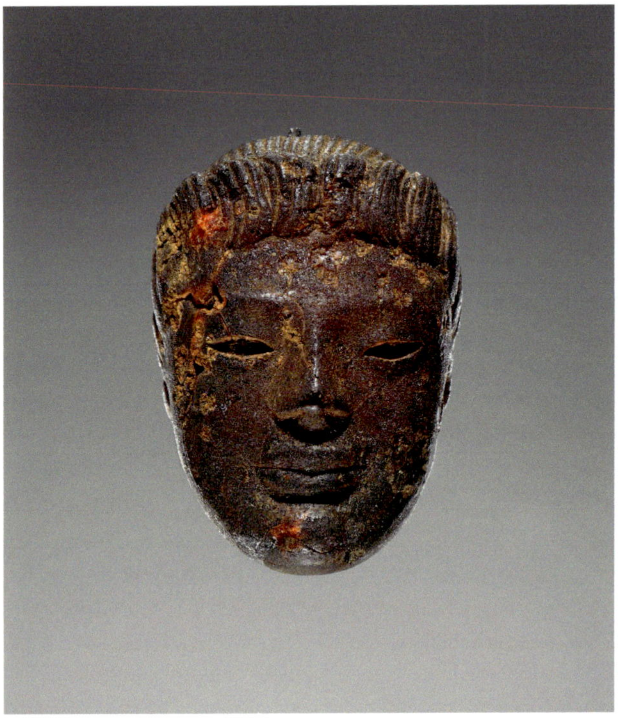

Accession Number	76.AO.79
Culture	Etruscan
Date	550–525 B.C.
Dimensions	Height: 34.5 mm; width: 24 mm; depth: 16 mm; Weight: 7.7 g
Subjects	Ionia, Greece (also Ionian, Greek); Jewelry; Sphinx

Provenance

–1976, Gordon McLendon (Dallas, TX), donated to the J. Paul Getty Museum, 1976.

Condition

There are no visible inclusions. The overall condition is good; the surface is firm and stable, although an extensive crack network is visible in transmitted light. There are small chips above the left eyebrow, near the left eye, and on the left side of the hair that join to form a fine diagonal crack; there are tiny losses on the right cheek and right ear. There are small chips and losses on the back of the object. It is red-brown and opaque in ambient light and bright orange in transmitted light. The metal insertions (wire or pins?) in the top of the head are broken off; one is rectangular in cross section. (They have not been scientifically analyzed but appear to be of silver.)

Description

The amber is convex on the obverse; the reverse is smooth and curved so that in profile to right, the back is C-shaped. Only the front of the head is depicted—that is, the face and the front part of the hair. The eyebrow ridges lie flat on the surface of the face; the upward curves are slight (the left somewhat more curved than the right), with a delicately suggested transition to the root of the nose. The shallow, feather-shaped eye sockets (which may have been intended for inlay) incline sharply upward at the upper corners. The lids are indented at the inner canthi, swelling above the eyes with the lids overhanging the lower rims. The area under the lower lid is sensitively modeled around the shape of the eye itself. The cheeks are full, especially around the mouth area.

The head's facial features are not precisely symmetrical: the right eye turns up more at the outer canthus, its inclination more manifest, while the nose lists slightly to the right, its right nostril higher. The smallish nose sits close to the frontal vertical plane of the face. Straight and tipped upward, the wings of the nose are somewhat fleshy, with the nares clearly defined. The mask has bow-shaped lips, the upper one overlapping the lower one, that are slightly pointed at the center, and the corners of the mouth are tucked in, to create the impression of a slight smile. The mentolabial sulcus is deep. The chin is wide and rounded, jutting forward to the plane of the brow. In full-face view, the ears are hardly noticeable, but in side view, they are clearly defined. They are large and placed high on the head. The antihelices are indicated, the antitraguses not indicated, and the lobes are separate.

The back is less polished than the front. There are traces of abrasion around the circumference and on the reverse.

The individual strands of hair were cut with a graver. There are the remains of three metal inserts (silver?) in the top of the head, the central one larger than the two flanking it, and the residue of a green (bronze) corrosion product on the reverse.

Discussion

76.AO.79 has no exact parallel. No conclusion may be drawn about the sex, function, or identity of the figure. It is not drilled for suspension, nor is there a mount; it has a curved back; and the green (bronze?) corrosion product on the reverse appears to be the result of direct contact with metal. The metal inserts in the top of the head are placed in the position of a crown, and although a suspension device could have been incorporated, there is no apparent parallel for this method.

76.AO.79 was accessioned as the head of a girl and published in 1993 by this author as that of a male divinity. If this latter identification holds true, the divinity could be Apollo. The sexing of the head now seems less sure; it may be that the original designation was the correct one. If 76.AO.79 indeed represents a female figure, the amber may represent a "brilliant" divinity such as Artemis or Aphrodite, *Potnia Theron*, or perhaps a beautiful sphinx.

In form, 76.AO.79 is related to the previous catalogue entry, 76.AO.85.1 and 76.AO.86 (cat. no. 10). They both have curved backs, although that of 76.AO.79 is curved on one plane only and is not rounded. The one other amber head-pendant that has a similar curved back is a female head-pendant in Munich.[1] It, too, has no visible boring for suspension; instead, the Munich head has a metal (bronze?) suspension loop (of uncertain date) inserted in the top of the amber.

The purpose of the three metal (silver?) inserts in 76.AO.79 may have been to suspend the head, but since there are three, two small and one large, it is more likely that they are the remains of a crown or other head decoration (silver would suit such an ornament).[2] Although unlikely, an incorporated suspension loop should not be ruled out. The only parallel for such added decoration is a head-pendant of a female figure (a siren?) in London (British Museum 60), said to be from Armento and probably dating to the mid-fifth century B.C. Metal remains (silver? wire or pin?) are in the lateral suspension boring, and further remains are in a larger vertical boring in the top of the head.[3]

The curved back of 76.AO.79 and the green residue on it make difficult a hypothetical reconstruction of any support. Contemporary mounted scarabs, amber scaraboids, and gemstones have characteristically flat backs. It is possible that the head was fitted with a gold bezel, comparable to the way that an Etruscan mounted three Egyptian faïence *pataikos* figures for a necklace.[4] However, the corrosion product on the back of 76.AO.79 suggests it was in direct contact with bronze. If the amber was set into bronze, the bronze support would have had to fit the curve of the amber's back. (Could it have been set into the base of a vessel handle?)[5] All other extant amber faces used as inlay are flat on the back, and all are mounted in amber, ivory, or bone.[6]

If the mounting of the Munich pendant is original, when suspended against a flat surface, the head would appear to look upward; however, if it were the single pendant of a carrier and worn high on the neck at the jugular notch, the amber would have fit snugly into the depression and the mask would have looked straight out. There are many Archaic illustrations of figures male and female wearing simple necklaces with one pendant (or a few) at this position or even higher up, as a choker. Might the same be true of 76.AO.79?

For the style of the head, the best comparison for 76.AO.79 among amber carvings is the Getty amber kore pendant 76.AO.77 (cat. no. 8), which is attributed to a South Ionian carver. Both compare well with a number of Archaic East Greek sculptures in terracotta and marble, especially those from the ambient of Miletos. 76.AO.79 shares the general stylistic qualities of the Milesian school—the frontality, the horizontality, the solidity, the "Massigkeit"—as outlined by François Croissant. He convincingly demonstrated that there are two great artistic tendencies in Milesian sculpture, a "style graphique" and a "style pictural."[7] It is within this latter trend, within the naturalistic vein of Milesian art, that 76.AO.79 seems to find its proper berth.

76.AO.79 can be linked to specific Milesian works, especially the terracotta protomes of Croissant's Type B3–B6 series.[8] They have in common a relatively flat face, the nose at a low angle relative to the cheeks, a short and overlapping upper lip, and a short, squarish chin. The amber head also compares well with a number of marbles from East Greece: the head of a draped woman from near Ephesus in London (British Museum B89); a siren in Copenhagen, from near Cyzicus; the head of a sphinx from Keramos, in İzmir; and the Louvre Dionysermos.[9] The Keramos marble is closest. The head shape, the plastic transitions from area to area, the sweet smile, the air of quiet self-confidence, and the composition of the features are very alike. The relation of the eyes to the eyebrow arcs is the same.

The eye shape is a distinguishing feature of the amber. The eyes of 76.AO.79 are akin to those of many of the terracottas in Croissant's Samian and Milesian groups, South Ionian marbles, and the Getty amber *Kore*, but the best parallel is the large ivory "Artemis" from the Halos deposit at Delphi.[10] Her distinctive eyes are feather-shaped, turned up at the outer corners, swelling in the center, with the lids stretched over the orbit and the upper lid overhanging the lower and drooping slightly. (Those of 76.AO.79 turn up more sharply, are spaced farther apart, and are even narrower and smaller in proportion to the face.) This affinity is noteworthy not only because it establishes a stylistic connection between the amber and ivory but also because it offers further evidence for the argument that amber carvers worked ivory and vice versa.

The hair of 76.AO.79 is elaborately dressed: the front sections are combed forward, apparently secured by a band, and then flipped back toward the crown, forming a series of curved bunches of hair framing the head. The hair sections just above the ears take the form of so-called Ionian wings. Although the hairstyle may have originated in Ionia, it was quickly absorbed in western Greece and in Etruria. Variants are worn by male and female figures in the Archaic period. Such hairstyles are found on the female figures (both heads and caryatids) of a group of terracotta lamps from Magna Graecia, a type that likely originated in the Achaean cities of the Ionian coast, but which was further developed in the colonies of Magna Graecia.[11] The hairstyle of 76.AO.79 is analogous to those worn by certain sirens and sphinxes. The hair of the marble siren from Cyzicus (Milesian work?) in Copenhagen, a number of Etruscan stone sphinxes from Vulci, and the sphinxes of a gold fibula from Vulci in Munich allow the possibility that 76.AO.79 represents an Etrusco-Ionian version of the fantastic creatures.

An exotic comparison is the *Tridacna squamosa* shell made into a female figure (siren?) by a Syro-Phoenician(?) artisan; an example in the British Museum excavated at Vulci.[12] The face carved into the shell's umbo is very like that of 76.AO.79.

The comparisons presented above point to a date for the head in the period of circa 550–520 B.C. and to a skilled carver from East Greece, perhaps in Etruria, where the impact of artisans from South Ionia and elsewhere in the east had such impact on art in the second half of the sixth century. Whether 76.AO.79 was originally the inlay (face) of a figure in another material or the centerpiece of a pendant, the amber itself determines that the subject of the head is a divine or heroic figure, a supernatural or a demonic being. The metal attachments, if they are part of a headdress and of silver, do not allow another interpretation.[13] As Brunilde Ridgway has argued, the very presence of elaborate head decorations in Archaic art "serves to indicate superhuman or divine beings."[14] As she concludes, "The metal attachments on the heads of Archaic statues should be read as part of elaborate headdresses functioning as attributes and helping in the identification of the figures."[15]

If the head represents a siren or a sphinx, 76.AO.79 would be another example in amber of the liminal creatures. Whether their images were worn in life or in the grave, the "work" of these creatures was established: they were effective in protection and aversion, and in the journey to the afterworld, both were reputed guides. Both have regenerative aspects. If 76.AO.79 is the head of a female divinity, it may be the image of Artemis or another deity with solar aspects. Then again, if it represents a youthful male divinity, the amber head may denote Apollo. The head, as a glittering object in the form of a divine or demonic being carved with the greatest skill and detail in high-status materials, amber, silver, and perhaps ivory for the eyes, was worthy of the attention of the gods; it is, it was, a marvel to behold. The materials, the image, the exquisite craftsmanship—all were worthy of the gods' attention. During funerary rituals and in the tomb, 76.AO.79 could have played a small but important role. A shining tear crafted into the form of a beautiful, youthful face was surely appropriate as a funerary gift, especially if the deceased had died young. If the figure is Apollo, it might have brought to mind the stories connecting the solar material, the solar deity, and the mourning by the sun-god for the premature deaths of Phaethon and Asclepius.[16]

NOTES

1. Munich, Antikensammlungen 15.003.
2. See cat. no. 10, n. 16 and n. 18, for discussion of amber pendants in metal mounts.
3. Strong 1966, p. 74, no. 60, pl. XXIV. Strong concludes that the figure is probably female, pointing out that while the hair on the forehead divides like that seen on males, it is bound by a fillet. How the metal element and the caplike hair covering might have functioned together is not apparent. In style, it is very like two of the flying figures from Sala Consilina in the Dutuit Collection, Petit Palais, Paris (see introduction, n. 219).
4. See cat. no. 10, n. 17.
5. The peculiar form of the back of 76.AO.79 would have allowed it to fit neatly into the indentation of the jugular notch, one of the

most vulnerable spots on the body and the place where many early single pendants are represented as hanging.

6. The amber pendant from Novi Pazar with an amber mounting is discussed in cat. no. 10, n. 16. Ivory and bone examples are the amber faces of the sphinxes of two Laconian relief plaques, one of ivory, the other of bone, from a *kline* dating to about 600 B.C., and the now-lost faces of a divinity and her two "acolytes" set into bone plaques from furniture excavated at Belmonte Piceno, dating to the early sixth century. For the bone plaques from Belmonte Piceno, see Rocco 1999, pp. 82–85, nos. 135–36, pls. XLIV–XLV, where there is fruitful discussion of the identity (*Potnia Theron*? Artemis?). For the two sphinx appliqués with amber faces from a couch in the Iron Age Celtic tomb at Grafenbühl, Asperg (Stuttgart, Würtembergisches Landesmuseum), see n. 17 in the "Pendants in the Form of the Human Head" introduction.
7. Croissant 1983, p. 181.
8. Despite the inherent generational transformations in the terracotta series, the profiles of the amber and the terracottas are markedly similar.
9. For British Museum B89, see Croissant 1983, p. 62 (which he finds to have a rapport with the Milesian school); for the Copenhagen siren (Ny Carlsberg Glyptotek 2817), Johansen 1994, pp. 44–45, no. 7; for the Keramos sphinx head (İzmir, Archaeology Museum), Akürgal 1961, figs. 229–32; Tuchelt 1970, p. 125 (L 53); and Croissant 1983, pp. 64, 67, pls. 15–16; for the Paris Dionysermos (Louvre Ma 3600), Hamiaux 1992, pp. 59–60, no. 51.
10. Delphi Museum 10413: Lapatin 2001, no. 33.
11. As M. Cipriani (*Magna Graecia* 2002, pp. 122–23) states: "The eclectic character of this production, as cleverly emphasized by Croissant, and the difficulty in isolating specific stylistic elements raise again the more general problem of defining colonial styles." See F. Croissant, "Sybaris: La production artistique," in *Sibari e la sibaritide: Atti del XXXII Convegno di studi sulla Magna Grecia*, ed. A. Stazio and S. Ceccoli (Taranto, 1993), p. 548. This position does not necessarily contradict the Ionian-origin hypothesis of C. Sabbione, "L'artigianato artistico a Crotone," in *Crotone: Atti del XXIII Convegno di studi sulla Magna Grecia*, ed. A. Stazio and S. Ceccoli (Taranto, 1983), p. 272.
12. For the London shell and further bibliography, see n. 8 in the "Pendants in the Form of the Human Head" introduction.
13. A crown of silver, which appears "brighter and more like daylight than gold" (Pliny, *Natural History* 33.19.9), may, like the amber, establish the figure as divine. The brilliance of the materials and the attention to detail no doubt added to its marvelousness as a work worthy to behold (see "Ancient Names for Amber" in the introduction).
14. Ridgway 1990. See also B. Ridgway, "Metal Attachments in Greek Marble Sculpture," in *Marble: Art Historical and Scientific Perspectives on Ancient Sculpture (Papers Delivered at a Symposium Organized by the Department of Antiquities and Antiquities Conservation and Held at the J. Paul Getty Museum, April 28–30, 1990)* (Malibu, 1991), pp. 485–508.
15. Ridgway 1990.
16. Apollonius (*Argonautica* 4.611–18) refers to a Celtic myth that drops of amber were tears shed by Apollo for the death of his son Asclepius when Apollo ventured north in a visit to the Hyperboreans. See "Ancient Literary Sources on the Origins of Amber" in the introduction.

12. Pendant: Satyr Head in Profile

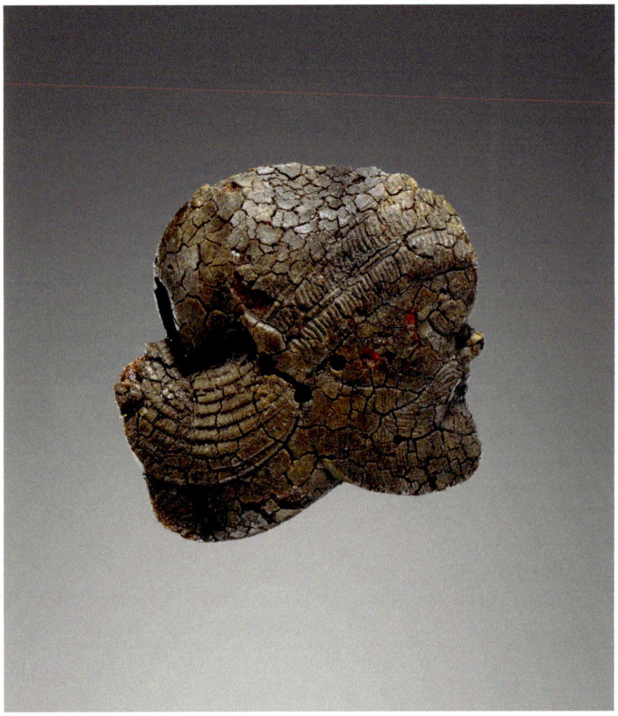

Accession Number	83.AO.202.1
Culture	Etruscan
Date	525–480 B.C.
Dimensions	Height: 65 mm; width: 68 mm; depth: 35 mm; Weight: 53.4 g
Subjects	Dionysos, cult of (also Satyr); Inclusions

Provenance

–1983, Antike Kunst Palladion (Basel, Switzerland); 1983, Vasek Polak, 1914–97 (Hermosa Beach, CA), donated to the J. Paul Getty Museum, 1983.

Condition

Small sections in the hair at the front top and reverse of the head are broken off. The amber surface is in poor condition, with an extensive network of cracking and loss, mostly on the reverse. The surface is fragile, with many loose flakes; some surface flake losses are 1 mm deep. Before the pendant entered the Getty Museum, the surface was treated with a surface consolidant, likely in an attempt to strengthen it. (This probably contributed to the overall cracking pattern, causing further surface deterioration and shrinkage.) In ambient light, the pendant is light to dark brown, with some areas of translucency. In the places where modern chips have exposed the interior, it is dark orange and translucent. In transmitted light, the amber is bright red-orange. There are various inclusions.

Description

The pendant is a large piece of amber, relatively flat on the reverse and rounded on the obverse. It represents the head and partial neck of a satyr in profile, facing right. The short, rounded forehead is modeled with a brow ridge and an eyefold in a manner close to the appearance of a human. The almond-shaped eye is plastically modeled, following the curve of the head. The eyeball area is recessed from the upper and lower lids. The lids are represented as filletlike lines and are of equal thickness. The upper lid line is more arched than the lower one, and overlaps it at the outer canthus. The short, upturned nose is deeply indented at the root, is rounded at the tip, and includes carefully modeled wings and nares. Rather small in proportion to the rest of the face, the long, teardrop-shaped ear is plastically modeled; the tip is acute.

A long mustache begins just under the nose. It is articulated with sweeping engraved lines, curving at the upper part and more horizontal below. The mustache overlaps the beard. The beard juts forward and ends in a rounded point. Closely engraved lines, straighter in front and more curved at the jaw, represent the beard hairs. A shallow ridge marks it off clearly from both the face and the neck. Just below the lower lip of the small mouth is carved a small patch of beard.

In front of the ear and surrounding the face are two rows of hair rolls, separated by a raised, filletlike line. At the forehead, the hair rolls bulge slightly, giving them the impression of volume. The lines defining the hairs in the bangs section are curved, the upper ones tilting slightly to the right and the lower ones slightly to the left. Behind the top row is another raised line, perhaps a fillet, defining the rolls from the crown portion of the hair. The cap of the hair is plain and follows the shape of the amber nodule to suggest the form of the skull. An engraved line separates it from the hair at the back of the head. The sweep of hair is modeled, with the midsection the most prominent. It is defined with nine horizontal rows separated by parallel curving grooves. Each of the rows is further described by short, finely engraved lines, giving the impression of coils of tightly curled hair pulled into a low-hanging bun. The neck is long and smooth and connects to a tiny section of the shoulder.

The idiosyncratic form of the pendant suggests that the carver closely followed the natural shape of the cleaned piece of amber. The head and neck are skillfully adapted to the surface's irregularities. On the obverse is an indentation at the top of the head and a shallow groove at the top of the beard line on the cheek, and on the reverse, a long smoothed groove at the back of the head. There are no visible tool marks because of the severe cupping of the surface.

In the break are the remains of a shallow, drilled groove, all that is left of the suspension perforation. There are four stopped bores: two on the obverse—one at the point where the beard meets the neck, another at the base of the ear—and two on the reverse, at the back of the head and at the neck. The two on the obverse retain their plugs (which are darker in color than the surrounding amber). When hung, the head would have fallen into position with the nose upward.

Discussion

This is one of the largest and most finely worked of all extant pre-Roman figured ambers. It has no exact parallel. Satyrs are the second most common subjects of head-pendants and date from the late sixth to the late fourth century. This profile head-pendant of a satyr and the other head-pendant of a satyr in the Getty collection (82.AO.161.1, cat. no. 13) are among the earliest portrayals of the subject in amber. They are also the two most common types, the profile head-pendant and the frontal face. Only a trio of satyr heads in a New York private collection is earlier.[1] The Getty and New York satyrs are of three different types and styles and reveal three different traditions of satyr illustration.

The two best comparisons for 83.AO.202.1 are the satyrs of two ambers in London: *Satyr and Maenad* (BM 35) and *Vintaging Satyr* (BM 36).[2] Both British Museum satyrs have small noses and mouths similar to those of the Getty head-pendant, and both 83.AO.202.1 and one of the British Museum satyrs (BM 35) sport luxuriant, long mustaches and extremely neat beards of rather short and straight hair (vertically delineated), barbered in two tiers. The two figures' coiffures are also similar, typified by a full bang that is longer in front of the ears. The mustache of BM 36 is less full and shorter, his beard more pointed, his hair shorter, and instead of a plastically rendered eye, his is hollowed out and flat, as for an inlay.

The reverse or main side of BM 35 is a figural group that includes a dancing pair, a nude bearded male and a draped female, with a young fawn leaping up between them; on the obverse is a large bearded snake, coiled and upright. The finial is a dolphin. BM 35 has been variously identified: Donald Strong called it a satyr and maenad (even if the male figure does not have pointed ears, hooves, or a tail), and it was formerly identified as Artemis and Zeus and as Artemis and a giant. In the view of this author, the main figures are better read as Bacchic revelers, with the male figure wearing the mask of Dionysos. The bearded snake may be a symbol of the spirit of the dead, or it may act as a chthonic symbol that refers to Dionysos (and perhaps Orphism). BM 36 represents a figure with normal ears (it is also hoofless and tailless), accompanied by a wineskin and grapes, crowded in the amber with a pointed amphora and a vine, and on the reverse, a coiled and rearing bearded serpent. The figure may represent a satyr, but it is more likely that he, too, is a Bacchic reveler wearing the mask of "the god, who does not appear." The bearded, coiled snake on the reverse may also refer to the chthonic Dionysos. For the initiated, the imagery of BM 36 may have conjured up Dionysos's mythic vineyard on Naxos, the wine of which was divine and held the promise of eternal life.[3] Reveling dancers are the subject of nine other carved amber pendants. Single male dancers ornament pendants in Boston and New York,[4] and single female dancers are the subject of many more pendants, including three from excavated graves in South Italy—one from Oliveto Citra and two from the Rutigliano-Purgatorio Necropolis (one is armed).[5] In addition to BM 35, a high-stepping couple is the subject of an amber pendant in the Louvre.[6]

The active pose of 83.AO.202.1 is demonstrated by the forward position of the head and neck as well as by the satyr's hair, streaming out behind him. Parallels for the head position are numerous; it is characteristic of many Attic black-figure painters and is a hallmark of the Etruscan black-figure vase painter known as the Micali Painter. The Attic examples include the active satyrs on an *aryballos* by Nearchos, three dancing satyrs on an amphora in the manner of the Lysippides Painter (an *aulos*-playing satyr on side A and a pair on side B, where one crushes grapes and the other attempts an abduction of a nymph), the dancers on an amphora by the Painter of Cambridge 47, and the ithyphallic harvester on an Amasis Painter amphora.[7] The beard and hair of 83.AO.202.1 are most similar to those of the satyrs of the Amasis Painter and the Micali Painter. The Etruscan master's rendering of the hair and beards of his fast-moving, ithyphallic satyrs—the dancing, *aulos*-playing, and running satyrs of Louvre CA 3185, the racers on his amphorae in Baltimore and Palermo, and those on the column krater in Berlin—are especially similar.[8]

Etruscan artisans working in bronze and in ivory refined the millennia-old manner of indicating fast movement—the stretching out of bodies with hair flying out behind. The rendering of the acrobats' hair on a tripod excavated at Vulci and the running figure of Perseus on a tripod foot from Orvieto are two outstanding parallels mirroring the speedy satyr of 83.AO.202.1. P. J. Riis has attributed these bronzes to his Group of the Mainz Censer, the earliest one of which he describes as "Late Ionizing (Late Ripe Archaic) or early Late Archaic."[9] The motif is current in ivory and bone carvings of male figures engaged in strenuous activities: driving chariots, wrestling, or fighting sea monsters.[10]

Many of the descriptive details of 83.AO.202.1—the very fine striations of the hair and beard, the tongue of hair beneath the lower lip, and the form of the eye—draw it close to the design and cold working of a group of bronzes thought to be Northern Etruscan, perhaps Chiusine, from about 500–480 B.C.: (1) a banqueter in London (British Museum GR 1831,1201.1); (2) the "Herakles" from Contarina (Rovigo); (3) the "Fufluns" in Modena; and (4) the Getty *Statuette of Tinia*.[11] 83.AO.202.1 is also very close to a group of Vulcian bronzes first brought together by Mario Del Chiaro (but which are for Riis part of his extended Mainz Censer group, as noted above).[12] Del Chiaro's group includes sirens, satyrs, and acrobats, in addition to the Cortona lamp.

83.AO.202.1 and the two Dionysiac ambers in London, BM 35 and BM 36, are strikingly similar to the earliest pseudo-scarabs made in Italy. Strong was the first to connect the figure of BM 36 with a cornelian in Boston,[13] on the back of which is a splayed, bearded Dionysos that "immediately argues a close stylistic connection with a whole group of S. Italian ambers. Other examples of such pseudo-scarabs, a number of which are in the British Museum, exhibit several of the same characteristic elements of style."[14] Strong thought the group to be South Italian, very probably Campanian, especially BM 459, which is made of sard, with a siren on the scarab side and Apollo(?) on the flat.[15] J. D. Beazley, writing earlier, was more cautious about the London sard: "The style of the stone may perhaps be called Etruscan rather than Greek. If it is Etruscan, it is one of the earliest Etruscan gems."[16] P. Zazoff, J. Boardman, and, most recently, J. Spier have noted the key role of the Master of the Boston Dionysos (as he is now called) in the story of gem engraving in Etruria.[17] Two silver objects from Lydian Usak have suggested to Spier a possible explanation for the immigrant master's training in a specific East Greek atelier.[18] Boardman's latest thoughts on the gem cutter are relevant for 83.AO.202.1:

> *Other gems by the same artist (Master of the Boston Dionysos) have their backs detailed in an "Etruscan" manner, and the style of the figures—stocky, big-headed, flat-footed in stance—is very close to that of peripheral Etruscan work in bronze and stone (as the Volterra stelai). This does not mean that the artist was not a Greek, since we may only be witnessing the establishment in Etruria of an immigrant and highly individual Greek style. But it is not one as yet well attested in the west so, whatever his nationality, the artist might be considered the first of the "Etruscan" gem engravers. There are later and still purely Greek works made in the west which contributed to the development of the local studios.*[19]

Was a gem engraver or bronze worker responsible for 83.AO.202.1, as might have been the case for earlier figured ambers? Was the head-pendant made in Vulci? Might it even have been made by the Master of the Boston Dionysos?

In spite of its state of preservation, this satyr is a striking object. The large size of the piece of amber and the exquisite artisanry of 83.AO.202.1 must have occasioned admiration from the day it was finished. In its function as an amulet, it may have been considered "homopoeic," meaning that through assimilation it would endow the wearer with the subject's characteristics, the swift and nimble satyr promising its owner fleetness of foot, the ability to speed away from danger or pain, as would a hare amulet. It may have been considered especially lucky, the magic and potency of the image being enhanced by the material. Satyrs were apotropaic images that could work on the "like banishes like" principle. An "active" satyr could call up the god Dionysos, the dance, and the ceremonies of sacrifice. Tied onto the body, in life or in death, such an amulet could place the wearer under the protection of Dionysos.

Since satyrs were not just useless hedonists but were also understood to be wise and to be participants in religious ritual as servants of Dionysos (Strabo),[20] "accomplished in dancing and in secret rites or initiations of Dionysos" (as described by Eustathios),[21] the pendant may have been involved in related activities. Etruscan funerary rites were not complete without the dances of reanimation, which took place at the cremation or burial place. Dancing satyrs are endlessly depicted on vases found in tombs throughout Italy, notably in Archaic Etruria. On the Micali Painter's famous vase of a funeral procession, satyrs figure prominently, dancing among veiled women, important insignia-bearing personages, and others. As Jean-René Jannot notes, the most archaic Etruscan illustrations of sacrifice "place very odd sacrificants in the scene: Sileni and satyrs, who impart a Dionysiac atmosphere, or at least invoke the savage character of the sacrificial act."[22]

If the head-pendant were sufficient to represent the whole and the satyr was understood as dancing, the subject may have alluded to the ceremonies of the funeral and reanimation through death. If the satyr were abstracted from an activity such as vintaging and the image understood to be a "quote," the Getty *Satyr Head in Profile* could allude to Dionysos's magic vineyard on Naxos. If the satyr were completed as ithyphallic, the pendant may have called up the phallic aspect of Dionysiac religion, and the head-pendant could have alluded to Dionysos Oipholios. Then again, 83.AO.202.1 may have conjured up a specific episode in Dionysian myth—the arrival of Dionysos by sea, or the union of Dionysos and Ariadne. Certainly, the amber's optical characteristics, recalling the rejuvenating sun or the color and sparkle of wine, would have been especially apt in underlining an association with the divinity for one of Dionysos's servants.

That 83.AO.202.1 is of amber, the golden "tear" of Phaethon's mourning sisters transformed into poplars, may be of special relevance for this head-pendant. If the gloss in the late lexicon of Harpocration is to be believed, that those who were initiated into the "Bacchic rites" (*Bakkhika*) were crowned with a wreath of white poplar "because the tree is chthonic, and chthonic also is Dionysos, the son of Persephone,"[23] what better material for an adherent to wear than Phaethonic amber?

NOTES

1. Unpublished.
2. Strong 1966, pp. 61–63, nos. 35–36, pls. XV–XVI. BM 36 is said to come from Canosa. The provenance of BM 35 is confused. Rival stories have it found at Ruvo, Armento, and Canosa. Ibid., p. 63, records the existence of another amber satyr, referring to Schultz (*Bullettino dell'Instituto di corrispondenza archeologica* 1843, p. 39), who "mentioned a fragment of a similar piece in the collection of Signor de Jorio, who also had an amber ram and a lion from the Basilicata."
3. Hedreen 1992, pp. 86–87, n. 149.
4. Boston, Museum of Fine Arts 02.2547 ("from Palestrina"); New York, Metropolitan Museum of Art 1992.11.4, Purchase, Renée and Robert A. Belfer Philanthropic Foundation, Patti Cadby Birch, and The Joseph Rosen Foundation Inc. Gifts, and Harris Brisbane Dick Fund, 1992.
5. For the Oliveto Citra dancer (Paestum, Museo Nazionale OC/00082), see Mastrocinque 1991, pp. 129, 133, fig. 84; and P. C. Sestieri, "Ambra intagliata da Oliveto Citra," ArchCl 4 (1952): 16, pl. 14. For the Rutigliano-Purgatorio Necropolis dancing figures, see Negroni Catacchio 1993, p. 199, fig. 7.6.
6. The Louvre amber of a couple, Bj 2253, is unpublished.
7. See the figures on an amphora in London (Fitzwilliam Museum GR.26.1864) by the Painter of Cambridge 47, and on an amphora by the Amasis Painter in Würzburg (Martin von Wagner Museum 265).
8. For Louvre CA 3185 (from Vulci), see Spivey 1987, p. 7, no. 3, fig. 1; for the Early II amphora in Baltimore (Walters Art Gallery 48.7, from Castel Campanile), ibid., p. 10, no. 27, fig. 5; for the Middle I amphora in Palermo (Museo Archeologico Regionale 1498, from Chiusi), ibid., p. 13, no. 55, fig. 11a; for the "late" column crater in Berlin (Staatliche Museen F4204), ibid., p. 28, no. 184, fig. 31a–b.
9. Riis 1998, pp. 42–52. Does this manner of painting derive from Laconian vase painting? Compare the flowing hair of the speeding harpy on the Boread Painter's name piece in the Museo Nazionale Etrusco di Villa Giulia (a vase found in Etruria) and that of the flying figure on a cup fragment by the Arkesilas Painter from the Samian Heraion (Berlin, Staatliche Museen 478x).
10. Compare, for example, the bone plaque of a charioteer in a *biga* pulled by winged horses from Tomb 15, the Crocifisso del Tufo necropolis, Orvieto (M. Bizzarri, *Le necropoli di Crocifisso del Tufo in Orvieto* [Orvieto, 1963], p. 85) and the small plaques from a chest in Paris (Louvre S. 2028: Martelli Cristofani 1985, p. 208, figs. 1–4; and A. Hus, *Les Etrusques: Peuple secret* [Paris, 1957], pp. 66–68).
11. The "Herakles" from Contarina (Rovigo) is Adria, Museo Archeologico Nazionale 9996; the "Fufluns" is Modena, Galleria Estense 12505; the *Statuette of Tinia* is Getty Museum 55.AB.12 (*J. Paul Getty Museum* 2002, p. 133; *J. Paul Getty Museum* 2010, 129; and Kozloff 1981, pp. 219–23).

12. M. Del Chiaro, *Etruscan Art from West Coast Collections* (Santa Barbara, 1976), pp. 12–13, no. 64. Close in spirit and technique is a small bronze in the Thorwaldsen Museum: T. Melander, *Thorvaldsens antikker—en temmelig udvagt samling* (Copenhagen, 1993), pp. 116–17, no. 93.

13. Boston, Museum of Fine Arts 21.197.

14. Strong 1966, p. 31.

15. Ibid.

16. J. D. Beazley, *The Lewes House Collection of Ancient Gems* (Oxford, 1920), p. 8.

17. See Boardman 2001, pp. 153, 420; J. Spier, "From East Greece to Etruria: A Late Sixth-Century B.C. Gem Workshop," in Tsetskhladze et al. 2000, pp. 333–35; Boardman 1968, pp. 162–63; Zazoff 1968, pp. 17–24; and Zazoff 1966, pp. 63–78. Beazley 1920 (in n. 16, above), pp. 31–33, was the first to note the connection between carved amber pendants and the pseudo-scarabs. See also Strong 1966, p. 31; and D'Ercole 1995, p. 286, n. 94. Bissing 1931 was the first to suggest that Early Etruscan amber scaraboids were carved by gem engravers.

18. Spier 2000 (in n. 17, above), p. 335.

19. Boardman 2001, p. 153.

20. Hedreen 1992, p. 168, nn. 83–84, refers to Strabo (10.3.11 [C 468]) and cites R. Seaford, "On the Origins of Satyric Drama," *Maia* 28 (1976): 214–15. Seaford draws the conclusion that the satyrs reflect initiatory practices.

21. Hedreen 1992, p. 168, referring to R. Seaford, *Euripides: Cyclops*, and to Eustathios on Homer's *Iliad* 1.311.25.

22. Jannot 2005, p. 41.

23. The reference to Harpocration (s.v. "leuke") comes from W. Burkert *Greek Religion*, trans. J. Raffan (Cambridge, MA, 1985), p. 294, n. 13. For the translation, see F. Graf, "Dionysian and Orphic Eschatology," in *Masks of Dionysus*, ed. T. H. Carpenter and C. Faraone (Ithaca, NY, and London, 1993), p. 244. In the *Suda*, s.v. "leuke" ("poplar"; Adler no. λ 319), "Those celebrating the Bacchic rites used to be crowned with white poplar because the plant is from the nether world and the Dionysos of Persephone, too, is from the nether world. He [Harpocration] says that the white poplar grew by the [river] Acheron, which is why in Homer it is called *acherois*."

13. Pendant: Satyr Head

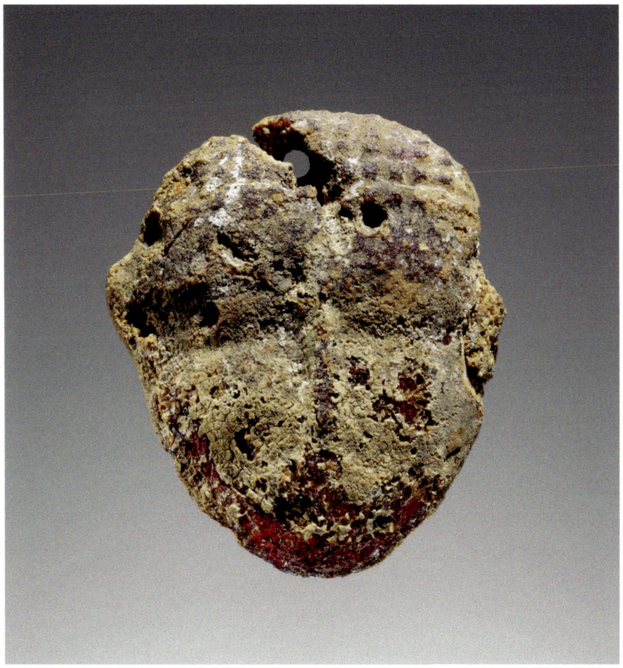

Accession Number	82.AO.161.1
Culture	Etruscan
Date	525–480 B.C.
Dimensions	Height: 53 mm; width: 48 mm; depth: 16 mm; Weight: 11 g
Subjects	Dionysos, cult of (also Satyr); Jewelry

Provenance

–1982, Jiří Frel, 1923–2006, and Faya Frel (Los Angeles, CA), donated to the J. Paul Getty Museum, 1982.

Condition

The piece is intact, although the surface condition is poor and the surface is covered by a thick, flaking yellow alteration crust. There are also many small chips on all sides. There is a large fissure on the top of the head and smaller fissures on the reverse side. A whitish encrustation covers some areas. The amber is yellowish brown and opaque in ambient light, translucent and orange where the interior is exposed by modern breaks, and bright orange and generally translucent in transmitted light. There are inclusions or deteriorated material visible in the fissures.

Description

This relatively large head is egg-shaped in front view and is like a rounded slab in profile view. It is slightly convex on the obverse, flat and plain on the reverse. Despite the poor condition of the piece, its subject is still legible. The hair is caplike, delineated by eleven rows of snail-like curls in even rows. The head is widest at the position of the ears. Traces of the right eyebrow remain; the brow itself is wide and smooth. The plastic, almond-shaped eyes are located equidistantly between the top of the head and the chin. The inner and outer corners (canthi) appear to be on the same line. Although broken, the ears are long, pointed, and prominent and are set high up on the head. The cheeks are wide and flat and the face long. The remains of the nose suggest that it was small and short. The mouth area is small and surrounded by a short mustache and low, close-cropped beard.

There are two suspension perforations, a narrow-gage lateral bore through the top of the head and a second, larger, rostrocaudal hole in the center of the forehead. All four exits show enlargement at the upper parts, abrasion troughs, and chipping. There is a stopped bore in the top left of the head. When suspended from the lateral bore, the head hangs with the brow tipped forward and the chin recessed. Suspended from the large hole, the head hangs perpendicular to the ground. If the large hole were used to secure the head to a support, its chin would have been back, the top of the head forward.

Discussion

The condition of this head—the evidence of pulling on the upper edges of the perforations, the frontal perforation that is likely secondary to the lateral bore, and the wear on the prominent surfaces of the head—suggests that it saw substantial use before it was buried. A number of pre-Roman figured ambers have a narrow-gage transverse perforation for stringing and one or more large front-to-back borings. Some have been found still attached to fibulae. Others retain only metal nails, or their traces (some bronze, others silver), perhaps used to attach the amber onto a wood (or other material) support. There are other significant examples of figured ambers with both narrow-bore lateral perforations and larger front-to-back borings.[1] They include a female head from Rutigliano still attached to a silver fibula, documented as from the early fifth century B.C.; a dancing figure from Oliveto Citra, which was first perforated with a transverse hole for suspension as a pendant and then with five front-to-back holes, one large one in the middle and four slightly smaller holes surrounding it; a fragmentary female dancer (probably once joined by a male figure) in a New York private collection, which has a transverse perforation and four large holes with the remains of bronze rivets; and two profile female head-pendants and a horse's head on the London art market (perhaps from the same findspot), whose frontal holes still retain silver nails. In each case, the large rostrocaudal holes are disfiguring and appear to be secondary to the lateral suspension bores, which are worn from pulling, with characteristic abrasion troughs on the upper inside edges of the exits.

The Getty *Satyr's Head* is illuminated by comparison with six female-subject head-pendants in the British Museum: BM 55, whose findspot is unknown, and five bequeathed by Sir William Temple, which are said to have come from Armento (BM 54, 56, 57, 58, and 60).[2] 82.AO.161.1 is most like BM 57, a head of a female figure wearing a feather crown.[3] The London heads and this amber satyr have a distinctive softness in the modeling, especially in the planar transitions, which must have been accomplished by abrasion. This is contrasted with the outlining of the eyes, probably done with a use of a graver, and with the description of the hair, perhaps accomplished with a carving tool such as those used for ivory or wood. The visual effect is more like stone carving and the best of ivory working or fine woodworking, and less like that of gem engraving. Donald Strong suggested that the London head-pendants, though said to have been found at Armento, were very likely made in Campania or "under the strong influence of Campanian art of the sixth century BC."[4] A comparison of the London pendants to selected Campanian *coroplastics* bears out Strong's observations.[5] Marked, too, are the East Greek aspects of 82.AO.161.1 and the London group; this is highlighted when they are compared to East Greek sculpted and molded works, and to the most East Greek–looking of Etruscan bronzes and painted vases.[6]

82.AO.161.1 has stylistic and iconographical features in common with 76.AO.85.1 and 76.AO.86 (cat. no. 10), attributed to an East Greek or East Greek–trained carver. Many of the comparanda important for 76.AO.85.1 and 76.AO.86 also elucidate 82.AO.161.1. In addition to these are coins and glyptics with frontal faces of Dionysos and his male followers. Three key comparisons are the "satyr" of an early electrum *hecte* from Cyzicus;[7] the device of a reclining satyr on an agate scarab in London, the name piece of the Master of the London Satyr, an East Greek carver working in Etruria;[8] and the Dionysos wrapped around the back of the cornelian pseudo-scarab in Boston, the name piece of the Master of the Boston Dionysos.[9] (The work of the Boston Master is very close to that of 83.AO.202.1, cat. no. 12.) A number of satyr heads on Attic black-figure vases are also important comparanda for the above-listed images of satyrs in amber: the large and staring eyes and the carefully groomed hair and beards are but two of the striking similarities.[10]

The only other amber satyr head related in format and size to 82.AO.161.1 is a well-preserved pendant from Tomb 106 in the necropolis at Braida di Vaglio (Basilicata).[11] Although the Braida amber differs in the satyr type, the face exhibits the same sober expression. The Braida satyr's hair is deeply waved around the brow, his beard long, and his large ears prominent. This is in contrast to the short beard, small ears, and curly hair of the Getty satyr. The context of the Braida satyr pendant is the early fifth century B.C., but it must have been carved earlier, perhaps as early as the third quarter of the mid-sixth century. It also shows considerable use wear and secondary working. The face is especially worn on the prominent surfaces, and the inserted suspension loop in the top of the head is likely secondary to the narrow-gage lateral boring. The same British Museum heads presented above as comparisons for 82.AO.161.1 are also instructive for the Braida di Vaglio satyr, especially BM 56.

For a discussion of the iconography of a satyr in amber, see the entry for *Satyr Head in Profile* (83.AO.202.1).

NOTES

1. This group is discussed in "The Working of Amber" in the introduction, n. 266.
2. BM 58, which is carved fully in the round, appears to be the earliest of the group: see Strong 1966, pp. 71–73, no. 58, pls. XXIII–XXIV.
3. Ibid., p. 73, no. 57, pl. XXIII, bears a resemblance to images of Hathor wearing a feathered crown, although her ears are human, not bovine. Strong describes the eyes of BM 57 as carefully worked and large in proportion to the other features. This head in London is also discussed in the entry for 76.AO.85.1 and 76.AO.86.

4. Ibid., p. 73.

5. See Riis 1981; Riis 1938; and W. Johanowsky, *Materiali di età arcaica dalla Campania* (Naples, 1983), pp. 72–73.

6. This is borne out by a comparison of British Museum 58 with a small marble head from Miletos in the Louvre (Ma 4546): Hamiaux 1992, no. 49 (circa 520–510 B.C.). The marble's squarish face and strong jaw, upward cant of the eyes, serious expression, waved hair at the brow, and crown, which encircles the head, are all similar to BM 58.

7. See Mottahedeh 1979, no. 5.

8. London, British Museum 465. For the Master of the London Satyr, see Boardman 2001, pp. 153, 181, 420 (with earlier bibl.).

9. See entry for 83.AO.202.1.

10. For the relevant satyr heads on Attic black-figure vases, see, for example, Hedreen 1992, pp. 169–70; Carpenter 1986, p. 97, n. 93; G. Ferrari, "Eye-Cup," RA 1986: 520; Mottahedeh 1979; and E. E. Bell, "Two Krokotos Mask Cups at San Simeon," *University of California Studies in Classical Antiquity* 10 (1977): 115. Carpenter cites the amphorae with heads listed in ABV 275 and examples of cups with satyr heads in the Group of Walters 48.42: see ABV, nos. 15, 206.

11. Potenza, Museo Archeologico Nazionale "Dinu Adamesteanu" 96684: *Magie d'ambra* 2005, p. 117; Bottini and Setari 2003, p. 66, no. 311 (satyr), fig. 37. The satyr head-pendant is discussed in the introduction and in the entry for 83.AO.202.1.

14. Pendant: Female Head in Profile

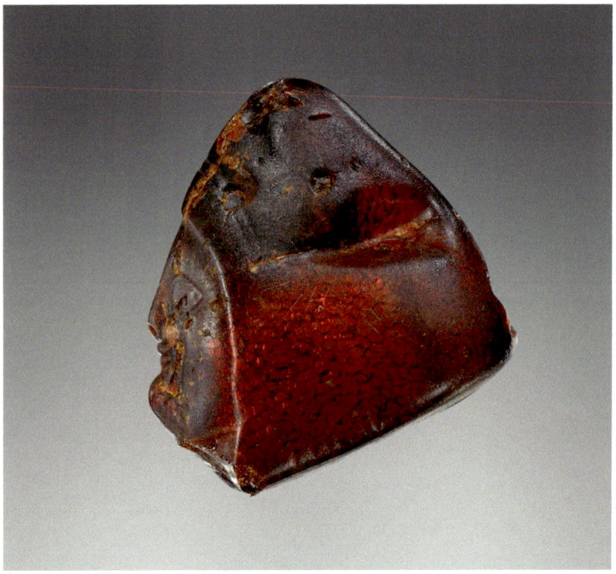

Accession Number	77.AO.81.4
Culture	Etruscan
Date	525–480 B.C.
Dimensions	Height: 57 mm; width: 56 mm; depth: 30 mm; Weight: 49.8 g
Subjects	Jewelry

Provenance

–1977, Gordon McLendon (Dallas, TX), donated to the J. Paul Getty Museum, 1977.

Condition

The piece is in a good state of preservation. The surface is smooth and firm, despite a number of recent as well as older (weathered) small breaks and chips. There are fissures extending from the apex of the headdress to the eye, and from under the eye to the corner of the mouth. All fissures and cracks contain yellow-ocher residue. The smoothing of the prominent surfaces (a wear pattern different from overall erosion); the patination covering some of the breaks, including on the tip of the nose, the lower edge of the eye, and several places on the cheeks; and wear on the suspension perforations are likely evidence of use wear. The amber is opaque in ambient light and reddish in color. With transmitted illumination, the interior appears bright orange and is transparent. There is a cloudy inclusion (possibly bubbles) at the top of the headdress.

Description

The shape of the pendant is nearly triangular. The obverse is rounded and the carving curves around the form of the amber except on the back, which is plain but uneven in surface. The figure is wearing a conical cap, and over it a veil and another overgarment. She also wears a circular earring. No hair is showing. The face is set off from her neck by an indentation. Her eye is large and almond-shaped, and is set off from the plane of the face by a continuous filletlike line, which represents the eyelids. The nose follows the same plane as the brow, with only a slight indention for the root. The upper lip area is short, and the barlike lips are pulled into a smile. The lower lip is wider than the upper. The horizontal sulcus is shallow and the chin prominent. The under-chin is fatty. The point of the chin extends forward, to the level of the tip of the nose. The neck is plump.

The uppermost layer of clothing, a shawl or the top portion of a cloak, covers a conical hat and veil. The grooves parallel to the edge of the overgarment are perhaps its border or the imprint of the edge of the hat beneath. The veil's front border is finished with a band, represented by fine parallel lines. Five soft, narrow grooves represent the fabric's folds as it hits the shoulder area. A shallow indentation in the overgarment, at about the jawline, indicates a break in the fall of the fabric.

When suspended, the head would have hung with the chin forward; an imaginary line drawn from the tip of the nose to the chin would have been perpendicular to the ground. The slope of the forehead and the slope of the chin would have been at approximately 60 and 120 degrees to the perpendicular.

On the reverse is a deep groove, probably a cleaned fissure. A 3 mm perforation for suspension extends from the flat area on the bottom side of the pendant and exits at the deep cleft on the side of the headdress. There are two bores stopped with amber plugs: one is located on the upper part of the headdress, and a second, 3 mm in diameter and about 11 mm deep, is under the chin.

Discussion

This head-pendant has no exact parallel. The pendant's triangular shape is unique. The dress of the figure, however, is similar to that worn by 76.AO.85.2 (cat. no. 15). Moreover, there are several parallels for the physiognomic type and the style of the pendant. 77.AO.81.4 is related to a number of images of women from Samos and other Ionian centers of the third quarter of the sixth century B.C. Although the comparison can be made only from an old photograph, 77.AO.81.4 is remarkably similar to a now lost (unfinished) head once in Berlin, which was uncovered in the Heraion at Samos.[1] The pendant also is akin to a family of East Greek terracottas. Other pertinent parallels include two mid-sixth-century alabastra in Paris and London, and a half-figure statuette of the second half of the century in Copenhagen.[2] Although the chin of 77.AO.81.4 is more prominent than those of the terracottas, their other morphological similarities are compelling.

Still closer comparisons are to be found in Etruscan art. For the dress of 77.AO.81.4, the most striking parallel is the Etruscan bronze votive in Florence (Museo Archeologico Nazionale 277), which belongs to Emeline Richardson's Tomba delle Barone group, one she characterizes by its Ionian associations. (Florence 277 is discussed in more detail in the entry for 76.AO.85.2). An amber head-pendant in London (British Museum 53) wears similar dress, although her veil is drawn more tightly in front and forms stronger horizontal folds.[3]

For the style of 77.AO.81.4, other Etruscan small bronzes and bronze reliefs are instrumental in situating the amber. Three figures (Peleus, Thetis, and a female companion) in one of the panels of the Loeb Cauldron C are very similar in physiognomy and expression, although their eyes are smaller and their hairstyles different.[4] Eight small bronzes are important comparisons for *both* the iconography and the style of 77.AO.81.4. They are an Etruscan votive in Paris (Bibliothèque nationale, Cabinet des Médailles 204); the four bronze corner figures from the box of a *carpentum*, now dispersed among Perugia, Munich, and Paris;[5] two "Latin" bronzes from Satricum in the Villa Giulia (Museo Nazionale Etrusco di Villa Giulia 10519 and 10922); and a "Latin" bronze in the Bibliothèque nationale, Paris (BN 205).[6] The Villa Giulia bronzes are distinguished by the fact that they wear the Ionian chiton properly understood (a rare feature; they are draped like the Samian sisters from the Genelaos dedication), and by the upright solar disk at the back of their heads. Richardson considers them images of the Mater Matuta, the goddess of daylight and a light bringer.[7] The figure of BN 204 is unusually pretty and her unusual dress uncommonly delicate: she is wearing an unbelted chiton, and her hairstyle is otherwise unknown in the Middle Archaic period.[8] The profiles of 77.AO.81.4 and BN 204 are so alike as to be from the same model, or to be products of the same ambient: the angles of the nose and the chin relative to the perpendicular are the same. BN 204 might even allow us to imagine how 77.AO.81.4 would have looked in full face: that is, wide through the temples, with eyes slightly far apart, and narrowing to the chin.

Who or what might 77.AO.81.4 represent? She wears a large earring and elaborate dress: cap, veil, and overgarment. Not one strand of hair is showing. At the least, this is the adornment of the elite. Does this indicate the figure's maturity or outdoor activity, two explanations offered for head coverings in Greek sculpture? On the other hand, does the dress indicate religious-political activity (for example, sacrifice or attendance at another ritual)? The material precludes that the representation is that of a mortal; it signals, rather, that the image is that of a heroic, divine, supernatural, or demonic being. If a divinity, then perhaps it is one whose powers were connected to the sun and to its regenerative powers. The Ionian dress and style of the head, too, may indicate something of its identity, as well as saying something about where, why, and by whom 77.AO.81.4 was carved and used. The style and dress point to South Ionia, Etruria, and "Latin" Etruscan imagery.

77.AO.81.4 could represent a divinity of light, similar to the Latin Juno Licina or the dawn goddess Mater Matuta, both of whom share some aspects of the Greek Artemis (perhaps in her guise as Artemis Phosphoros), notably midwifery. Since the dawn is also a potent symbol of new life, this may be a South Ionian Eos or the Etruscan Thesan. Two other alternatives are Artumes and the Latin Juno Gabina; Richardson records that bronzes of a "Latin" type, some of them wearing sun disks, have been found at Gabii, near the so-called temple of Juno Gabina.[9] If 77.AO.81.4 does indeed represent a light and life divinity, one that might be called upon with some regularity for aid in childbirth, in protection, or in the care of young children, such an identity may help explain the use wear on the face of the pendant.

NOTES

1. Berlin, Staatliche Museen 1875: Richter 1968, no. 157, figs. 504–5; Freyer-Schauenburg 1974, p. 41, no. 18, pl. 10.
2. Paris, Louvre MC681; London, British Museum 60.44.57; Copenhagen, Ny Carlsberg Glyptotek 949.
3. Strong 1966, p. 71, no. 53, pl. XXII.
4. See Höckmann 1982.
5. See discussion by S. Bruni in Torelli 2000, p. 581.
6. Richardson 1983, pp. 21–23, 26–67, 361.
7. Ibid., p. 265, pls. 605–6.
8. E. Richardson, "Moonéd Ashteroth?," in *In Memoriam Otto J. Brendel: Essays in Archaeology and the Humanities* (Mainz, 1976), pp. 21–24. See also Waarsenburg 1995, pp. 461–62.
9. Richardson 1983, p. 265, pls. 605–6.

15. Pendant: Winged Female Head in Profile

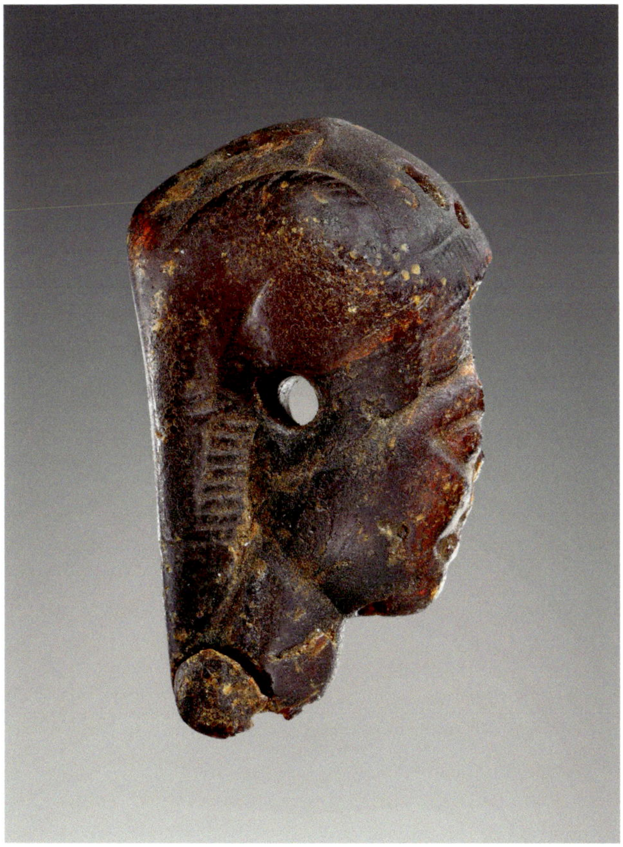

Accession Number	76.AO.85.2
Culture	Etruscan
Date	525–480 B.C.
Dimensions	Height: 79 mm; width: 49 mm; depth: 25 mm; Weight: 51.3 g
Subjects	Inclusions

Provenance

–1976, Gordon McLendon (Dallas, TX), donated to the J. Paul Getty Museum, 1976.

Condition

A piece broken off at the neck in modern times was reattached with an adhesive before the donor acquired it. There is evidence of ancient use wear on the prominent parts (cheeks, nose, head, chin, and wing) and abrasion troughs on the uppermost parts of the suspension perforations at the top of the head. (The weight of the pendant, gravity, movement, pulling, or other abrasive action caused the carrier to deepen, or "saw," the holes.) Modern damage includes pitting on the neck and cheek and large fracture losses at the right profile edge, at the forehead, and in the front section of the hair. The surface of the amber is firm and stable although abraded slightly and crazed overall. Several large fissures (with inclusions and degraded material inside) run throughout the piece, and there are other small areas of inclusions throughout the pendant. In ambient light, the surface of the amber is dark red-brown and opaque; in transmitted light, the pendant is dark red and translucent.

Description

The pendant depicts the head and neck of a female figure in profile facing to the right. It is rounded and decorated on the obverse, flat and plain on the reverse. The area around the eye is subtly modeled. Brow ridge, eye fold, and eye bag are all indicated. The spherical shape of the eye is suggested and set off from lids that are represented as raised lines. The cheeks are full, raised, and puffed; the face is smiling widely. The chin is full and rounded, and the jawline is set off from the neck by a groove as well as modeling. The neck is plump.

The hair is visible at the brow and at the right side. At least five overlapping scalloped waves once framed the face; the remaining ones lie close to the head and are defined with finely engraved parallel curving lines. Horizontal indentations behind the ear area and a curved line at the bottom indicate the shape of the hair fall. The shape and bulk of the head covering shows that over a cap (the conical projection at the back of the head is its apex) is worn a veil and, over it, a shawl. The outermost layer of the headdress is banded with a pattern of finely engraved parallel lines, perhaps a decorative woven edge, or an addition.

A striking feature of this piece is a large volutelike wing, projecting from the right top side of the head and curving back to the neck. Diagonal engraved lines at the top and back indicate secondary feathers. In front of and below the wing are sections of hair, engraved with horizontal parallel lines; the mass of the hair is curved at the bottom.

The smoothed grooves and craters are likely the result of the clearing away of surface inclusions or fissures in the amber blank. The object incorporates the natural undulations of the amber, especially at the back and in the wing area.

The pendant has two through-bores for suspension. The first is a lateral perforation at the crown 2.5 mm in diameter, with a hole in the front of the head, at the border of the head covering, and another in the top of the head. At the position of the ear is the second through-bore, 9 mm in diameter, which appears to have been made after the head-pendant was carved. There are also five stopped bores; the longest, 22 mm in length by 4 mm in diameter, begins under the chin, ascends vertically to the hairline, runs parallel to a long internal fissure, and does not retain a plug. Two other stopped bores, one of which retains its plug, are located on the forehead. The two others, at the back of the head and on the reverse, are without plugs. It is not possible to determine how this pendant would have hung, because of the multiple through-bores.

Discussion

One of the largest of the extant amber head-pendants, 76.AO.85.2 has no close counterpart. It is also significant within the corpus of head-pendants because of its style, iconography, evidence of use wear, and quality.

The figure wears several elements of dress that cover her head: a cap, veil, and shawl or mantle. In this, the combination is comparable to the dress worn by 77.AO.81.4 (cat. no. 14). For the dress, the best parallel is an Etruscan bronze in Florence (Museo Archeologico Nazionale 277). The outer garments worn by Florence 277 and 76.AO.85.2 both display a decorated band on the front edge. Emeline Richardson noted the unusual style of the bronze figure, including her unique banded scarf, and compared the work to a group of Greek-Ionian-looking bronzes, her Tomba del Barone group, named after the eponymous Tarquinian tomb. Richardson dated Florence 277 to the end of the sixth century B.C.[1] The amber's hairstyle, however, is different from that worn by most other Archaic Etruscan human-form figures. It is one found on a number of Greek marble korai, sirens, and sphinxes, and is comparable to that of a unique image, possibly Etruscan, possibly Magna Graecian, of a bronze of a crowned and curiously robed figure from Villa Ruffi, Covignano (Rimini), now in Copenhagen.[2]

There is a distinct "Greekness" about the physiognomy and style of 76.AO.85.2. This is brought out by comparison to three different marbles: a small head of a sphinx from Aegina in Athens,[3] a kore head from Lindos in Copenhagen,[4] and a small kouros (unfinished) from Paros in Paris.[5] There are especially striking correspondences between the Paris kouros and 76.AO.85.2: the full cheeks, which extend all the way from the angles of the mouth to the edge of the jaw, the form and depth of the nasolabial furrows, the placement of the eyes, and the hair framing the face. These Archaic marbles all date to the second half of the sixth century, with the Paros kouros close to the mid-century mark and the female heads in the later sixth century.[6]

The hairstyle of 76.AO.85.2, a complex fashioning of overlapping rounded waves, is an ancient one, first found in the Near East for gods and kings and refashioned in Early Archaic art for male and female figures, including sphinxes.

76.AO.85.2 may be the earliest extant representation of a winged amber head-pendant. The wing—growing or attached behind the ear—implies the existence of a symmetrically placed ear on the other side of the head. This is not the only amber head-pendant to have a wing: a number of heads have a wing or wings. 76.AO.85.2 offers the opportunity to consider how an amber carver might address the concept of the head and wings of a winged divinity.

The choice to represent one wing or a pair must have depended on the form of the amber blank. For obvious technical reasons, wings projecting upward from the shoulders would be difficult to represent convincingly. Certain forms inherent in the undulating forms of the raw material might even have provided inspiration for the images: the rounded volute wing of 76.AO.85.2 may have started out as a natural high point on the resin. Other winged amber head-pendants are a pair from the Tomb of the Ambers, Ruvo (Naples);[7] a head-pendant in profile to right, with one clearly represented wing, from an amber-rich grave of the Rutigliano-Purgatorio Necropolis;[8] a fragmentary head-pendant of a frontal female head with one wing in New York;[9] two others in the Getty collection, 77.AO.81.5 (cat. no. 23) and 77.AO.81.29 (cat. no. 16); a large frontal head with two wings in the Steinhardt collection, New York;[10] and three unpublished examples in a London private collection. Many of these winged heads are closely related in artistic style and in manner of dress to other, wingless amber head-pendants.

The wing might well be an attribute of the head, acting as *pars pro toto* of a divinity, demigoddess, or demon; the winged head might be considered the sculptural compression of a complete winged figure.

The large and disfiguring frontal hole in 76.AO.85.2 was bored through the ear and surrounding area. As noted in the introduction and in the "Pendants in the Form of the Human Head" opening, such large holes are found on a number of other figured ambers, of both male and female subjects, and of complete figures as well as both frontal and profile head-pendants, including satyr head-pendants, female head-pendants, a dancing female, a pair of sirens, a horse head–pendant, and a head of Herakles in a lionskin headdress.[11] Was this pendant formerly attached to a fibula, as is the case with a head-pendant excavated at Villalfonsina (Chieti) and the two pins with amber heads threaded on them from Rutigliano, a satyr head on a silver pin and a female head on a bronze pin?[12]

76.AO.85.2, like the large profile head-pendant of a satyr, 83.AO.202.1 (cat. no. 12), has four stopped bores, a plug retained in one of them. Were they part of the original production or made later? All the physical evidence—the smoothed prominent surfaces[13] (this is not the typical breakdown of the cortex from oxidation), the multiple through-bores, the abrasion troughs, and the central hole—indicates that this pendant must have been used over a period of time before its final interment. The inherent evidence of 76.AO.85.2 suggests that it served as an important amulet-ornament for the living long before it was finally placed in a grave, perhaps pinned to the funereal dress of a girl or woman, in protection and in "mourning attending the death of the young."[14] The wing, the old-fashioned coiffure, the Etruscan dress, and the style suggest that it was made in an Etruscan ambient where a Greek (perhaps Parian or Greco-Etruscan) artisanal tradition was significant.

NOTES

1. Richardson 1983, pp. 285–86, pl. 198, figs. 669–71.
2. Copenhagen, National Museum 4203. The bronze likely records an older image type from Ionia or elsewhere in East Greece, one that looks back to older Oriental models. It is possibly Etruscan and has been compared to the korai of northeastern Etruria; B. Bundgaard Rasmussen in Torelli 2000, p. 622, no. 278, makes a convincing suggestion that it is Magna Graecian, perhaps from southern Sicily.
3. Athens, National Archaeological Museum 1939.
4. Copenhagen, National Museum 12199.
5. Paris, Louvre Ma 3101 (circa 540 B.C.): Hamiaux 1992, pp. 80–82, no. 73. Two heads comparable to the kouros, which *are* finished but are worn, are in the Liebieghaus, Frankfurt (H. von Steuben, *Kopf eines Kuros*, Liebieghaus Monographie 7 [Frankfurt am Main, 1980]), and in the Iolas collection (C. Rolley, "Tête de kouros Parien," BCH 103 [1978]: 41–50, figs. 1–7).
6. Croissant 1983, pp. 60, 77, 98, places the Louvre kouros in his Paros group of the mid-sixth century and the Aegina head with his Chios group, dating it to the later sixth century, and associates the Lindos head with Knidian coinage of the years around 500 B.C. Gisela Richter placed the Lindos and Aegina heads ("kore or sphinx?") in her Samian Cheramyes Genelaos group, dated to the second quarter of the sixth century, and the Louvre kouros in her Melos group, with a date of about 550 B.C.: Richter 1968, no. 66, figs. 214–16 (Athens head); no. 77, figs 244–47 (Lindos head); no. 116, figs. 356–58 (Paros kouros).
7. Naples, Museo Archeologico Nazionale 113644, 113646 (excavated in 1876). No. 113644 has symmetrical wings; no. 113646 is by another hand and shows a different approach to the subject (it has a single small wing on the right side of the head). These winged heads are two of the five surviving figured carved ambers from the Ruvo tomb. The others are an elongated female head (fragmentary) by still another hand; a piece with an indeterminate subject; and a large pendant of a kneeling warrior, armored with helmet, shield, and sword (no known inv. nos.) and attended by a crow. G. Prisco in *I Greci in Occidente: La Magna Grecia nelle collezioni del Museo Archeologico di Napoli*, exh. cat. (Naples, 1996), pp. 114–16, figs. 10.5–9, identifies the warrior pendant as Achilles, and dates it to the end of the sixth or beginning of the fifth century B.C. For a possibly relevant interpretation of Achilles in relationship to the cult of Apollo, see Simon 1998. An amber Achilles pendant would be a powerful *apotropaion*, the color of the amber reinforcing the potency of the martial subject.
8. Taranto, Museo Archeologico Nazionale 138144 (from the male Tomb 9, fifth century B.C.).
9. New York, Metropolitan Museum of Art 23.160.97: Richter 1940, p. 32, fig. 100.
10. New York, collection of Michael and Judy Steinhardt: Grimaldi 1996, p. 151; and Negroni Catacchio 1999, pp. 289–90, fig. 5.
11. Tomb 9, Rutigliano-Purgatorio Necropolis (Taranto, Museo Archeologico Nazionale 138144: *Ornarsi d'ambra: Tombe principesche da Rutigliano*, ed. L. Masiello and A. Damato [Rutigliano, 2004]; Mastrocinque 1991, p. 131, n. 408; and G. Lo Porto in *Locri Epizefirii: Atti del XVI Convegno di studi sulla Magna Grecia* [Naples, 1977], pl. CXV). On the subject of pendants with large secondary holes see "The Working of Amber," in the introduction, n. 266.

12. The head-pendant from Villalfonsina, pierced through the top of the hair, dangles from the pin of the fibula: Negroni Catacchio 1975, pp. 314–15; R. Papi, "Materiali archeologici da Villalfonsina (Chieti)," ArchCl 31 (1979): 83–85, 91. The head-pendants from Tomb 9 (satyr) and Tomb 10 (female) from the Rutigliano-Purgatorio Necropolis are discussed in the introduction.

13. The touching, rubbing, and kissing of faces in prayer, adoration, propitiation, warding away danger or evil, and communication of other kinds are the subject of an enormous body of literature. See the introduction, particularly n. 176.

14. Diodorus Siculus, *Library of History* 5.23–24: amber "is commonly used in connection with the mourning attending the death of the young."

16. Pendant: Winged Female Head

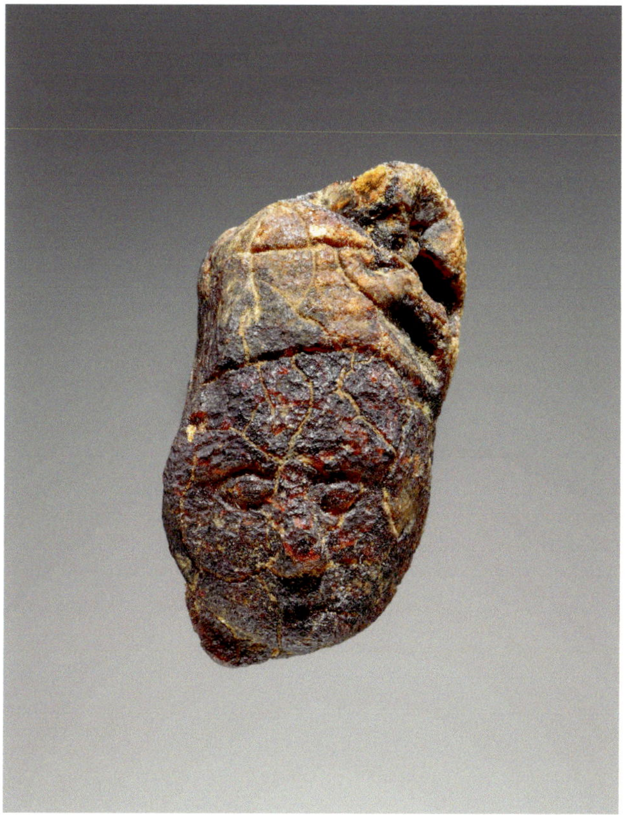

Accession Number	77.AO.81.29
Culture	Etruscan
Date	525–480 B.C.
Dimensions	Height: 40 mm; width: 20 mm; depth: 18 mm; Diameter of suspension hole: 6 mm; Weight: 8.7 g
Subjects	Etruscan culture

Provenance

–1977, Gordon McLendon (Dallas, TX), donated to the J. Paul Getty Museum, 1977.

Condition

The piece is largely intact, except for a large loss at the top proper right of the pendant. The lateral suspension perforation passes through two fissures and is broken off at the top of the left exit. There is wide-interval cracking and crazing over the entire surface. Degraded amber residue, yellow-ocher in color, covers parts of the surface and is inside the cracks. In ambient light, the piece is opaque and dark reddish brown; in transmitted light, it is translucent and red. There are no visible inclusions.

Description

This head-pendant of a female figure includes the head and neck, which are contorted and abbreviated to fit the natural shape of the amber blank. The pendant is rounded on the obverse and nearly flat on the reverse, which is undecorated. The composition is asymmetrical: the figure is more deformed on her right side and more naturalistic on her left. The woman's face is round but relatively lean. The high forehead is flat and smooth, and her brow ridge is modeled. She has full, wide cheekbones, shallow cheeks, and a small, pointed chin. The eyes are small and deep-set, with the orbits plastically rendered. The eyelids are indicated by engraved lines. The outer corners of the eyes turn up slightly. The nose is indented at the bridge, and its tip is missing. The lips curve up slightly; below them is an indentation representing the mentolabial sulcus. The area below the chin is flat. The junction of the neck is higher than the point of the chin.

The headdress, a conical hat with veil, or perhaps a cloth-wrapped hat, sits directly on the forehead and is separated from the smooth brow by an engraved line. No hair shows on the forehead. Four horizontal grooves and a finer horizontal engraved line suggest the layers or wrapping of the headdress. On the upper left side of the figure's head is a volute-shaped wing. At the side of the neck is a long, narrow segment of undifferentiated amber, which is set off from the neck by a long groove. It probably represents a long, single hank of hair.

The suspension perforation is drilled in the upper part of the headdress and is 6 mm in diameter at its widest point.

Discussion

Both the iconography and the style of this amber are Etruscan. The rounded, almost egg-shaped head is that of Emeline Richardson's Middle Archaic korai, which she dates to the third quarter of the sixth century B.C. What Richardson has proposed about this series of bronzes might also describe 77.AO.81.29: "The longest axis [is] from crown to chin, at an angle to the vertical."[1] The best parallels for 77.AO.81.29 among this Richardson grouping are bronzes in Florence from the votive deposit of Fonte Veneziana, Arezzo, including Museo Archeologico Nazionale 258 and 265–67.[2]

77.AO.81.29 belongs to a family of amber head-pendants that are characterized by their sharp features—wide cheekbones, thin cheeks, pointed chins—vestigial necks, single locks of hair hanging down, and high headdresses. 77.AO.81.29 is a relative of the oldest-looking head from Tomb 106 at Banzi. The latter is one of a group of thirteen pendants, of different types and styles, from a grave dated to the second half of the fifth century B.C.[3]

NOTES

1. Richardson 1983, pp. 258–59.
2. Ibid., pp. 259–62, figs. 583–84 (Florence 258), figs. 590–91 (Florence 265), figs. 588–89 (Florence 266), and figs. 590–91 (Florence 267). Florence 266 is the earliest bronze to wear the diagonal Ionian mantle.
3. Melfi, Museo Archeologico Nazionale del Melfese "Dinu Adamesteanu" 116846: Bottini 1990, pp. 59–61, no. 8, fig. 2.8. The Getty amber may also be compared to two head-pendants from two different tombs at Banzi, one from the early-fifth-century Tomb 106 (illustrated in *Magie d'ambra* 2005, p. 50) and the other from the end-of-the-fifth-century Tomb 428 (ibid., p. 125).

17. Pendant: Female Head in Profile

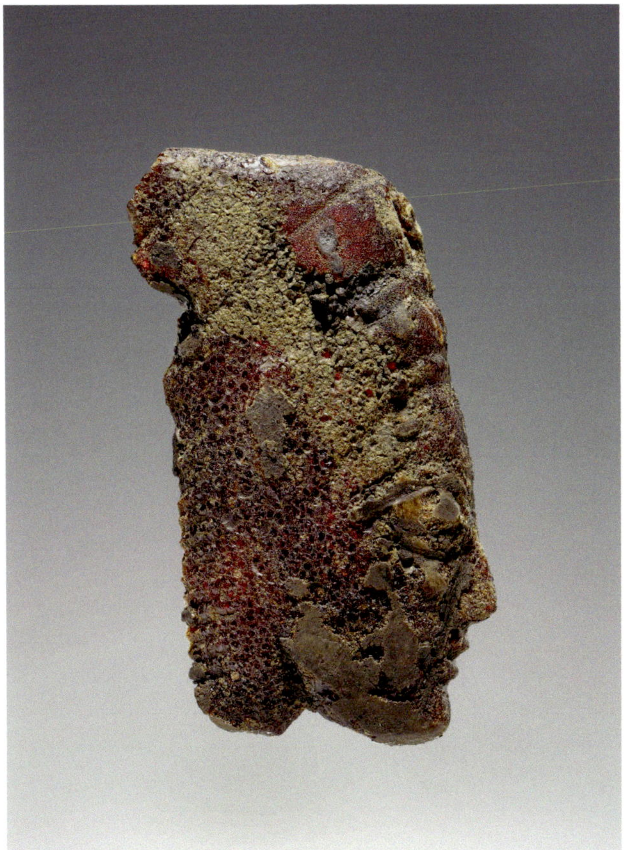

Accession Number	82.AO.161.3
Culture	Italic
Date	500–450 B.C.
Dimensions	Height: 46 mm; width: 45 mm; depth: 11 mm; Weight: 3 g

Provenance

–1982, Jiří Frel, 1923–2006, and Faya Frel (Los Angeles, CA), donated to the J. Paul Getty Museum, 1982.

Condition

The pendant is intact, but there are losses on the head and at the suspension perforation exit on the back of the head. The front perforation exit is smoothly worn on its upper edge, suggesting pulling. Details are smoothed at the tip of the cap on the obverse and at the middle of the pendant's back more than elsewhere, probably through use. The opaque brown surface reflects the original surface level, visible on the cheek and jaw. The surface is crazed, and extensive flaking and cracking all over reveal a reddish, sugary, translucent sublayer. In ambient light, the amber is yellow-brown. At the modern breaks, the interior is translucent and bright orange. In transmitted light, the pendant is translucent and dark red. There are no visible inclusions.

Description

This pendant is oblong, its obverse slightly convex, and its reverse flat and plain. 82.AO.161.3 represents the head and neck of a female in profile. The forehead is broad and rounded. The large, almond-shaped eye is plastic, bulging out almost as in nature; it is recessed at the corners. The lids are rendered as shallow grooves with slightly curved upper and lower rims of varying thickness. Above the eye is the eyebrow ridge, which continues into the temple area. A longer and deeper groove extends from the inner canthus of the eye to the chin, a combination of the nasolabial line and the mouth furrow. The nose is indented at the root. The ears are smooth, rounded shapes at the angle of the jaw; the right ear is more legible than the left. The area between the nose and the upper lip is short. The upper lip slightly overhangs the lower. The mentolabial sulcus is a shallow curve. Although the chin is prominent, it is recessed from the line of the brow. The jaw intersects the neck, forming a small, recessed triangular area.

The figure wears a conical cap covered with a wrapping of cloth (showing three distinct sections), and over it a crown. No hair shows on the forehead. At the back of the head is a fall of hair (a ponytail-like section?) marked with diagonal grooves, except for the plain terminal section.

The carver has taken advantage of natural protrusions in the original amber blank for the modeling of the face. The long groove that runs from the temple to the top of the cap is plausibly a trace of fissure removal. The suspension perforation runs laterally through the top of the headdress, causing the pendant to hang slightly to its left, with the forehead tilted forward, the chin backward.

Discussion

This head-pendant is similar to a large and varied group of female head-pendants that includes works documented as coming from various sites in peninsular Italy and Corsica.[1] 82.AO.161.3 is especially close to two head-pendants from a sporadic find, a tomb group at Roscigno–Monte Pruno (Salerno, Museo Provinciale), and with two head-pendants from the Spina–Valle Lege Tomb 740B.[2] They may all be of comparable date and reflect a common sculptural prototype. The Roscigno tomb is dated to the end of the fifth or beginning of the fourth century, and the *terminus ante quem* for the burial of the Spina head-pendants is no later than the mid-fifth century B.C.

NOTES

1. La Genière 1961. See also Losi et al. 1993, pp. 205–6.
2. For the Salerno pair, see La Genière 1961, pp. 75–88. For the pendants from Tomb 740B at Spina (Museo Archeologico di Ferrara), see *Due donne* 1993, pp. 42–47, nos. 20–21. For additional discussion of the type, see D'Ercole 2008, pp. 69–75.

18. Pendant: Female Head

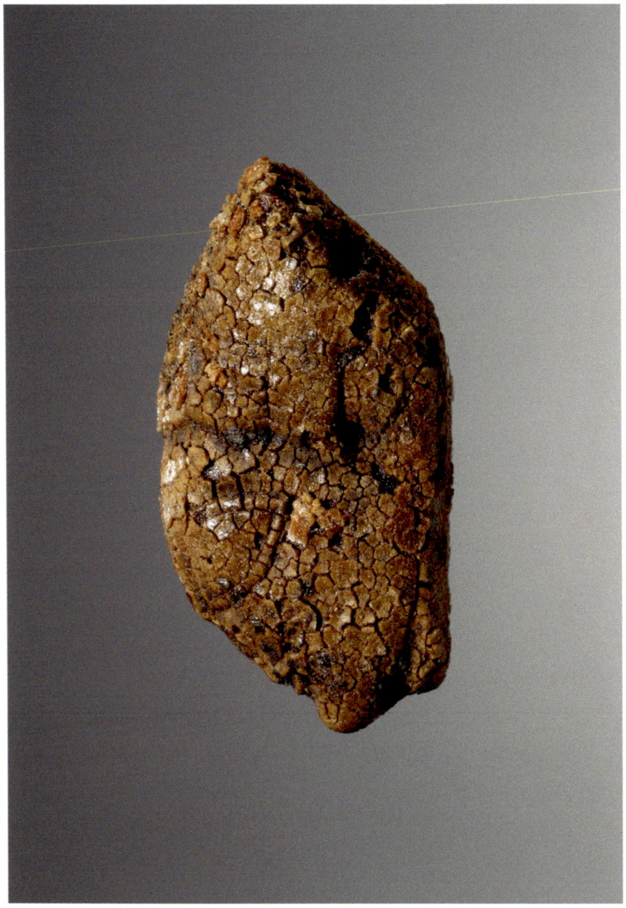

Accession Number	83.AO.202.4
Culture	Italic
Date	425–400 B.C.
Dimensions	Height: 40 mm; width: 23 mm; depth: 16 mm; Diameter of suspension hole: 2 mm; Weight: 4 g
Subjects	Etruscan culture

Provenance

–1983, Antike Kunst Palladion (Basel, Switzerland); 1983, Vasek Polak, 1914–97 (Hermosa Beach, CA), donated to the J. Paul Getty Museum, 1983.

Condition

The pendant is intact, but the nose is broken off. The surface is in poor condition and flaking, there are small losses over the entire piece, and it has a network of cracking. The surface is slightly glossy, suggesting an applied consolidant coating. The entire surface of the pendant is covered in a weathered yellow-ocher crust. In ambient light, the amber is yellowish brown and opaque; in transmitted light, it is dark red and clear. There are no visible inclusions.

Description

The contortion of the form and the asymmetry of the headdress suggest that the original amber blank influenced the form of the head. The pendant represents the frontal head and a small section of the neck of a female. The reverse is flat and plain, the obverse much more rounded and figured. Although the piece is worn and looks almost inorganic, the anatomy of the narrow face is modeled, including the eyeballs. The transitions from plane to plane are smooth. The brow is high and smooth, with the edge of the hair set well back. Above the eyes, the brow ridge swells slightly. The large, almond-shaped eyes meet at the root of the nose and tilt up sharply, the outer canthi higher than the inner, and both the upper and lower lid lines curve. The eyelids are unusual, flat and circumscribed by even, filletlike raised lines. The upper parts of the ears are not discernible. Where the lobe of the ear would be is a small rounded area: is it the lobe or an earring? The mouth is small and curves upward. The lips are separated by a groove. The lower lip is wider than the upper one. The sulcus is curved, leading to the sharply pointed chin. The under-chin area is flat and angles backward to the jaw. The juncture of the head and neck is indicated by a groove. More of the neck vestige is visible on the left side. Framing the face at the brow are scalloplike waves, the individual strands indicated with curving parallel lines. Above the hair is a headdress that is wider than the brow and ends in a soft point. At its base is a slightly rounded, raised area bordered by two parallel engraved lines, which probably represent either the turnup of a felt(?) hat or the bottom of the hat and the edge of the veil.

Alternatively, they may represent just the edge of the veil. There is no evidence of hair at the back or sides of the head.

A 2 mm suspension perforation extends through the headdress near the top, but both holes are on the left side. This would have made the frontal head hang in profile. When suspended, the head tilts forward, with the back perpendicular to the ground.

Discussion

This head-pendant was in the same donation group as four other female head-pendants: 83.AO.202.5 (cat. no. 19), 83.AO.202.6 (cat. no. 20), 83.AO.202.12 (cat. no. 21), and 83.AO.202.18 (cat. no. 22). All five are alike in size, condition, and general typology. All are asymmetrical and have off-center suspension perforations that cause them to hang crookedly. Although they are different in style and details of dress, it is possible that they come from the same original context. Figured ambers of different style, date, and type are often found in the same burial.

83.AO.202.4 is dressed in a high conical hat covered with a veil. 83.AO.202.5 and 83.AO.202.6 each wear a conical cap that is set off from the straight bangs by a pair of spaced engraved lines, perhaps representing the turnup or roll-up of the cap (perhaps made from felt). The flange of amber at the back of the head of 83.AO.202.5 likely represents a wing.

The combination of cap, earrings, and veil of 83.AO.202.4 is the same combination worn by the figure of another Getty head-pendant (77.AO.81.4, cat. no. 14) and by one of the British Museum heads in-the-round (BM 45).[1] The Getty head-pendant 77.AO.81.30 (cat. no. 25) is dressed in a cap (but has no veil). A much larger head-pendant in Cleveland has her hair dressed in scalloped waves at the brow like 83.AO.202.4, but like 77.AO.81.25 (cat. no. 26), the Cleveland figure wears a crown in addition to the cap.[2] The straight bangs topped by the cap of 83.AO.202.5 and 83.AO.202.6 is a more common fashion than that of 83.AO.202.4—which is also worn by the imposing frontal head-pendants of a heavy-faced woman in London (British Museum 48–51).[3] Two unprovenanced profile head-pendants in a London private collection represent a figure of the same physical type. They each wear a cap and have hair coifed like that of 83.AO.202.4. The profile, but not the garments, of 83.AO.202.4 is comparable to the faces of a pair of legless sirens excavated from Tomb 43 at Melfi, to one of the heads from the Melfi Tomb 48 burial, and to one of the heads from Tomb 428 at Banzi.[4] The profile is also not far from that of a head-pendant in London (British Museum 52).[5]

This group of female head-pendants are characterized by their elaborate hair fashions and head coverings—hats, veils, and crowns—and occasionally by their earrings. The dress depends on Etruscan fashion, and the style can be traced to Etruscan inventions. Despite their schematic depiction, these tiny amber pendants clearly reflect the art of the votive bronzes placed by Emeline Richardson in her Late Archaic Series, many of them in her Group A, Ionians.[6]

NOTES

1. Strong 1966, p. 68, no. 45, pl. XX.
2. Cleveland Museum of Art 1992.61 (Andrew R. and Martha Holden Jennings Fund).
3. Strong 1966, pp. 69–71, nos. 48–51, pls. XXI–XXII.
4. For illustrations of the comparable head-pendants, see *Magie d'ambra* 2005, pp. 70 (siren), 121 (head from Melfi), 122, 125, 127 (head from Banzi).
5. Strong 1966, p. 71, no. 52, pl. XXII. Strong compares this head to one found at Populonia (NSc ser. 6, 2 [1926]: 326), to a head in Bari (inv. 6598), and to the Valle Pega heads from Tomb 640B and Tomb 514A. I know the heads from Populonia and Bari only from photographs. The Valle Pega head from Tomb 514A looks less like this Getty head-pendant than does that from Tomb 640B.
6. Richardson 1983, pp. 271–302.

19. Pendant: Female Head

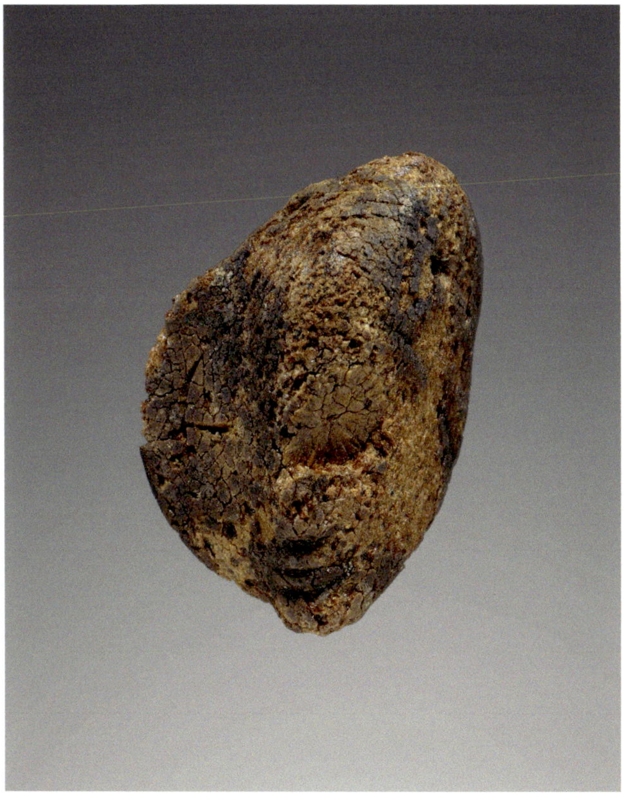

Accession Number	83.AO.202.5
Culture	Italic
Date	425–400 B.C.
Dimensions	Height: 29 mm; width: 38 mm; depth: 14 mm; Diameter of suspension hole: 2 mm; Weight: 10.5 g

Provenance

–1983, Antike Kunst Palladion (Basel, Switzerland); 1983, Vasek Polak, 1914–97 (Hermosa Beach, CA), donated to the J. Paul Getty Museum, 1983.

Condition

The pendant is in poor condition, and most detailed features are lost. The nose and the left eye are broken off. The surface is chipped and flaking, with many minute losses. The outermost, brown alteration layer is flaking; the layer below is more compact and stable. Opaque and yellow-brown in ambient light, the pendant is translucent and dark orange in transmitted illumination. There are no visible inclusions.

Description

The pendant was carved from a lobed piece of amber. One section was used for the face, and a small spur for a section of the neck—this is located to the back of the mouth. A third lobe, at the back of the head and flangelike in shape, is plausibly a wing. It does not have any engraved lines.

83.AO.202.5 represents the head and a section of the neck of a female. Although the piece is fragmentary, it is evident that the head was wide across the brow and tapered at the chin, much like that of 83.AO.202.4 (cat. no. 18). The left eye and the bridge of the nose are broken off. The nose was clearly triangular in form, although much of it is missing. The upper lip area is short. The mouth is almost straight, but the bottom lip turns up slightly. The engraved line that separates the lips is curved upward slightly in a smile. The sulcus is short and shallow, and the chin wide and full, with a prominent chin boss. Above the brow is a fringe of bangs, the strands marked by vertical striations. Behind the bangs is a smooth, rounded-top cap. The two engraved lines separating the hair from the hat depict the rim, which is either a flat turnup or a rounded roll-up.

The suspension perforation runs through the top of the head. When worn, the head would have been seen in profile to the left, the forehead tilted forward, casting the eyes downward.

Discussion

From the front, the lips, mouth, and chin resemble those of two heads, one from Tomb 164 and one from Tomb 428 at Banzi.[1] The lower part of the face bears a familial likeness to a seated figure from Tolve.[2] For further discussion, see 83.AO.202.4.

NOTES

1. For the ambers from Banzi, see Bottini 1987. Illustrated in *Magie d'ambra* 2005, pp. 122, 125.
2. For the Tolve figure, see *Magie d'ambra* 2005, p. 114.

20. Pendant: Female Head in Profile

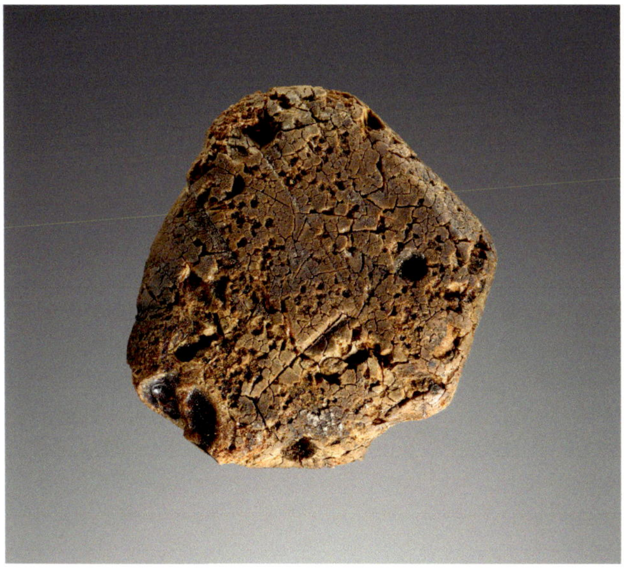

Accession Number	83.AO.202.6
Culture	Italic
Date	425–400 B.C.
Dimensions	Height: 51 mm; width: 51 mm; depth: 15.5 mm; Weight: 23.3 g

Provenance

–1983, Antike Kunst Palladion (Basel, Switzerland); 1983, Vasek Polak, 1914–97 (Hermosa Beach, CA), donated to the J. Paul Getty Museum, 1983.

Condition

The pendant is in extremely poor condition. The lower parts of the face (and possibly the neck) are missing. A break, perhaps ancient, starts at the middle of the lower eyelid, and losses include the nose, mouth, and chin. There are many other large losses and chips all over the pendant. The surface is friable, with deeply cracked and pitted weathering layers. A thick yellow-ocher crust covers much of the surface. The pendant is opaque and yellow-brown in ambient light except at the breaks, where the amber is red. In transmitted light, it is translucent and red. There are no observable inclusions.

Description

This fragmentary pendant represents a female head in profile to the left. On the obverse, the relatively flat piece of amber is figured, and on the reverse, plain. Despite the condition of the piece, many engraved lines are still visible. A large hole in the area of the jaw is likely a stopped bore. Some features remain legible. The brow is wide and smooth. Below this, a large, almond-shaped eye is set off by two engraved lines indicating lids. At the top are bangs described with parallel vertical striations. There is a rounded protrusion at the area of the ear, perhaps an ear and earring.

The two parallel engraved lines at the front of the hat represent the rim of the hat, either a flat turnup or a rounded roll-up. The cap is rounded at the top and more oblong at its back.

Discussion

See 83.AO.202.4 (cat. no. 18) and 83.AO.202.5 (cat. no. 19).

21. Pendant: Female Head

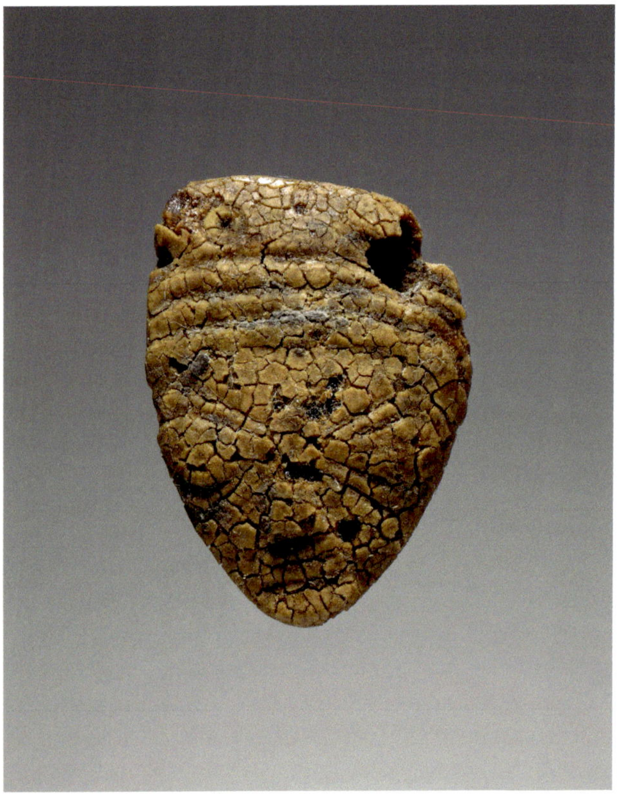

Accession Number	83.AO.202.12
Culture	Italic or Campanian
Date	500–480 B.C.
Dimensions	Height: 30 mm; width: 26 mm; depth: 4 mm; Weight: 2.4 g
Subjects	Magic

Provenance

–1983, Antike Kunst Palladion (Basel, Switzerland); 1983, Vasek Polak, 1914–97 (Hermosa Beach, CA), donated to the J. Paul Getty Museum, 1983.

Condition

The pendant is intact but in poor condition, with a friable, flaking surface. The surface has an overall crack network. A large modern chip is on the left side of the head, and there are many other, smaller modern losses over the entire surface: these reveal the dark red-brown translucent amber beneath. In ambient light, the pendant is opaque and light yellow-tan; in transmitted light, it is dark red and translucent. There are no visible inclusions.

Description

This head is worked from a small, flattish piece of amber. There is no indication of a neck. The reverse is uncorked and flat, and the obverse is slightly more rounded and carved. The face covers half of the pendant, while the other half is devoted to the hair and headdress. The face is widest at the eyes, narrower at the top of the head, and tapered at the chin. The brow is flat and smooth. The large, diamond-shaped eyes wrap around the head, extending from the front plane of the face to the sides of the head. The eyeballs themselves bulge slightly. Both top and bottom rims, indicated by narrow fillets, are angular at the midsection. The eyelid fillets taper in slightly at both of the canthi; they meet at the outer corners but not at the inner ones. The apexes of the upper eyelids nearly meet the lowest horizontal section of the headdress.

The flat, triangle-shaped nose is set off from the cheeks by a long groove that continues to the jawline, incorporating the nasolabial and mouth angle furrows. The mouth is almost straight, with a groove separating the barlike lips. The mentolabial sulcus is also grooved. The small chin projects forward to the level of the lips. Above the brow is a bandlike section of the headdress, composed of three rounded horizontal sections separated by grooved lines. Above the horizontal sections is a tapered, plain section of the headdress that is squared off on top.

There is a stopped bore (1 mm in diameter) at the left side of the head, which still contains its plug. A perforation extends laterally across the head at the headdress. When suspended, the head would have tilted forward, the chin back.

Discussion

This head and 83.AO.202.18 (cat. no. 22) have many features in common with 83.AO.202.4 (cat. no. 18),

83.AO.202.5 (cat. no. 19), and 83.AO.202.6 (cat. no. 20). (For a discussion of the common features of the group, see 83.AO.202.4.) This pendant and 83.AO.202.18 depict the frontal face rather than the complete head. They are both much flatter and more schematic in conception than 83.AO.202.4 and seem to represent an older female than the other three. Most importantly, the headdress worn by 83.AO.202.12 and 83.AO.202.18 is distinctly different: it is characterized by horizontal bands on its lower edge (83.AO.202.12 has three bands; 83.AO.202.18 has four). The headdress of 83.AO.202.12 sits directly on the brow, with no hair showing. The headdress of 83.AO.202.18 sits farther back on the head, behind the straight bangs.

The headdress of 83.AO.202.12 and 83.AO.202.18 is a variant of the commonly represented head wraps of amber head-pendants. A striking comparison is an Etruscan terracotta votive figure in Kansas City. She wears a comparable head covering, about which Stephen Smithers noted that the three bands incised into the lower third of her cylindrical *polos* probably represent the winding of a cloth headdress.[1] 83.AO.202.18 has straight bangs but no hair showing in back, 83.AO.202.12 has no hair showing at all, and the Kansas City terracotta head has hair showing in front: it is parted in the middle at the brow, sweeps to each side in waves, and is short in back.

The almost insectlike eyes of 83.AO.202.12 and 83.AO.202.18 (and of the amber head-pendants cited as comparisons above) are similar to those of bronze korai in Emeline Richardson's Late Archaic Series C, Group 3A ("Perugians"), such as those in Florence (Museo Archeologico Nazionale 261) and Leiden (Rijksmuseum van Oudheden C.06).[2] These bronzes are sophisticated in design and mannered in details, and some have old-fashioned features. Although Richardson does not single out for mention the thickly outlined, wide-open eyes which wrap around the face from the front to the side planes, this manner of representing the eyes must have recalled to its viewers the most ancient of hieratic female images to be found in Italy. So, too, must be the case with the amber head-pendants with huge and staring eyes.

83.AO.202.12 and 83.AO.202.18 gain from a comparison with four other groups of amber heads. All have features stretched and skewed, all are asymmetrical, and all would have hung crookedly. All are related, even if the workmanship of 83.AO.202.12 and 83.AO.202.18 is more precise. The Getty pieces show how more was accomplished with a fine graving tool.

When the Getty faces are compared to the small, late-sixth-century B.C. faces from Sabine Eretum,[3] the commonalties in sculptural approach are apparent: they are all small, flat, and schematic and represent a severe, mature figure. The morphological differences between the Getty faces and those from Eretum are less marked than those between the Getty faces and a pair from a larger group of amber pendants from the earlier-discussed tomb at Roscigno–Monte Pruno,[4] perhaps to be dated to the early fifth century. While 83.AO.202.12 and 83.AO.202.18 are closer to the Eretum and Roscigno–Monte Pruno sets, they also have features in common with two grave groups of amber head-pendants from female tombs of the first half of the fourth century. The examples are from Tombs 2 and 3 at Melfi, Cappucini.[5] The three amber head-pendants of Tomb 2 were found with other figured amber pendants—a strange siren, a schematic "Achelous," and three faces.

How might this continuity be explained? These rather ugly objects demand answers to this, one of the most compelling questions concerning the working of amber in pre-Roman Italy. Certainly, the makers of the head-pendants are part of the explanation. Another part lies in the origin of or prototype(s) behind the type, which must date to the earlier sixth century, if not before, in Etruria. That all of the heads were perforated to hang crookedly may be evidence of their use and identity. Is this meant to convey that they are reverted, and thus specifically refer to death, guardianship, and other magical workings?

NOTES

1. Smithers 1988, pp. 214–15, compares the votive to a marble kore in Athens (Acropolis Museum Akr. 688) dated circa 480 B.C.
2. Richardson 1983, pp. 314, 744–46. These same insectlike eyes are seen on a number of amber heads. The earliest is the late-eighth-to-early-seventh-century head from Stephens Tomb 108, Cumae (E. Gabrici, "Cuma," MonAnt XII [1913]: col. 609, fig. 220). For two now-lost heads from Cumae, two from Oliveto Citra, one from Canosa, and three from Termoli (Molise), see Losi et al. 1993, p. 210, n. 20.
3. For the ambers from Eretum, Colle del Forno, Tomb XIII, see, for example, Losi et al. 1993.
4. See cat. no. 17. For a recent discussion of this group, see Losi et al. 1993, p. 203.
5. Bottini 1990, pp. 61–63, no. 5 (22ab, 23ab).

22. Pendant: Female Head

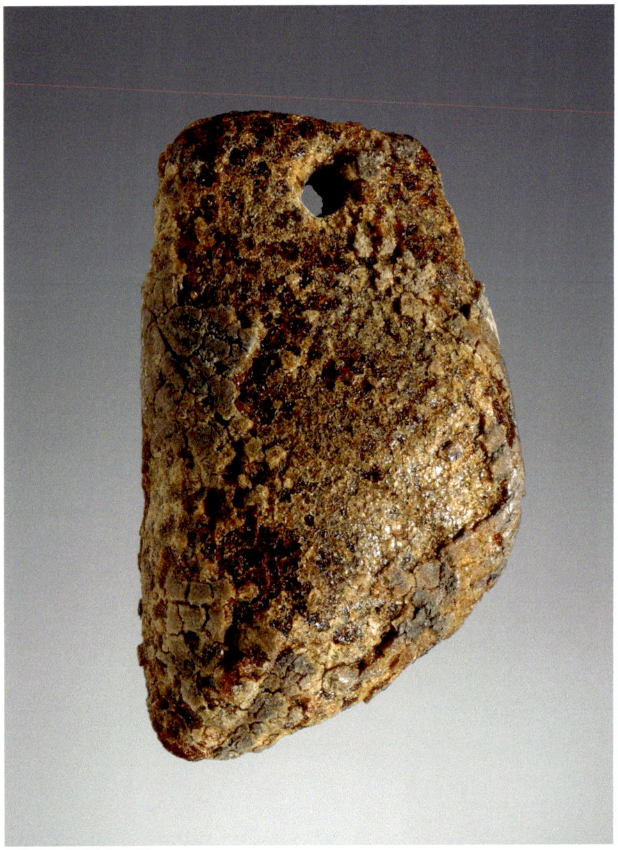

Accession Number	83.AO.202.18
Culture	Italic
Date	480–450 B.C.
Dimensions	Height: 30 mm; width: 21 mm; depth: 9 mm; Weight: 4.3 g

Provenance

–1983, Antike Kunst Palladion (Basel, Switzerland); 1983, Vasek Polak, 1914–97 (Hermosa Beach, CA), donated to the J. Paul Getty Museum, 1983.

Condition

The pendant is intact but in poor condition. The surface is friable and has suffered many small chips and losses. The tip of the nose is missing. On the reverse, the amber's surface-layer flaking is particularly severe. A yellow-ocher crust covers most of the surface. Areas with crust losses reveal a more compact and less altered amber beneath, and under transmitted light, additional subsurface cracking is visible. The pendant is opaque and yellow-brown in color in ambient light. Under transmitted illumination, it is translucent and dark red-brown. There are no visible inclusions.

Description

The original amber nodule seems to have determined the form of the pendant. The obverse is more convex, the highest point at the bridge of the nose. The reverse is plain and flat. The asymmetrical, triangle-shaped face tapers from a wide brow to a small, pointed chin. The hair and headdress are wide at the base but narrow at the chin. The holes for suspension are located in the top of the headdress on obverse and reverse. When hung, the head exposes its right profile. The brow is flat above the enormous bulbous eyes. Each eyelid is rendered with two filletlike rims that are the same width both above and below the eyes. They meet at both inner and outer canthi. The eyeballs are on the same plane as the eyelids and bulge as in nature. The glabella is wide. Despite the losses to the nose, its form can be discerned as being short and narrow. The upper lip area is also short, and the mouth is small. The lips are rendered as two bars, the upper one set off from the nose by a short groove. Two short vertical grooves, which run from the wings of the nose to the chin line, demarcate the mouth angle furrows. A slight ridge sets off the thick bangs from the brow. The bangs continue from the obverse onto the narrow sides; the pattern of the hair is indicated by fine diagonal grooves fanning out from the center. There are neither ears nor neck indicated on this figure. Atop the head is a headdress separated from the hair by a slight ridge. This headdress is made up of four rounded horizontal fillets.

Discussion

This pendant has many features in common with 83.AO.202.4 (cat. no. 18), 83.AO.202.5 (cat. no. 19), 83.AO.202.6 (cat. no. 20), and 83.AO.202.12 (cat. no. 21). For a discussion of 83.AO.202.18 in relation to related face and complete head-pendants, see the entries for 83.AO.202.4 and 83.AO.202.12.

23. Pendant: Winged Female Head

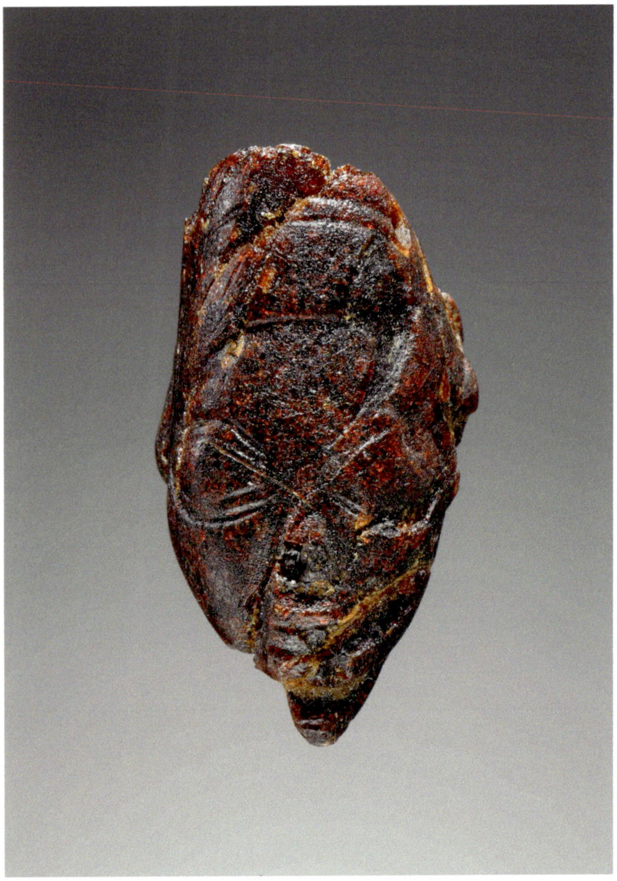

Accession Number	77.AO.81.5
Culture	Italic
Date	500–480 B.C.
Dimensions	Height: 42 mm; width: 23.5 mm; depth: 32 mm; Diameter of suspension hole: 2 mm; Weight: 11.9 g
Subjects	Magic

Provenance

–1977, Gordon McLendon (Dallas, TX), donated to the J. Paul Getty Museum, 1977.

Condition

The surface is stable, firm, and clean, but it is worn, especially on the front of the face. There is a large fracture loss on the reverse at the top; large losses behind the left ear, above the right ear, and at the top front of the head around the cracked suspension hole; and small chips under the right eye and below the jaw. Fissures run downward along the right side of the nose to the chin as well as along the right temple. There are cracks on the face, beneath the chin on the left side, on the temple, and running from the crown to the right ear. There are traces of yellowish degradation residue in the cracks and grooves. The surface is slightly glossy, suggesting an applied consolidant coating. The amber is translucent and dark red-brown in ambient light. Under transmitted light, the pendant is transparent and red. In the large fissure extending from the top, near the eye, to the chin is an inclusion.

Description

77.AO.81.5 represents the head and neck of a female figure. On the back of the head is a wing. The pendant is carved fully in the round. The eyes are large and lozenge-shaped, with the proximal corners nearly joined over the nose. The eyes curve around from the front to the side planes. The lids are indicated by parallel grooves, roughly carved and relatively angular. What remains of the nose suggests that it was a flattish triangular bar. The mouth is small and turned up slightly, as if in a smile. The lips are rendered as curved bars. The nasolabial furrow continues to the jawline, giving the face a jowly look. The ears are small, flat, semicircular nubs. They are located high on the head and overlap the edge of the headdress. The chin is small and the under-chin area fleshy. The neck is set off from the face and from the hair by two deep grooves.

The figure wears a conical hat with a high rounded crown. It is cloth-wrapped. None of her hair is showing in the front. However, at the back of the neck, a section carved with finely spaced engraved lines must represent a long tress. The secondary feathers of the wing on the back of the head are indicated with upward-directional engraved lines.

The shape of the pendant suggests that the original form of the amber was teardrop-shaped. The long smoothed grooves are the result of a precarving cleaning of surface fissures or inclusions. The vestigial neck of the figure is formed from a spur of amber. A 2 mm perforation for suspension passes laterally through the top of the headdress. When suspended, the chin was recessed close to the wearer's neck, making the eyes look downward.

Discussion

The smoothing on the prominent surfaces—nose, eyes, mouth, and chin—is evidence of considerable preburial use wear. For the figure's dress, see the entry for 77.AO.81.25 (cat. no. 26); for a discussion of wings on amber head-pendants, see the entry for 76.AO.85.2 (cat. no. 15). In frontal view, this head has an uncanny similarity to a unique and odd Etruscan votive bronze of a togate male figure in Naples, who wears a close-fitting hat or net incised with a scale pattern. (Or is this the hair?) Hans Jucker places this togate figure in his Populonia group; Emeline Richardson places it in her Late Archaic grouping, Type 2A, Ionians.[1] Although there is no close parallel for 77.AO.81.5, it corresponds in dress and type to three other amber head-pendants. One was found with thirteen other head-pendants in a "princely" tomb at Roscigno–Monte Pruno, the context dated to the early fifth century B.C.[2] The second is a head-pendant from Tomb 106 at Banzi, from a context dated to the first quarter of the fifth century.[3] The third is a much-worn head-pendant in the Getty, 77.AO.81.25. The schematic carving of the eyes, nose, and mouth of 77.AO.81.5 is comparable to that of one of the heads from Tomb 428 at Banzi.[4]

NOTES

1. Naples, Museo Archeologico Nazionale 5534, found on Elba: H. Jucker, "Etruskische Bronzen," AA (1967): 620–21, figs. 5–6; H. Jucker, "Etruscan Bronzes from Populonia," in *Art and Technology: A Symposium on Classical Bronzes*, ed. S. Doeringer, D. G. Mitten, and A. Steinberg (Cambridge, MA, 1970), pp. 199–203, figs. 8a–f; and Richardson 1983, p. 232, figs. 522–24.
2. For a recent discussion of the find, see Losi et al. 1993, p. 206, n. 20.
3. See Bottini 1990, p. 59, no. 4, fig. 2.4.
4. For an illustration of the head-pendant from Tomb 428 at Banzi, see *Magie d'ambra* 2005, p. 125; for discussion of the ambers from the tomb, see Bottini 1987.

24. Pendant: Female Head

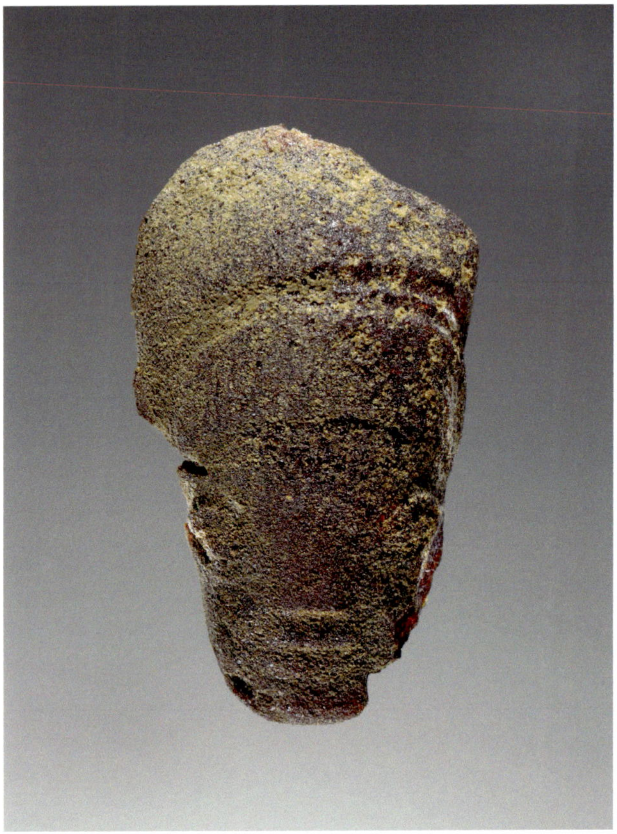

Accession Number	82.AO.161.7
Culture	Italic
Date	500–480 B.C.
Dimensions	Height: 48 mm; width: 28 mm; depth: 12 mm; Weight: 5 g

Provenance

–1982, Jiří Frel, 1923–2006, and Faya Frel (Los Angeles, CA), donated to the J. Paul Getty Museum, 1982.

Condition

The pendant's top proper right quarter was broken and reglued in modern times from three fragments. On the obverse are a number of both old and modern small chips: at the perforation, on the left and right sides of the face, and at the top and right sides of the head. On the reverse is a fissure. The lack of detail on the prominent surfaces of the face strongly suggests use wear. There is a scattered, spotty, yellow-ocher degradation overall, with some associated surface pitting. The amber is friable in the vicinity of breaks. The amber is opaque and yellow-orange in ambient light. Under transmitted light, it is translucent and dull red. With the exception of the material in the fissure, there are no visible inclusions.

Description

This frontal face of a female figure is concave on the obverse and flat on the reverse. The figure is wearing a high soft hat and a veil. The visible section of the brow is short and broad. There are no indications of ears. The widely spaced large eyes are flat and almond-shaped and are bounded by rims composed of thick fillets. The nose is broad and snubbed. All that remains of the mouth are two grooves. Nevertheless, it is clear that the upper lip hung over the lower one. The jaw is squarish and the chin pointed. Above the brow are straight bangs with vestiges of vertical striations. The softly rounded high hat has a turnup, indicated by two parallel engraved lines. The veil covers the hat and ears but not the bangs.

Because of the surface erosion of this piece, no tool marks are visible. A suspension perforation was drilled laterally through the top of the piece. When it was suspended, the back would be perpendicular to the ground, with the head tilting forward, the chin close to the neck, and the eyes appearing downcast.

Discussion

82.AO.161.7 wears a costume similar to that of many other amber head-pendants; however, her hat is rather more rounded on top than most. In its form and its manner of hanging, the frontal face is akin to a face from a tomb at Kompolje, Croatia. In style, it is similar to the three well-preserved heads from the tomb.[1] The profile relates it to a head-pendant from Tomb 2 at Tolve.[2]

NOTES

1. For the ambers from the Kompolje tomb, see R. Bižiž-Drechsler in *Vjesnik Arheološkog Muzeja u Zagrebu*, ser. 2, no. 2 (1961): 109–13, pls. XXVIII–XXIX; and Mastrocinque 1991, p. 134, fig. 82, pl. VII.14.

2. For the Tolve head-pendant, see *Magie d'ambra* 2005, p. 115.

25. Pendant: Female Head in Profile

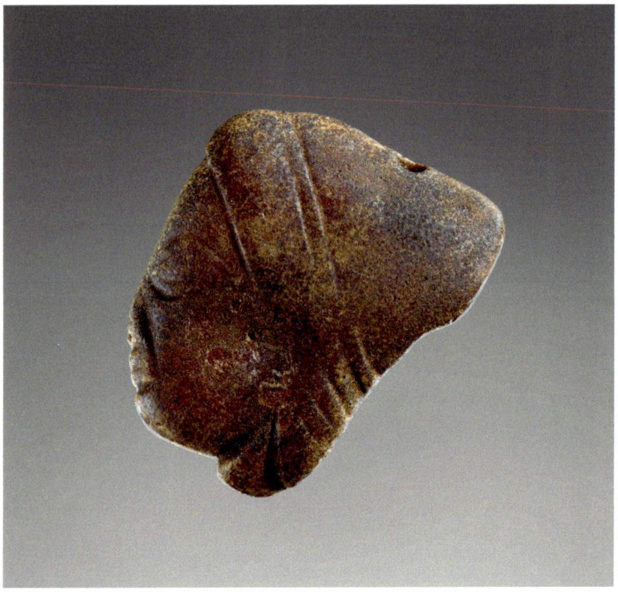

Accession Number	77.AO.81.30
Culture	Italic
Date	500–480 B.C.
Dimensions	Height: 44 mm; width: 38 mm; depth: 16 mm; Diameter of suspension holes: 3.5 mm and 4 mm; Weight: 15.1 g

Provenance

–1977, Gordon McLendon (Dallas, TX), donated to the J. Paul Getty Museum, 1977.

Condition

The surface of the amber is smooth and solid, but shows an uneven pattern of abrasion and loss of detail on the most prominent areas, suggesting use wear. Before purchase by the donor, the pendant was treated with a surface consolidant. There are several ancient breaks (they have the same degraded surface as the unbroken areas), including at the bridge of the nose and in a section on the reverse. There are small chips on the cheek. A more recent break at the back was reglued before the object's arrival at the Getty Museum. On the reverse are three large depressions, possibly from the removal of flaws or fissures by the carver, or weathering losses. In ambient light, the piece is somewhat translucent and dark reddish brown; in transmitted light, it is red and fully translucent. There are no visible inclusions.

Description

The pendant is composed of the head and a section of the neck of a female, carved from a relatively flat piece of amber. The figuration is on the obverse and the two sides. The back is plain, but uneven because of the shape of the original amber piece. The shallow crevices are the result of preparation of the amber nodule, by the removal of fissures or inclusions.

The figure's brow is smooth and on the same plane as the mouth and nose. There is a groove above the large almond-shaped eye, and the eye's outside corner is higher than the inner corner. The lids curve smoothly, taper slightly at the corners, and are outlined with a raised fillet. The nose is set off from the cheeks and upper lip area by grooves. The area above the upper lip is short. The lips are large bar-shaped forms that curve from the obverse to the front edge. They are separated from each other by a groove. The mentolabial sulcus is a deeper groove. The receding chin is rounded and the under-chin area full. The triangular section of neck is described by two grooves. The area of the ear is indistinct, not only because the surface is much worn but also because there is a natural concavity there. However, a slight swelling at the bottom of the bangs may indicate the ear.

Above the brow is a fringe of bangs, the strands of which are indicated by eight short diagonal incisions. Behind these is a cap set off from the hair by a raised filletlike form, which probably represents the rim of the cap. The cap is a squared conical shape. At the back of the head is a curved chignonlike section of hair extending to the nape of the neck. It is engraved with three lines perpendicular to those in the bangs.

The pendant was hung from a V-shaped suspension system, drilled from two sides through the top of the cap. The front hole is 3.5 mm wide, and the rear hole 4 mm.

The pendant would have hung with the chin recessed, the brow forward.

Discussion

77.AO.81.30 is generally similar in style and dress to 77.AO.81.25 (cat. no. 26). It is closer, however, especially in the overall format and in the hairstyle, to two well-preserved pendants from Tomb 164 at Banzi.[1] The styling of the hair at the back of the heads is remarkably similar in all three works: it is short and curled under, with the lower edge forming a soft curve. The horizontal waves are rendered by fine engraved lines. However, there are differences in the hair treatment. 77.AO.81.30 has straight bangs, while one of the Banzi head-pendants has horizontal waves at the brow and the other scallops. Both Tomb 164 figures also have a raised fillet at the neck edge, perhaps a necklace, perhaps a mark evoking the form of the decorative protome—as is characteristic of amber horse's-head pendants. 77.AO.81.30 does not have such a fillet.

Angelo Bottini compared the Banzi amber head-pendants to coin types of Syracuse and Athens of the late sixth and early fifth centuries and dated Tomb 164 "to the first quarter of the fifth century, at the latest."[2] Bottini made a strong case for a local manufacture of the Tomb 164 head-pendants but at the same time argued for a more careful consideration of the territory and its cultural relations with the larger world.

The Banzi Tomb 164 head-pendants and 77.AO.81.30 correspond in general type to Emeline Richardson's series of Late Archaic bronze korai. A comparison with her Group 5D, especially with the group's name piece, Naples, Museo Archeologico Nazionale 5532 (provenance unknown), brings out their common features. They look to be of the same physiognomic type and are dressed in a similar manner.[3] As Richardson establishes, Naples 5532 is an old-fashioned work, one that looks back to earlier conceptions of the kore type in Etruria, especially to the Ionian series of the late sixth and early fifth centuries.

NOTES

1. Melfi, Museo Archeologico Nazionale del Melfese 51436: Bottini 1990, pp. 61–62, no. 17, fig. 4.17a–b. Related is the pendant from Tomb 55 at Banzi: ibid., p. 60, no. 5, fig. 2.5. For discussion of the Banzi heads, see ibid., pp. 59–63, nos. 1–2; Bottini 1987, pp. 1–16; and Losi et al. 1993. Bottini 1990 left open the question of whether the parallel striations at the back of the heads represent coifed short hair or a *kekryphalos*.

2. Bottini 1990, p. 62.

3. For the Late Archaic bronze korai, see Richardson 1983, pp. 271–332. For Naples 5532, see ibid., pp. 323–25, figs. 770–71. Naples 5532 wears a short necklace and disk earrings. Unlike the amber heads, she wears a low diadem and a veil. See also the discussion of Naples 5532 in the entry for 77.AO.81.25.

26. Pendant: Female Head

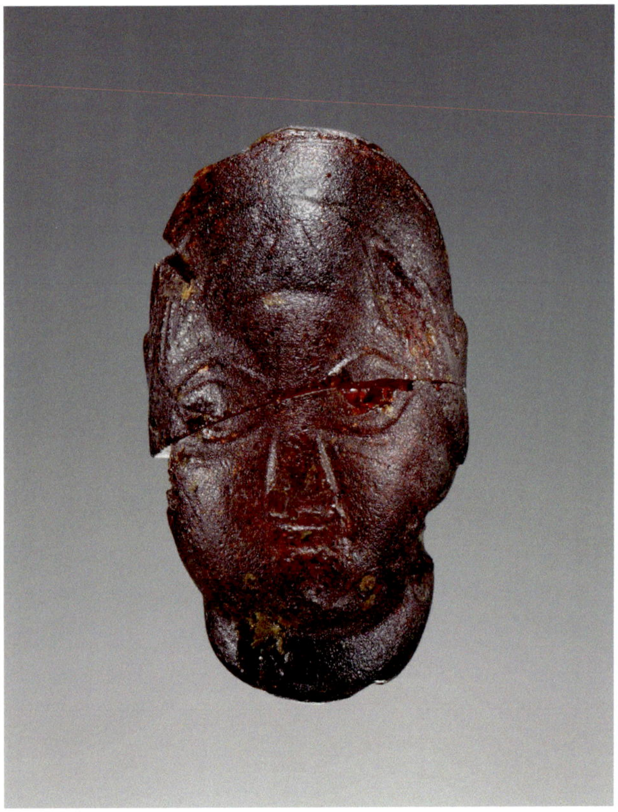

Accession Number	77.AO.81.25
Culture	Italic or Etruscan
Date	500–480 B.C.
Dimensions	Height: 42 mm; width (across face): 24 mm; depth: 29 mm; Diameter of suspension hole: 3.5 mm; Weight: 14.9 g
Subjects	Etruscan culture; Ionia, Greece (also Ionian, Greek)

Provenance

–1977, Gordon McLendon (Dallas, TX), donated to the J. Paul Getty Museum, 1977.

Condition

The pendant was joined from two pieces along a slight oblique plane at the level of the eyes. Chips are missing at the breaks, on the eyes, at the area of the hair on the left side, behind the left ear, at the corner of the left jaw, along the right side of the neck, and in the region of the left eye. There is a large fissure along the left side of the head below the suspension perforation. Another fissure crosses laterally immediately above the perforation. The surface of the piece is crazed overall, and subsurface cracking is visible in transmitted light. There is a pocket of flaky yellowish residue under the left side of the chin. A cloudy inclusion at the center of the piece is visible at the break. In ambient light, the pendant is opaque and dark reddish brown in color; in transmitted light, it is translucent and a lighter red-brown.

Description

The ovoid head-pendant is composed of the head and a portion of the neck of a female figure. The pendant seems to follow the convolutions of the natural lump of amber from which it was worked: the face, neck, and headdress are on the more rounded obverse, while the flatter reverse is smoothed but not figured. A groove on the back and a crater above the ear may be evidence of the removal of inclusions. The suspension perforation is located in the head: it is a lateral bore 3.5 mm in diameter that passes behind the crown on each side of the head. Both exits are abraded on the upper inside edges, no doubt from friction from the carrier. This and the wear on the prominent surfaces are evidence of ancient wear.

Despite the shallow carving and wear to the object, its features are still legible. The face is full and rounded. The eyes are large and almond-shaped, with thick and cordlike eyebrows. The outside corners of the eyes are upturned slightly, the left more than the right. The brow ridges rise slightly above the eyes. There is only a slight indentation for the root of the nose, which is now flat. Grooves running from the inner canthi of the eyes to the corner of the lips on each side incorporate the sides of the nose and the nasolabial furrows. The lips are almost straight, semicylindrical bars. The mouth angle furrows are indicated by shallow grooves, the mentolabial sulcus by an indentation. The chin is full and the jawline rounded. The figure has large ears and wears round earrings. The neck is cylindrical, with a distinct fullness.

Above the smooth brow are bangs that divide in the middle of the forehead and are brushed diagonally to each side; the strands of hair are individuated by diagonal grooves. The figure wears a conical hat, a veil, and a crown. The veil covers only the hair, cap, and ears. The veil has straight pleated folds indicated by modeling on the back of the head. The tiaralike crown is decorated with a grooved border at all of the edges. The lower edge of the crown is set off from the bangs by an engraved line. The crown's upper edge stands above the head. The sides of the crown stop just above and in front of the ears.

Discussion

There is a parallel for this head-pendant in Munich, one also modeled fully in the round.[1] The Munich amber is drilled with a lateral suspension hole in the top of the head in the same position as that of 77.AO.81.25. In addition, the Munich head has a metal loop (date uncertain) inserted in the top of the head. The two figures differ in dress: although the Munich head has bangs, a cap, and a veil, it does not have a crown or earrings.

Variants of this head-pendant's elements of adornment—hat, crown, earrings, veil, and styled hair—are characteristic of many other head-pendants worn by Etruscan Archaic bronze korai, the earliest of which are in Emeline Richardson's Middle Archaic series.[2] The best Etruscan bronze parallels for the amber are found in Richardson's groupings of Late Archaic korai, especially Late Archaic Series A, Ionians,[3] and a related group of bronzes, Group 5D, Naples, Museo Archeologico Nazionale 5532.[4] Two bronzes from Series A, one in Vienna (Kunsthistorisches Museum 71) and one in Copenhagen (Ny Carlsberg Glyptotek H224), are a set of important comparanda for 77.AO.81.25.[5] The Getty amber head-pendant and the two Series A bronzes are all recognizably Ionian in style, as the name given to the bronzes by Richardson underlines. Richardson placed Vienna 71 and Copenhagen H224 into different subgroups of Series A: Vienna 71 in her Group 2, Tomba del Barone, and Copenhagen H224 in her Group 6, Late Korai. The physiognomy of 77.AO.81.25 is most like that of Copenhagen H224 (which is bareheaded), but the dress is more like that of Vienna 71—which, too, is capped, crowned, earringed, and veiled.

A comparison to Naples 5532 brings out additional Ionian and Etruscan aspects of 77.AO.81.25. This bronze kore represents a physical type very much like 77.AO.81.25; they both have the same "big head and long face."[6] The votive also provides a model for how 77.AO.81.25 would look if the amber were a complete figure. Richardson describes the draping of Naples 5532's veil as "pulled over the cap so that two ends fall on the shoulders [with] the rest [hanging] in a long panel down the back."[7] This is the same South Ionian fashion of veil worn by the Getty amber *Kore*, 76.AO.77 (cat. no. 8).

NOTES

1. Munich, Antikensammlungen 15.003.
2. Richardson 1983, pp. 258–70.
3. Ibid., pp. 275–76.
4. Ibid., p. 323.
5. For Vienna 71, see ibid., pp. 284–85, fig. 663; for Copenhagen H224 (perhaps made at Capua), ibid., pp. 295–96.
6. Ibid., p. 323. See also the entry for 77.AO.81.30 (cat. no. 25) for a discussion of Richardson's Group 5D and its relevance for the amber head-pendants.
7. Richardson 1983, p. 323.

Animals

27. Roundel: Animal

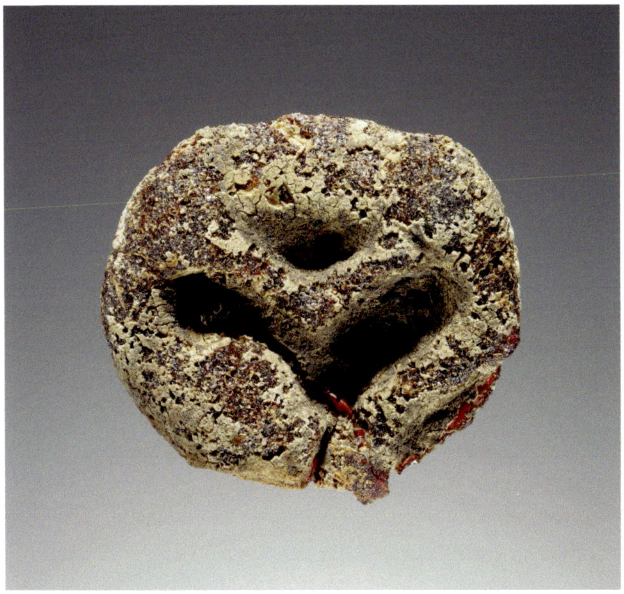

Accession Number	82.AO.161.2
Culture	Italic or Etruscan
Date	700–600 B.C.
Dimensions	Diameter: 44–50 mm; depth: 16 mm; Weight: 6 g
Subjects	Animals; Artemis; Dionysos, cult of (also Satyr); Funerary use of amber (also Burial)

Provenance

–1982, Jiří Frel, 1923–2006, and Faya Frel (Los Angeles, CA), donated to the J. Paul Getty Museum, 1982.

Condition

The piece has a severely degraded and friable surface. A large fragment on the lower edge of the animal's head has been reattached. The surface has altered to an opaque, light tan degradation layer that is flaked and chipped overall, and there is a corresponding loss of surface detail. In ambient light, the pendant is yellow-orange. It is not translucent, and there are no visible inclusions.

Description

The disk-shaped amber is carved in high relief on the obverse and is plain and completely flat on the reverse. The sides are tapered inward slightly from the bottom. The animal's head is depicted in top view, with the chin, throat, and neck ventrally flush with the base. The left flank of the body is presented in profile view, with only one each of the fore- and hindlegs shown. The large head is paddle-shaped, rather flat on top, and wide through the ear area. The lower jaw is narrow and flat, the mouth an engraved line that extends as far as the area of the eye. The neck is thin. The bulges and indentations on each side of the head at the point of its greatest width must represent the ears. There is no legible evidence of eyes. The long tongue is extended, touching the rear hoof. Flush with the curve of the pendant's edge, the animal's back is rounded, with a slight indentation just above and before the curve of the haunch, and just behind it is the tiniest indication of a tail. The chest and abdomen areas are each approximately the same size as the head. The front leg is long and thin from knee to hoof and is bent at the ankle, as in nature. The powerful back leg and haunch curve forward. The nonfigured area in the middle of the roundel is recessed below the animal. There are two sets of holes: a perforation between the rear hoof and the tongue, and a lateral perforation from one side of the animal's neck to the other. Along the dorsal ridge, equally spaced from the neck to the ankle, are six tapered, stopped bores 3.5 mm deep.

Discussion

Unique in form and subject, the roundel is unlike any other amber object. The round, thin form, with its flat reverse, beveled edge, perforations, and stopped bores, suggests that 82.AO.161.2 was originally the lid of a small, round pyxis. If that is the case, the bores, the perforation in the neck area, and the hole between the foot and the tongue may have been used for attaching a lid to a container. Alternatively, the lateral bore through the top of the animal may have been drilled to allow for hanging, perhaps as an ornament. This may have been a secondary use.[1]

One of the best parallels for the roundel is not an amber object but an ivory lid from the Idaean Cave, Crete, published as North Syrian by J. A. Sakellerakis.[2] This beveled-edge lid is decorated with an overall geometric pattern and has a similar system of stopped bores, or mortises, on its edge. 82.AO.161.2 might also be compared to a group of Roman-period amber pyxis (or perfume pot?) lids, three of them in the British Museum: a nonfigured lid (BM 115), turned with a series of convex and concave moldings, engraved lines, and narrow fillets (very close in size to 82.AO.161.2),[3] and two slightly larger figured lids, one of a sleeping swan with putti on his back (BM 117) and the other, a satyr face (BM 118).[4]

The placing of an animal, resting or in movement, within a circular format is age-old. A contorted animal within a tondo is a distinct subset of the schema.[5] As John Boardman notes, compositions with contorted animals whose form is characterized by the dislocation of the legs or another portion of the body imply movement and allow the circular field to be filled more symmetrically.[6] Even though the contortion in 82.AO.161.2 extends only to the twisting of the animal's head into top view and the body and legs into profile, the composition still calls to mind the whirling compositions of Cretan stone seals, which

> *express the old Minoan feeling for torsion and for spreading designs which own no top or bottom or sides. But these contorted animals are not simple essays in the grotesque, as they are often described, but the artist's rendering of a novel but natural viewpoint, top three-quarter of a reclining animal with his legs before him, his hindquarters twisted to one side.*[7]

A group of stone seals from the Greek islands, dating to the second half of the seventh century B.C., appear to be the only comparable post–Bronze Age Mediterranean objects decorated with the "old Minoan" type of contorted animals. Boardman considers a group from Melos to be dependent on actual Bronze Age seals found on the island, noting that they "are of an importance and interest far beyond their intrinsic merit, because they show us how artists could be influenced by the arts and artifacts of a past civilization, otherwise remembered only by the poets."[8]

It may be that 82.AO.161.2 is a comparable seventh-century response—although there is no Bronze Age object with a comparable representation of a quadruped. The tongue extension, too, is unusual. Does the animal lick its hind leg, or is the tongue extended in exertion?[9]

The condition of the amber and the schematic depiction of the animal do not allow for a sure classification of the quadruped. However, the salient physiognomic characteristics and the position of the tongue lead me to think that it is a fawn. This identity is posited despite the lack of a close comparison and despite some resemblance to a number of seventh-century B.C. ivory and amber dogs.[10] However, the feet of 82.AO.161.2 are entirely different from the wide, multitoed feet of these dogs: they are tiny and undifferentiated.

An isolated fawn is an infrequent subject in ancient art, uncommon as the subject of a pendant, and exceptional in amber.[11] The morphological characteristics of the animal depicted in 82.AO.161.2 compare well with those of a number of Greek Late Geometric fawns, does, and groups of a doe and her suckling fawn. Two bronze statuettes and a pair of bronze amphora handles assure the identity of the amber animal. The bronze of a standing fawn on a rectangular base in the Harvard collections has a similarly paddle-shaped head and nubs of ears set far back on the head.[12] A standing fawn in the Menil Collection, Houston, which has a shorter, blunter head, huge ears, and a dappled coat suggested by tiny concentric marks, is another schematic representation.[13] More naturalistic are the early-fifth-century pairs of fawns (deer?) of two nearly identical Etruscan (Vulcian?) bronze amphora handles, one in Boston and the other in a private collection.[14] On each handle, at the base, squats a syrinx-playing satyr. The connection between a satyr playing panpipes and sleeping fawns may be relevant for the amber roundel.

The subject of an animated fawn in amber, especially if it were an ornament, calls to mind the most famous brooch of ancient literature, a *daidalon*, the cunningly fashioned gold pin worn by the disguised Odysseus:

> *Godlike Odysseus wore a purple cloak of wool, double thick; but on it was fashioned a pin of gold with double clasps, with a* daidalon *in front: a hound was holding in its forepaws a dappled fawn, preying on it while it struggled. All were marveling at it, how though they were [of] gold, the one preyed on the fawn throttling it, but the other struggled with its feet as it tried to flee.* (Odyssey *19.225–31)*

Because 82.AO.161.2 may have been an ornament, a rare image of a figure wearing a fawn pendant should be recalled: this is the bronze youth wearing a fawn's head in the Guglielmi Collection of the Vatican.[15]

The only other amber fawn known to me is in the center of a large pendant in London (British Museum 35), a representation of Bacchic revelers.[16] From between the two dancers leaps up a fawn, a scene from the Dionysian *thiasos*.[17] The fawns and satyr of the Etruscan bronze amphora handles link the vessel, wine, and Dionysos. The fawn of 82.AO.161.2 may have been intended to refer to a nature divinity other than Dionysos. In the Geometric period in Greece, deer (and fawns) were associated with the Olympians Hera, Athena, and Apollo. In Archaic and Classical Greece, deer and fawns were most commonly associated with the children of Leto, although they were important in depictions of Dionysos, Herakles, and hunting generally. There appears to have been a special association of deer with weddings and cultic activity in Attica.[18] Fawns are held in the arms of many Archaic terracotta images of Artemis or her votaries, and images of fawns are found in sanctuaries of Artemis.[19]

Not only did young girls imitate she-bears for Artemis at the Attic site of Brauron, but in Thessaly, girls performed a ritual in which they played the part of fawns. Both rituals were considered preparatory for pregnancy and childbirth.[20] By the end of the sixth century B.C., the fawn is pictured on Attic vases as a love gift between older and younger men, a custom that introduces Aphrodite into the picture.[21]

Whether 82.AO.161.2 was lid or pendant, dog or deer, its ultimate use was funerary. The animated creature embodied in the amber tondo, eternally circling in a whirling composition, a design without beginning or end, might signify the cycle of life. Here the idea of regeneration would be perfectly synthesized in material, subject, and form.

NOTES

1. Theoretically, a filament strung through the through-bore in the neck area would cause the pendant to hang perpendicular to the ground, with the animal head downward. When suspended from both perforations, the animal would have hung with head upward, as if trussed for carrying.

2. J. A. Sakellerakis, "The Idean Cave Ivories," in Fitton 1992, p. 114, pl. 8.

3. Strong 1966, p. 93, no. 115, pl. XLI.

4. Ibid., p. 94, nos. 117–18, pls. XLII–XLIII. Might the dormant swan be a direct connection to the Ligurian prince Cygnus, who mourned Phaethon? See "Ancient Literary Sources on the Origins of Amber" in the Introduction.

5. For an illuminating study of whirling animals in Early Etruscan art, see L. Donati, "Rappresentazioni etrusche della capra e del cervo di tipo 'sciita,'" in Staccioli et al. 1991, pp. 919–38.

6. Boardman 2001, p. 24.

7. Boardman 1967, p. 34.

8. Ibid., p. 105.

9. If the animal of 82.AO.161.2 is a running dog whose tongue is extended, the subject finds many comparisons in ancient art from the Bronze Age onward. The running dog may specifically refer to the hunt. If the animal is licking its hind foot, this action may have been considered a medico-magical technique, as it was in Mesopotamia and Egypt. As Ritner 1993, p. 933, notes, "The magical transfer of health or blessing by saliva reflects such naturally observable phenomena as the licking of wounds. As a magical technique, licking represents a ritualized extension of such instinctive acts." Licking could be equated with a solar blessing and solar "licking" with the rays of the early dawn, as ibid., p. 94, points out. For licking imagery in Mesopotamian magic, Ritner refers to J. Westenholz and A. Westenholz, "Help for Rejected Suitors: The Old Akkadian Love Incantation MAD V 8," *Orientalia* 46 (1977): 215 ("Then the ewe licked [literally 'took good care of'] her lamb"). The composition might also be compared to the motif of a trussed ibex licking its hind foot on Egyptian toilet articles in the shape of ibexes.

10. These include the single dog from Tomb VI at Satricum (see Waarsenburg 1995, pp. 452–53 and passim) and three from Narce (a pair from the Monte Cerreto necropolis, Tomb 103, and a singleton from the Pizzo Piede necropolis; for references, see Negroni Catacchio 1999, p. 283, n. 16). Three unpublished dogs that also have collars are Louvre Bj 2124–26. Bj 2124 has large circular inlays, Bj 2125 is plain, and Bj 2126 is complete and retains fragments of a gold collar. These dogs are all close in style and conception and are very like the ivory dogs from Cameiros, also noted by Waarsenburg 1995, p. 452, citing D. G. Hogarth, *Excavations at Ephesus: The Archaic Artemisia of Ephesus* (London, 1908), pl. 30. The Italian amber dogs are from the same family as the confronted pairs embellishing identical ivory roundels from the Tomb of the Ivories at Marsiliana d'Albegna near Grosseto: see M. Benzi, "Gli avori della Marsiliana d'Albegna," *Atti dell'Accademia Nazionale dei Lincei, Rendiconti* 21 (1966): 253–92. The importance of the dog in early Etruria is indicated by the trios of dogs on the ritual tripod-basins from the Barberini and Bernardini (Praeneste) tombs. However, if 82.AO.161.2 does represent a dog, the iconography would match well with the roundel's material, since the dog is associated with female goddesses, including Hekate, Artemis, and Eileithyia: see Waarsenburg 1995, pp. 452–53, nn. 1268–73. Waarsenburg underlines the importance of the dog in Italy, indicating key comparanda in art, and shows that it "played a key role in religious and magical-superstitious beliefs. Popular beliefs connected the dog with ghosts, death, fertility, and childbirth.... As was expressly declared by Pliny NH.39.58, and evidenced by the nature of the mentioned rites, the dog's sacral association was very old ... and the meaning of the rites related to dogs was not understood anymore, not even by the priest performing them." The origin of the amber dog must lie not in Egypt but in the ancient Near East. The early Italian amber and ivory dogs are probably associated with healing and specific

goddesses. As Black and Green 1992, p. 71, summarize: "The sitting dog first occurs as a divine symbol in the Old Babylonian Period and continues through to the Neo-Babylonian. Inscriptions on kudurrus identify it as the symbol of Gula, goddess of healing.... King Nebuchadnezzar II records the placing of statuettes of gold, silver and bronze dogs as deposits in the gates of Gula's temple at Babylon.... In the Neo-Assyrian and Neo-Babylonian Periods, the dog, sitting and standing, was also used as a magically protective figure, not attached to any individual deity." On dogs as large-scale talismans in the ancient world, see Faraone 1992. See also Waarsenburg 1995, n. 1099, for a list of relevant examples.

11. Although it represents a deer instead of a fawn, the unique amber seal from the Vetulonian Tomb of the Trident should be noted: see Massaro 1943, pl. XXVI, 1ab.

12. Harvard University Art Museums, Arthur M. Sackler Museum, David M. Robinson Fund 1966.108: Langdon 1993, pp. 21–67, no. 86.

13. Boeotian, Late Geometric bronze fawn, formerly in the collections of Captain E. G. Spencer-Churchill and George Ortiz: ibid., pp. 21–45, no. 85.

14. Boston, Museum of Fine Arts, H. L. Pierce Fund 99.464, E. P. Warren Collection: Comstock and Vermeule 1971, pp. 360–61, no. 507 (with reference to the related handle in a Haverford, PA, private collection; includes bibl.).

15. J. D. Beazley and F. Magi, *La Raccolta Benedetto Guglielmi nel Museo Gregoriano Etrusco*, vol. 2 (Vatican City, 1941), pls. 47–49 (cited by Brown 1960, p. 106).

16. Strong 1966, pp. 61–62, no. 35, pl. XV.

17. As is noted in "The Archaic and Afterward" in the introduction, many other pre-Roman ambers from southern Italy, especially those of fifth-century date, are also carved with Dionysian subjects, such as satyrs, vintagers, and maenads. Roman amber objects—rings, perfume containers, boxes, small figures, and small figural groups—are often carved with Dionysian subjects. All of these survived because they were ultimately grave furnishings. On Roman amber pyxides (or perfume pots?), Dionysian themes are the most common of subjects. There are many reasons for the tie between amber and Dionysos, among them amber's age-old association with fecundity, regeneration, and healing, its role in averting danger, its chthonic importance, and its winelike optical properties. An Etruscan stone sarcophagus lid of an older woman who is joined by a fawn suggests the presence of Dionysos.

18. See S. Klinger, "An Attic Black-Figure Pyxis in Athens and Some Observations on Deer Escorting Chariots," AA (2003): 23–44 (with earlier bibl., including S. Klinger, "A Terracotta Statuette of Artemis with a Deer at the Israel Museum," *Israel Exploration Journal* 51 [2001]: 208–24). See also Y. Morizot, "Autour d'un char d'Artemis," in *Agathos daimon*, ed. P. Linant de Bellefonds (Paris, 2000), pp. 383–91; and Bevan 1986, pp. 389–93, where she records the deer remains found in sanctuaries of Artemis.

19. Bevan 1986, pp. 389–93.

20. Cole 1998, p. 33, cites inscriptions from Demetrias (Pagasoi) in Thessaly and three from near Larisa of women who had served Artemis or "played the fawn" for Artemis; Barringer 2001, p. 246, n. 110, cites P. Clement ("New Evidence for the Origin of the Iphigeneia Legend," *L'antiquite classique* 3 [1934]: 393–409) on the cult of Artemis Pagasitis in Thessaly, in which young girls of marriageable age were identified with deer rather than bears, as they were at Brauron.

    Scanlon 2002 (see cat. no. 4, n. 15) discusses girls' running in the Brauronia and the Munichia, concluding that these races were chases based on the analogy of the hunt. See also C. A. Faraone, "Playing the Bear and Fawn for Artemis: Female Initiation or Substitute Sacrifice?," in *Initiation in Ancient Greek Rituals and Narratives: New Critical Perspectives*, ed. D. Dodds and C. A. Faraone (London, 2003).

21. The fawn's vulnerability (and corresponding need for protection) may have been inherent in the image. For the fawn as a love gift, see Barringer 2001, esp. pp. 88–98. She emphasizes that ancient Greek authors use the hunted fawn as a metaphor for the pursued *eromenos* in the amatory activity of an *erastes* and shows how the fawn is also a metaphor for unwitting prey in contexts expressive of betrayal and entrapment (pp. 54–55).

28. Plaque: Addorsed Sphinxes

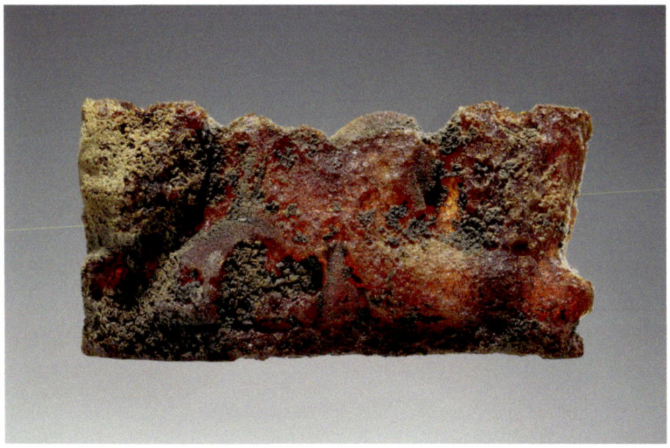

Accession Number	78.AO.286.2
Culture	Campanian
Date	575–550 B.C.
Dimensions	Height: 29 mm; width: 62 mm; depth: 15 mm; Weight: 16.7 g
Subjects	Color; Egypt; Jewelry; Sphinx

Provenance

–1978, Gordon McLendon (Dallas, TX), donated to the J. Paul Getty Museum, 1978.

Condition

The plaque is intact except for two losses: a large chip on the reverse, at the bottom of the right-hand perforation, and a portion of the obverse wall of the left-hand perforation, just below the jaw. The surface of the flat bottom is much less degraded than that of the rest of the object, which is in poor condition, with a uniformly degraded top layer that is granular and friable. Dirt and white-gray encrustations adhere to the surface on all sides, especially inside the top of the left-side bore. No inclusions are visible. Where the surface is intact, the amber's color varies from sandy yellow to gray-brown; in the numerous areas where the internal material is exposed, it is rich brown. Held against the light, the amber is orange.

Description

Addorsed sphinxes, in repose and *reguardant*, decorate the rectilinear plaque, which is carved in the round. The thin slab of amber is modeled in low, flat relief on the obverse, reverse, and sides; it is smoothed almost flat on the bottom.

Despite the poor condition of the pendant, the salient features of the two sphinxes are still decipherable. Engraved lines in the lower central part of the composition are the remains of divisions between the vanes of the primary feathers. On the top side, the hatted heads (the hats may be flat *petasoi*) and the details of the wing tips are clearly visible. On the lateral sides, indistinct details of the heads and the indentations for the necks, the breasts, and the legs remain. On the reverse, the backs of the hair, the contours of the upper sectios of the wings, the lower bodies, and the legs are still visible. The hair of each is fashioned in a shoulder-length cut, the *Etagenperüke*, or stepped wig, and it is visible on the obverse, the short sides, and the reverse. The large frontal faces are in higher relief than the bodies. The heads are wide and flat at the top and are large in proportion to the bodies and wings: each head measures over one-quarter of the width and over half the height of the composition.

The sphinxes have broad, flat brows, the suggestion of long, prominent noses, high and wide cheeks, wide jaws, narrow but prominent chins, and small mouths. The distance from the parting of the lips to the apex of the chin is short. The better preserved of the heads (at right) is carved with bulbous, slanted eyes, the outer canthi just slightly higher than the inner ones.

Little can be determined about the manufacture due to the poor state of conservation. The plaque appears to reflect the shape of the original piece of amber, for the level of relief varies over the obverse, the reverse, and the sides. The obverse has a greater depth of relief than the other sides. The upper contours are asymmetrical, and the bottom is uneven.

Two large tapered perforations located 7 mm from either end were bored on the same vertical line as the medial line of the creatures' heads. At the top, the holes are 4.5

mm in diameter, and at the bottom, 6 mm. The fragment of a bronze(?) tube, constructed from a roll of thin sheeting (0.3 mm thick), remains in one perforation, and the residue of a metallic corrosion product is found in the other. This strongly suggests that both holes were similarly lined. In the depression between the wings at the top is a stopped bore, 5 mm deep and 3 mm in diameter (for an addition?).

Discussion

This amber plaque has no parallel in subject or form. The sphinxes, while related to various Greek and Etruscan types, have no close analogues. When the piece was new, the natural translucency and color of the amber would have been augmented by the internal reflections of the metallic sheaths lining the two vertical perforations. These perforations must each have carried a filament, and because of this it does not seem likely that a suspension device was inserted into the stopped bore on the top edge. An additional decorative element (of amber or another material) was almost certainly placed there instead. The direction of the through-bores and the figural design imply that the plaque functioned in a horizontal position, perhaps as part of a larger object, hanging from a pin, head ornament, or necklace (in front, or in back as a counterweight).

Metal-lined perforations are not found on any other figured amber objects but are common on many fibulae and on many other kinds of amber objects. The technique is an old one and is known in eighth-century B.C. Greece, with examples from Lefkandi, Tekke, and Salamis.[1] Seventh-century examples are the fibulae and other kinds of amber objects from the opulent burials of Verucchio and Cumae.[2] The metal linings not only offered protection from abrasion and breakage, but also added to the brilliance of the amber, exploiting its natural clarity, brilliance, and luminosity.

The sphinxes of the plaque exhibit many conventions of seventh-century representations: large triangular faces projecting from small squat bodies, a stepped hairstyle, and low, flat hats. The style of 78.AO.286.2 is a complex blend, one that shows connections to earlier North Syrian objects[3] and to even earlier Mycenaean ivory-carving traditions. The latter is demonstrated by comparison with early sphinx-subject ivories from Mycenae and Athens.[4] The low-relief carving and squared-up sphinxes of 78.AO.286.2 are similar to a number of small seventh-century B.C. ivories from Greece and Italy, from Ephesus, Sparta, and Perachora, and from Comeana.[5] However, perhaps the best analogues are the relief metope-sphinxes of certain Cretan amphorae and terracotta *pinakes* of the seventh century.[6] There is also an unmistakable relationship between 78.AO.286.2 and some Archaic Campanian sphinxes of later date: the bicorporal sphinxes of horseshoe-shaped terracotta plaques from Capua (of circa 575–550 B.C.; a notable example is in Copenhagen)[7] and three small bronze sphinxes (one in Boston and two in Baltimore, possibly from the same bowl) thought to come from Cumae.[8] These similarities may be taken as evidence that the maker and/or model of 78.AO.286.2 had a lasting influence in Campania. A sixth-century date for 78.AO.286.2 can be extrapolated from the other amber objects said to have been found with it, including the Getty *Hippocamp*, 78.AO.286.1 (cat. no. 29).[9]

A double sphinx is a nonnarrative subject, unlike the single sphinx, the devourer, which might call up the story of Oedipus. The double sphinxes may have held special force, since they look backward and forward, left and right, perhaps doubling the power of a single sphinx. The motif of a double sphinx had a venerable history as a potent subject in the circum-Mediterranean area. Generally, repetition is an age-old formula for increasing the potency of any amulet, spell, or curse.

Generally related in form to 78.AO.286.2 are the Egyptian amulets in the form of addorsed lions and of back-to-back foreparts of bulls and rams, the latter having a full moon with a crescent nestling between their backs.[10] Contemporary with 78.AO.286.2 is one variant of the Egyptian addorsed lions amulet type. The suspension loop was placed between the animals' backs in such a way that it resembles a sun disk, suggesting an underlying connection with Rwty, "over whose back the sun rose each day."[11] Might the stopped bore on the top of 78.AO.286.2 have once supported an added image (perhaps a device in the form of a solar disk or other symbolic element)?

There is no evidence for the preburial function of the amber (the condition prevents any conjecture about signs of wear). If it were worn in life, 78.AO.286.2 may have functioned in a way parallel to that of the Egyptian addorsed animal and sphinx amulets. The amber may have brought to the wearer the underlying savagery of the lion, as the sphinx subject did in Egypt. (Since at least the Middle Kingdom, the single-sphinx amulet had been understood to link the wearer with the pharaoh's protective power and authority and the lion's power.) The doubled sphinxes could have invited the protective and propitious powers of the composite creature. Double sphinxes of amber might double the curse of "fighting fire with fire." Certainly, the inherent potency was magnified and focused, and the object more efficacious, when amber was carved with such a time-honored potent subject. In the grave, 78.AO.286.2 would have played a powerful guardian role: the sphinx who escorts the dead, the watchdog who punishes those who disturb the dead, could also protect the "house" of the grave.[12]

NOTES

1. For the technique and its eighth-century B.C. history in Greece, see introduction, n. 48 and n. 50.

2. For the Verucchio material, see *Verucchio* 1994. For the amber objects from Cumae, see Strong 1966, pp. 23, 32. The relevant fibulae and ring pendants that I have studied firsthand include those in the Getty Museum, the British Museum, the Metropolitan Museum of Art, and a Geneva private collection. I owe both the discovery of the metal tubes and observations about their probable original effects to John Tucker.

3. Compare, for example, eighth-century B.C. North Syrian harness attachments ornamented with a frontal nude female under a sun disk: see J. J. Orchard, *Equestrian Bridle-Harness Ornaments: Ivories from Nimrud* I, 2 (Aberdeen, 1967), pls. XX–XXI, XXVIII–XXXII.

4. For the ivory sphinx from the Athenian Acropolis (Athens, National Archaeological Museum 2486), see Poursat 1977, no. 493, pl. 53. For the Mycenaean sphinxes on the lid from the House of the Sphinxes (Athens, National Archaeological Museum 7525), see ibid., no. 138, pl. 12.

5. For the ivory sphinx from the Sanctuary of Hera Limenia, Perachora (Athens, National Archaeological Museum 16519), see T. J. Dunbabin et al., *Perachora II* (Oxford, 1962), p. 403, pl. 171; for the sphinx from Ephesus, Akürgal 1961, pp. 194ff., figs. 135, 154; and D. G. Hogarth, *Excavations at Ephesus: The Archaic Artemisia of Ephesus* (London, 1908). For the sphinx from the sanctuary of Artemis Orthia at Sparta, see Dawkins 1929; and Marangou 1969. For a recent discussion of Laconian and Laconian-influenced sphinxes, see S. Descamps-Lequime, "Une sphinx en bronze: Élément de décor d'un trône archaïque?," in Clark and Gaunt 2002, p. 116. For a miniature sphinx from Poggio Civitate (Antiquarium 71-198), see Phillips 1993, pp. 75, 78, n. 219; and N. Spivey, *Etruscan Art* (London, 1997), p. 27, no. 28. For the fragmentary ivory sphinx from Comeana, see Bartoloni et al. 2000, p. 262, no. 328.

6. Compare the metopes with pairs of sphinxes on Cretan amphorae. For the relief amphorae, see W. Hornböstel, "Kretische Reliefamphoren," in *Dädalische Kunst auf Kreta im 7. Jahrhundert v. Chr.: Museum für Kunst und Gewerbe Hamburg* (Hamburg, 1970), pp. 56–59; and J. Schäfer, *Studien zu den griechischen Reliefpithoi des 8.–6. Jahrhunderts v. Chr. aus Kreta, Rhodos, Tenos und Boiotien* (Kallmünz, 1957); for the sphinxes, P. Müller, *Löwen und Mischwesen in der archaischen griechischen Kunst* (Zurich, 1978). For the Cretan terracotta *pinakes*, see P. Blome in *Orient und frühes Griechenland: Kunstwerke der Sammlung H. und T. Bosshard* (Basel, 1990), p. 50, no. 78.

7. Copenhagen, Ny Carlsberg Glyptotek HIN 157 (from ancient Capua): T. Fischer-Hansen et al., *Campania, South Italy and Sicily: Ny Carlsberg Glyptotek Catalogue* (Copenhagen, 1992), p. 196, no. 148.

8. Boston, Museum of Fine Arts 51.2469, Frederick Brown Fund (circa 540 B.C.): Comstock and Vermeule 1971, p. 37, no. 35, with reference to the two bronze sphinxes in Baltimore (see also D. K. Hill, *Catalogue of Classical Bronze Sculpture in the Walters Art Gallery* [Baltimore, 1949], p. 122, nos. 280–81, pl. 54).

9. Many other beads, pendants, and fragments, all unpublished, were part of this donation.

10. For the relevant amulet types and their functions, see Andrews 1994, pp. 78–79, 89–90.

11. Ibid., p. 90.

12. Vermeule 1979, pp. 69–70, 171.

29. Pendant: Hippocamp

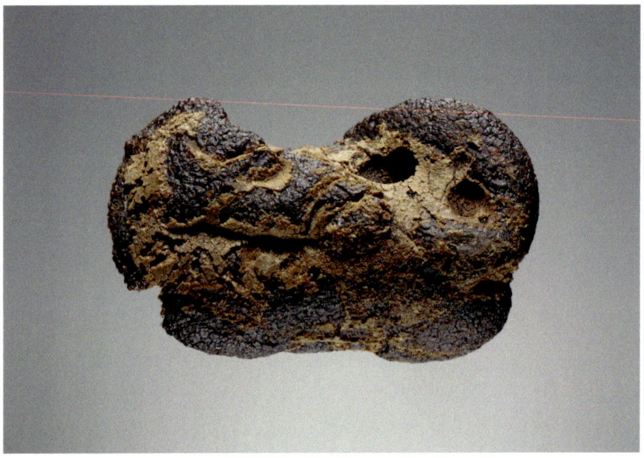

Accession Number	78.AO.286.1
Culture	Etruscan
Date	575–550 B.C.
Dimensions	Height (i.e., length along major axis): 70 mm; width: 43 mm; depth: 27 mm; Weight: 39.6 g
Subjects	Animals; Etruscan culture; Inclusions

Provenance

–1978, Gordon McLendon (Dallas, TX), donated to the J. Paul Getty Museum, 1978.

Condition

Although the surface of the pendant is in poor condition (with an extensive loss of surface detail), it is intact minus a triangular break in the mane. There is extensive flaking on the surface. A surface coating may have been applied before it entered the Getty Museum. This piece is opaque, but part of its surface retains some of the original integument, which is tan-brown; the rest of the surface is darker reddish brown. Illuminated with transmitted light, the amber is translucent and orange-red. Several areas have small inclusions.

Description

The pendant, carved fully in the round, represents a dormant hippocamp. The head is resting on the coiled body and tail. On the obverse are the head, the neck, the right lower leg, and, by implication, the upper body of the sea horse and a section of its tail, which loops in a counterclockwise fashion. On the reverse are the left front leg and the tail end of the coils, which curl in a clockwise direction. The ruffled edge on the tail must be a section of the dorsal fin.

In profile view, the head is long, and it has a straight muzzle and jawline, a large, almond-shaped eye (in profile), and a full, rounded nose and mouth. The top of the upright, pointed ear is visible at the poll. On the forehead lies a petal-shaped lock of hair. Behind it is an even ridge of short-cut mane that surmounts the arched neck, the hair depicted by regular, straight striations.

The obverse is convex, the reverse slightly concave. On the obverse at the lower right are two smoothed craters, apparently resulting from the removal of faults. One is hemispherical, 7 mm in diameter, and the other oblong, 10 mm across and 8 mm deep at its widest point.

The object has two sets of perforations: one, 3 mm in diameter, passes through the piece from the top of the forelock to the middle of the mane. The second, 2 mm in diameter, has exits in the mane (near the first perforation) and near the coronet of the leg. A short perforation, 2 mm in diameter, intersects with the latter one from a hole near the pastern. When the pendant was suspended, the nose would have been nearly vertical. The perforation holes would have forced the pendant to face obverse side out when suspended.

Discussion

This pendant is one of four amber pendants of hippocamps. A pendant in London (British Museum 73)[1] shares with 78.AO.286.1 the same basic configuration of a coiled monster, but its style is different. BM 73 does not have the strong forelegs of the Getty Museum hippocamp. It is more snakelike and less subtly modeled, described instead with engraved lines. Both the London and Getty pendants have two sets of suspension perforations at the rostral end. When the pendants are suspended, the creatures' heads are in a vertical position.[2] An amber hippocamp pendant in the Metropolitan Museum of Art is perforated so that it hangs in a sejant position, much like that of a seahorse in motion.[3] One of six amber pendants, now lost, which was excavated at Marzabotto in the nineteenth century, was thought to be a hippocamp.[4] A hippocamp also may be represented on one side of an exceptionally large amber pendant in London (BM 38), the subject(s) of which remains unexplained. On the main side is a charioteer driving a four-horse chariot. The head of the distant horse on one side corresponds with that of the hippocamp(?) on the other. On the reverse is a male figure (perhaps nude) who is similar to the charioteer, perhaps struggling with a hippocamp, perhaps pulling in the creature, using his legs and feet to secure the slack in a thick rope, or, more likely, half-astride the creature, holding on for a wild ride.[5] The subjects must be related, since when the amber was newly carved, the image on one side would have been visible on the other, and there do not appear to be pre-Roman ambers with unrelated subjects. Do both sides show aspects of the cycle of the sun, the infant sun drawn by the charioteer (Apollo?) on the main side, and the nighttime passage of the sun beneath the ocean, drawn by the hippocamp, on the reverse? The use of amber, a solar material with a waterborne phase in its formation, augments the iconography and magic of the image; this may help to explain the existence of other hippocampic carved ambers.

The horse part of 78.AO.286.1 is generally similar in style and type to horses painted on Middle and Late Corinthian vases and Early Attic black-figure ware. They share common features: heavy forelock, large profile eye with carefully outlined rims, full, rounded nose and mouth, large nostril, and diagonal grooves above the nose. The amber mane is shorter, however, than those of the horses on the painted examples. Illustrative sculpted parallels that bring out the Etruscanness of 78.AO.286.1 are the protomes (with perhaps hippocamp rather than equine forepart?) on various bronze, *bucchero*, and terracotta works. From the same family are the horse protomes of two bronze braziers from the Vulcian "Isis Tomb," British Museum 436 and 437, dated by P. J. Riis to the second quarter of the sixth century,[6] and other Vulcian bronze horse protomes, among them attachments to the tripod in Cap d'Agde.[7] Comparison can also be drawn with the protomes of various *bucchero* vessels and offering trays (*focolare*)[8] and with a terracotta of a horse head in Basel.[9] The Getty *Hippocamp* shares with the protome type of BM 437 the same rounded, curving neck, the short mane worked with fine striations, and a comparable form of eye. The head shape of 78.AO.286.1 is closer to that of BM 436. When suspended, the London and Getty ambers have a position identical to that of the Vulcian brazier protomes.

The hippocamp first appeared in Etruria during the Orientalizing period and gained greater popularity in the Archaic.[10] The monster as found in ancient art must owe something to the appearance of the tiny fish *Hippocampus antiquorum*, common in warm seas. There are far more representations of the hippocamp in the art of Etruria and South Italy than in that of the mainland and East Greece. Throughout the sixth century and into the fifth, the hippocamp appears in tomb paintings and on bronze vessels and stands, cinerary urns, and sarcophagi. It joins other sea monsters on Caeretan ceramics of the mid-sixth century. At first, marine creatures like the hippocamp and the sea dragon are dreadful and powerful monsters, inhabiting an unknown and dangerous place. Only heroes like Herakles or Perseus can defeat such marine monsters. This changes during the sixth century, when the hippocamp becomes an ally of humans and gods. In later times, the hippocamp is associated with Thetis, the daughter of Nereus, mother of Achilles, and wife of Peleus. In Italian Hellenistic-period representations, the hippocamp is Thetis's mount when she descends into the oceans to get the armor of Achilles. (Is there a connection between Thetis's steed and safe travel across the ocean to the Blessed Isles?)

The significant part played by the hippocamp in Etruscan and other Italian sepulchral symbolism is underlined by its frequent appearance on objects made purely for funerary purposes. The outstanding examples in Etruscan art date from the middle of the sixth century to the middle of the fifth and include the Etruscan stone figures of hippocamp riders, such as the famous example from Vulci in the Villa Giulia, of about 550;[11] the hippocamps of Etruscan black-figure amphorae; and the images of hippocamps in Etruscan painted tombs. The best visual explanation of the amber hippocamps' meaning may be found in the Tomb of the Bulls' pedimental, heraldic composition of hippocamp riders racing toward a central island.[12] In the fifth century in Etruria and elsewhere on the peninsula, hippocampic imagery is joined by that of other hybrid beings of the sea. The message is the same, too, in fourth-century South Italian red-figure vases, Canosan vases,[13] and the gilded terracotta reliefs once ornamenting Tarantine sarcophagi.

For the Etruscans, the hippocamp ferries the dead to the afterworld beyond the ocean. In tomb paintings, the hippocampic scenes are not simple allusions to the sea, as Jean-René Jannot emphasizes: "Painters used dolphins to evoke that element.... It is highly probable that the hippocamps ... served the function in Etruscan imagery of *psychopompoi*, guides for the dead.... These representations show death as a voyage toward an island Afterworld. The *hinthial* [soul, shade] rides the sea monsters across the sea (or ocean) toward a land where he will dwell."[14]

Why a pendant in the form of a hippocamp? The subject's Bronze Age antecedents (Mycenaean rather than Minoan) show how important the subject was for adornment and for seals—and for the grave. "The sea-horse is frequently found in the Mediterranean, and is mentioned by several ancient authors because of medicinal (or poisonous) properties that were imputed to it," as Campbell Bonner writes.[15] This must account in part for its presence on representations with (unexplained) magical significance. It might be assumed that the lore of the sea horse predated writing about it by Dioscorides and Galen. Whether worn in life or in death, a hippocamp carved from amber would bring together two aqueous and magical subjects. Amber, being naturally buoyant in ordinary water, could float in saltwater, and the hippocamp's home was the salty sea. No matter which of the amber origin stories were current in sixth-century southern Etruria (where 78.AO.286.1 was likely carved, or where it was buried), the amber was believed to have been produced by the action of solar powers combined with falling into a river or the ocean before being carried onward. Amber, before carving, had already experienced a successful watery voyage.

The undulating forms of both the Getty and British Museum hippocamp pendants recall both the natural form of the fossil resin and the movement of water and suggest that the carver incorporated the original shape of the raw amber into the carving. The subject may even have originated in the appearance of the nodule. The New York hippocamp, in contrast, is much more like contemporary sculpted and painted examples. When newly worked, 78.AO.286.1 would have been an arresting sight: because of the waterlike translucency of the amber, the head, front legs, and long tail would have been visible all at the same time. Worn on the body, a couchant hippocamp, its eyes open in quiet watchfulness, might have conjured up the sea, its myriad dangers, and the necessity of a guardian monster. The hippocamp, as a demonic creature, was a protective, danger-averting subject that could work on the amuletic principle of "like banishes like." Additionally, the hippocampic pendants might have functioned by assimilation—that is, they were worn with the hope of acquiring the nature of the hippocamp and thus gaining access to its characteristics and particular powers. As a sea creature, the hippocamp was likely believed to have the gift of prophecy, as Emily Vermeule reminds us:

> *Almost all sea-creatures have the gift of prophecy. It may be minor and limited, but some who were born with the beginning of the world, older than the Olympian gods (Hesiod,* Theogony *131, 233), had vast aboriginal experience combined with knowledge of the intense constant changes of the sea under wind and sun, and their prophetic power had an authority which land oracles and newer gods could not rival. It was a knowledge of multiple possibilities, of transformations, mutations, and grandeur because it was not limited to the simple affairs of men on land.*[16]

NOTES

1. Strong 1966, pp. 79–80, no. 73 ("Snake [?]"), pl. XXIX. Strong found the interpretation of the amber problematic. This author interprets the grooves and hatching of the top edge as the dorsal fin. The form and pattern of the skinny neck and the thin, closely striated mane hairs are similar to the finely incised manes of a number of seventh-century B.C. ivory horse heads from the Temple of Artemis Orthia at Sparta (Dawkins 1929, pl. CXLIX, 1–2) and to a mid-seventh-century Etruscan ivory arm ring from Tivoli in Oxford (Ashmolean Museum Pr. 323: Brown 1960, p. 33, pl. III).

2. Strong 1966, p. 71, refers to G. Gozzadini, *Di ulteriori scoperte nell'antica necropoli a Marzabotto nel Bolognese* (Bologna, 1870), pl. 15. The amber hippocamp(?) was found with amber rams' heads and frontal and profile female head-pendants.

3. New York, Metropolitan Museum of Art 1992.11.23, Purchase, Renée and Robert A. Belfer Philanthropic Fund, Patti Cadby Birch, and The Joseph Rosen Foundation Inc. Gifts, and Harris Brisbane Dick Fund, 1992: *The Metropolitan Museum of Art Annual Report* (1991–92), p. 37.

4. Strong 1966, p. 64.

5. Ibid., no. 38, pl. XVIII. Strong thought the youth on the reverse side to be nude, riding the hippocamp, and unrelated to the subject of the main side. He considered the style to be "closely connected with Etruscan work and this may be explained ... if we suppose it was made in Campania or under the influence of Etrusco-Campanian art." For this author, the amber reveals a close link with the art of Southern Etruria, and is characteristic of "Ionian" works from the last quarter of the sixth century B.C. The hippocamp tamer evokes both Iolaos grasping one of the Hydra's necks on the Getty Caeretan *hydria* (83.AE.346) and some of the cavorting *comasts* on the Getty "Campana" Group *dinos* (83.AE.249), both wares "probably the product of East Greek (Ionian) artists working in southern Etruria": R. De Puma,

CVA, *United States of America*, fasc. 31, *The J. Paul Getty Museum, Malibu*, fasc. 6 (Malibu, 1996), p. 31.

6. Riis 1998, pp. 22–25, 121 (with bibl.). As Riis has argued, relatives of the two London protome types are found on other bronzework produced in the same active Vulcian workshop.

7. Cap d'Agde, Musée de l'Éphèbe ME 1171: O. Bérard-Azzouz, *Les bronzes antiques du musée de l'Éphèbe: Collections sous-marines* (Agde, 1997), pp. 40–42.

8. For the *bucchero* horse protomes, see CVA, *Getty* 6 (in n. 5, above), pl. 304 (with extensive bibl.).

9. A. Bignasca, in her catalogue entry for the publication of the horse protome in Basel (Antikenmuseum, Collection Ludwig BO 153, *Orient und frühes Griechenland: Kunstwerke der Sammlung H. und T. Bosshard*, ed. P. Blome [Basel, 1990], pp. 115–16, no. 172), establishes links with South Italian works such as the Grumento horse and rider (see cat. no. 55, n. 7). Bignasca's comparisons are significant also for this amber, 78.AO.286.1.

10. For the hippocamp, see, for example, M. Boosen, *Etruskische Meermischwesen: Untersuchen zu Typologie und Bedeutung* (Rome, 1986); and K. Shepherd, *The Fish-Tailed Monster in Greek and Etruscan Art* (New York, 1940). The significance in Etruscan funerary art of the hippocamp in the afterworld is outlined in F. Roncalli, "Iconographie funéraire et topographie de l'au-delà en Étrurie," in Briquel and Gaultier 1997, pp. 37–54; and Jannot 2005, esp. chap. 4.

11. For the stone rider, see A. Hus, *Vulci étrusque et étrusco-romaine* (Paris, 1971), p. 76, pl. 4.

12. For the Tomb of the Bulls, see Steingräber 2006, passim.

13. See, for example, A. Rinuy, F. van der Wielen, P. Hartmann, and F. Schweitzer, "Céramique insolite de l'Italie du Sud: Les vases hellénistiques de Canosa," *Genava* 26 (1978): 141–69.

14. Jannot 2005, pp. 60–61.

15. Bonner 1954, p. 142.

16. Vermeule 1979, p. 190, n. 16.

30. Pendant: Cowrie Shell / Hare

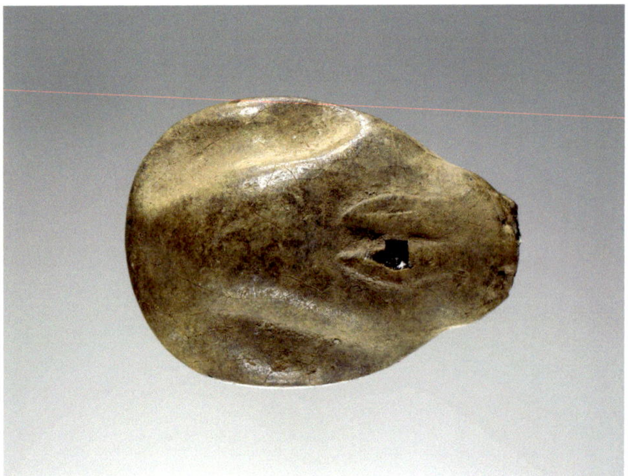

Accession Number	79.AO.75.28
Culture	Italic or Etruscan
Date	600–500 B.C.
Dimensions	Height: 37 mm; width: 26 mm; depth: 14 mm; Weight: 3 g
Subjects	Color; Egypt; Fertility; Hare; Transparency

Provenance

–1979, Stanley Silverman (Huntington Beach, CA), donated to the J. Paul Getty Museum, 1979.

Condition

The pendant is intact, with the exception of a large fracture loss to the nose and chips between the ears. The surface is smooth and firm. Minute surface crack patterns are visible on the dorsal surface and the ventral side. Before acquisition by the donor, the pendant was consolidated with a coating material that likely cemented together the consolidant, the yellow-ocher degraded amber, and soil, probably unintentionally. The amber is opaque and brownish yellow in ambient light, except at the break, where the rich red translucency of the interior is exposed. There are no visible inclusions.

Description

This teardrop-shaped pendant is an amalgamation of two sculptural forms, a crouching hare and a cowrie (Cypraeidae) modeled on the adult shell of the mollusk. The body of the hare is elevated from the ventral surface by a sort of platform. This is the edge of the cowrie shell. The rounded form of the pendant is a conflation of the hare's back and the swelling of the cowrie's dorsal surface. The hare is wide at the shoulders and narrow at the rump. Neither the front nor the hind legs are indicated. The hare's ears are long, point straight backward, and lie flat on the animal's head. Its almond-shaped eyes are carefully incised, tapering to a point at the outer canthi. The ventral surface is smoothed but not flat; it curves gently upward at the base of the cowrie's anterior canal and at the edge. The aperture is represented as a long groove.

The pendant retains evidence of the prepared amber blank from which it was carved. The indentation on the left side of the hare's body and the three declivities in the ventral surface (each approximately 6 mm wide by 7 mm in length) are likely the result of the prefiguration removal of imperfections. The amber's natural form, perhaps originally a large drop, may have directed the figuration. Its shape may also have conditioned the position and shape of the hare's eyes. The use of a graver is seen in the working of the eyes and the articulation of the long groove of the ventral surface and the groove separating the body from the base.

The pendant was suspended from a perforation that passes laterally through the cheeks of the hare; the exit holes are each 2 mm in diameter. When suspended, the animal would have faced upward, its back facing the viewer.

Discussion

This pendant is an extraordinary combination of four things: amber, a recumbent hare, a cowrie shell (the ventral side mimics the aperture), and, when hanging, the shape (in silhouette) of the Egyptian sign of the East, the flame of the new sun's light. The lustrousness of the fossil resin echoes that of the shell, and its transparency the light of the sun. Like amber, the cowrie was highly valued for its rarity, distinctive colors, and luster; like the cowrie, amber had a marine aspect: amber was made, found, or transported by water, and some specimens may still have included encrustation of shells. Amber is like the color of the sun and flame.

"Cowrie" is the common name applied to marine gastropods belonging to Cypraeidae, a large family of marine snails abundant in the Indian Ocean, particularly in the East Indies and the Maldive Islands. (Cowries are also found off the coast of Southern California.) They characteristically have massive, smooth, shiny shells with striking patterns and colors. As is the case today, only some species of cowrie were highly prized (and imitated) in the ancient world; the most sought after then as now were those deemed "exotic," that is, particularly rare, lustrous, or distinctly colored.[1] Cowries were traded to Egypt as early as the fifth millennium B.C., and cowrie-shaped pendants used as ornaments are documented as early as the late Old Kingdom period in Egypt. The much-valued cowrie shell inspired various kinds of jewelry, direct transfers in hollow electrum, gold, and silver,[2] as well as the cowroid, a mixture of the scarab and the cowrie, in which the scarab back was replaced with the cowrie shell and the underside engraved.[3]

Cowries are among the most popular of all pendant shapes in pre-Roman Italy and are among the first subjects to appear in amber, the earliest dating to the late eighth century B.C. Amber cowries gained popularity in the seventh century in Etruria, the mid-Adriatic, and the south of Italy and remained popular until the end of the fifth century B.C., when the shell was imitated in silver and in gold.[4] Amber cowrie-shaped pendants come in all sizes, from 20 to 90 mm in length, and range in their degree of naturalism. Some pendants are schematic; others include greater morphological detail, as is the case with 79.AO.75.28. The evidence from controlled excavations of Orientalizing-period sites in South Italy shows that amber cowroids are documented exclusively in female graves, where they were the primary elements of necklaces and girdles. Other uses are documented, however. A unique pectoral with cowries and tiny female figures was buried with a woman in a grave at Ascoli Piceno.[5] A late-seventh-to-early-sixth-century grave (Tomb 315) at Alianello-Cazzaiola documents another unusual use: the deceased woman was buried with an elaborate headband composed of small disk-shaped beads of ivory, bone, amber, and faïence, as well as with Egyptian faïence scarabs, real cowries, and a small amber imitation cowrie.[6]

The Getty cowrie-hare is one of six amber cowroid pendants known to me. A unique combination is the pendant from Tursi, dated to the first half of the eighth century B.C., which incorporates a cowrie and a scarab (but differently than in Egypt): one side of the Italic amber is a scarab and the other side that of a cowrie.[7] The others are a cowrie-hare pendant in the Metropolitan Museum of Art,[8] a schematically rendered cowrie / crouching hare, teardrop-shaped and with a plain underside, in a London private collection,[9] and three hare-subject pendants in a German private collection.[10] The Getty, New York, and London hare-cowrie pendants are all relatively close in size, although each is different in style and shows variations in form. The New York cowrie-hare is the most naturalistic: the raised base at the anterior canal of the cowrie, the distinctive caudal elongation of the hare's head, and the rangy muscularity of the hare's long body are faithfully captured.

This trio is related to the amber frog-cowrie pendants found at Vetulonia and to an example (now in the British Museum) said to come from Armento.[11] A similar pendant was once in the Stroganoff collection,[12] and another (allegedly from Metaponto) is recorded as once being on the art market.[13] As with the scarab-cowrie combination, these hare-cowrie and frog-cowrie pendants bring together powerful subjects with age-old fertility, protection, and regenerative significance in amber, a material with the same properties.

The silhouette of this pendant, and of cowrie-shaped pendants in general, forms the shape of the Egyptian flame of sunlight, the sign of the East. The Egyptian flame accompanies the infant sun, the East, the direction of the sunrise and rebirth, an important step in the formation of the world.[14]

In Egypt, from the late Old Kingdom through the Late Period, both males and females wore hare and frog amulets, but the cowrie seems to have been worn exclusively by women.[15] In Italy, the hare as a subject of adornment may have had resonance for both men and women; it might have symbolized a specific divinity: Dionysos, Artemis, or another female goddess of nature or the hunt. The hare and the frog were both associated with fertility, but a hare pendant might have had an additional significance in the tomb, offering its wearer special sight even in the dark, speed (away from danger and for fast passage through the afterworld), and the possibility of rebirth.[16] The symbolic meaning of the hare in the iconographic tradition of Syria (the ultimate source of Phoenician hare images?) "grew out of its biotope (animal of the open fields and desert) on the one hand, and out of its stunning reproductive capacity (the superfecundatio was known since antiquity) on the other hand."[17] "Hare amulets may have stood for the vital forces connected with fertility ..., alert quickness, and swiftness of the animal, and are to be viewed as life-giving symbols."[18] (See *Divinity Holding Hares*, 77.AO.82, cat. no. 4, for additional discussion of the hare as a symbol.)

The mature cowrie has been likened in appearance to the human eye and female genitalia, both powerful danger-averting subjects.[19] The cowrie shell has been used to "replace" the eyes of the deceased and in Egypt and Italy is especially important for women. The long and narrow aperture on the underside of the cowrie recalls the external appearance of the vulva, and with the animal emerging from it, birth itself. The overall shape of the shell may have been thought to intimate the shape of the womb, "so when beads of its shape formed an element of a woman's girdle, they were in exactly the right place to ward off evil influences from the relevant bodily part of the wearer, especially if she were pregnant."[20]

The function of a cowrie amulet would have been to enhance the particular bodily functions of the organs it represented (the eyes or the genitalia) and to act as a substitute for them in the afterworld. The amber, itself magic and regenerative, no doubt enhanced the fertility properties of the cowrie, as did the image of the proverbially fertile hare. That the hare and the cowrie are two animals in which the females are larger than the males must have added to the amuletic aspects of the pendant. In the tomb, the combination of hare and cowrie in a beautiful ornament-amulet might have been especially valuable for protection in rebirth or for the journey to the afterworld. In the beliefs of some in ancient Italy, this was a complicated voyage, part by land, part by sea, and part submarine.[21]

NOTES

1. White 1992, p. 549, theorizing a source-distance gradient for Aurignacian-period personal ornaments, underlines that they were usually made of materials exotic to the regions in which they were found. This continues to hold true for the prehistoric period and the early historic period in Italy for amuletic and exotic materials, such as amber, ivory, and cowries. Shennan 1993, pp. 62–66, discusses amber's value, especially in light of its acquisition by political-religious elites living spatially distant from amber sources, and cites Helms 1988, p. 114: "Many exchanged items have inherent magical or religious significance as 'power-charged' treasures acquired from extraordinary realms outside their own heartland." Indeed, as Helms 1988 concludes on p. 130, "the ultimate goal of those seeking [shields or shell or stones or holy incense, or amber] may well be directed towards obtaining (maintaining) access to material manifestations of the power and potency that imbues their cosmos, thereby continuing their close association and inclusion with the dynamics of the universe of which they are an integral part." This idea is more fully developed in M. W. Helms's *Access to Origins: Affines, Ancestors and Aristocrats* (Austin, 1998).

   The cowrie was associated with fertility and the sea, whose tides are controlled by the moon, the effect of which on women was well established. For this reason, among others, the moon is associated with many female fertility divinities in the Mediterranean, among them Artemis (with whom amber is also closely linked). An amber cowrie could be seen as a material manifestation of the dynamics of the universe.

2. An early outstanding example is the girdle formed of gold cowries belonging to Queen Mereret (Dynasty 12) in the Cairo Museum: see, for example, Andrews 1994, p. 42.

3. Ibid. Amber scarabs were among the first carved amber objects to appear in Italy in the Orientalizing period, with many excavated examples. Although many demonic subjects of Egyptian and Near Eastern origin appear to have lost or changed their meaning as the visual forms were adapted in Greece, Etruria, and elsewhere in Italy, the scarab seems never to have lost its original associations with the sun, life, and regeneration. The early amber scarabs and scaraboids are evidence of this. See A. F. Gorton, *Egyptian and Egyptianizing Scarabs: A Typology of Steatite, Faience and Paste Scarabs from Punic and Other Mediterranean Sites* (Oxford, 1996), p. 158, n. 65.

4. For example, the precious metal cowries buried with the woman in Tomb 419 at Banzi (along with three amber profile head-pendants): *Magie d'ambra* 2005, p. 126.

5. I. Dall'Osso, *Guida illustrata del Museo Nazionale di Ancona* (Ancona, 1915), p. 303 (referred to by Waarsenburg 1995, p. 454, n. 1276).

6. For the headband from Tomb 315 at Alianello-Cazzaiola (Heraclea, Museo Nazionale della Siritide 209862), see *I Greci in Occidente: Greci, Enotri e Lucani nella Basilicata meridionale* (Naples, 1996), pp. 152–53, no. 2.12; and *Ornamenti e lusso* 2000, p. 17, fig. 10.

7. *Magie d'ambra* 2005, pp. 87–88 (discussed by S. Bianco). The amber was excavated from a woman's tomb in the necropolis of Santa Maria di Anglona–Valle Sorigliano.

8. Metropolitan Museum of Art 1992.11.13, Purchase, Renée and Robert A. Belfer Philanthropic Fund, Patti Cadby Birch, and The Joseph Rosen Foundation Inc. Gifts, and Harris Brisbane Dick Fund, 1992: *Art of the Classical World* 2007, pp. 295, 473, no. 343.

9. Unpublished.

10. These appear, from the photographs, to be similar to the London hare pendant: see K. A. Neugebauer, *Antiken in deutschem Privatbesitz* (Berlin, 1938), no. 255.

11. For the amber in the British Museum, see Strong 1966, p. 79, no. 72 ("Frog or toad"), pl. XXVIII. The effluent fertility of both frogs and toads (rarely differentiated in Egypt, according to Andrews) apparently led to their association with fertility generally and the renewal of life after death. The frog was seen as a chthonic animal that alluded to the forces which brought life into being. The Egyptian goddess Heket, who manifested herself in the shape of a frog or as a woman with a frog's head, was also associated with childbirth. The goddess had "participated in the creation of the divine child, crouched beside the potter's wheel on which Khnum shaped the small naked immortal, [and thus] all frog amulets might be intended to represent the goddess in her animal manifestation": Andrews 1994, p. 63. Representations of frogs are to be found on apotropaic wands, objects associated with the protection of new mothers and their babies: see E. Sullivan in *The Quest for Immortality: Treasures of Ancient Egypt*, ed. E. Hornung and B. M. Bryan (Washington, DC, 2002), p. 157, no. 71; and Andrews 1994, p. 63. The frog amulet first appears in burials of predynastic date in Egypt, and similar amulets are known from Mesopotamia and Susa (for example, the lapis lazuli frog bead from a find at Early Dynastic IIIB Susa: *First Cities* 2003, pp. 303–4, no. 202c).

12. Pollak and Muñoz 1912, vol. 1, pl. XL, 3.

13. Strong 1966, p. 79, records that "a piece much nearer the style of this one was part of a find at Metapontum." This find, which Strong refers to on p. 30, was on the London market in 1953; the photographs of the group were in the British Museum at the time of his writing, but are now lost.

14. A depiction of this initial stage in the formation of the world is found on a Ptolemaic granite sarcophagus, illustrated in Desroches-Noblecourt 2006, p. 24.

15. For the frog/toad, see Andrews 1994, p. 63; for the cowrie, ibid., p. 42. Both of the latter subjects are known in third-millennium amulets from Sumer and ancient Susa.

16. For the hare in Egypt, see S. Schroer, "Hare," in *Iconography of Deities and Demons: Electronic Pre-publication* (last revision November 15, 2006), www.religionswissenschaft.uzh.ch/idd/prepublication_3.php.

17. Schroer (see n. 16), p. 2, col. 1.

18. Ibid., col. 2.

19. Pinch 1994, p. 107, notes that cowries have been used against the evil eye in many cultures. In addition to its rarity and lustrous beauty, other reasons why the cowrie accrued amuletic powers include the facts that the females of the Cypraeidae family are larger than the males; the shells secrete a purple staining liquid; and the life cycle of the cowrie includes a development metaphorically similar to that of female genitalia. Some cowries have coloration that looks like the skin of spotted cats, some have "eyes," and some of the rarest cowrie species are thought to resemble female genitalia. Culled from R. J. Griffiths, "Size and Sex in *Cypraeidae*," *Proceedings of the Malacological Society of London* 34, no. 6 (1961): 322; and F. Lorenz and A. Hubert, *A Guide to Worldwide Cowrie*, 2nd ed. (Hackenheim, 2000).

20. Andrews 1994, p. 42. Pinch 1994, p. 107, notes the similarity of Egyptian representations of cowrie-girdle wearers to the contemporary custom of women wearing cowries on girdles in some parts of the Sudan. See also G. Clark, *Symbols of Excellence: Precious Materials as Expressions of Status* (Cambridge, 1986), pp. 23–26, for further parallels in contemporary societies.

21. See, for example, Jannot 2005.

31. Pendant: Lion

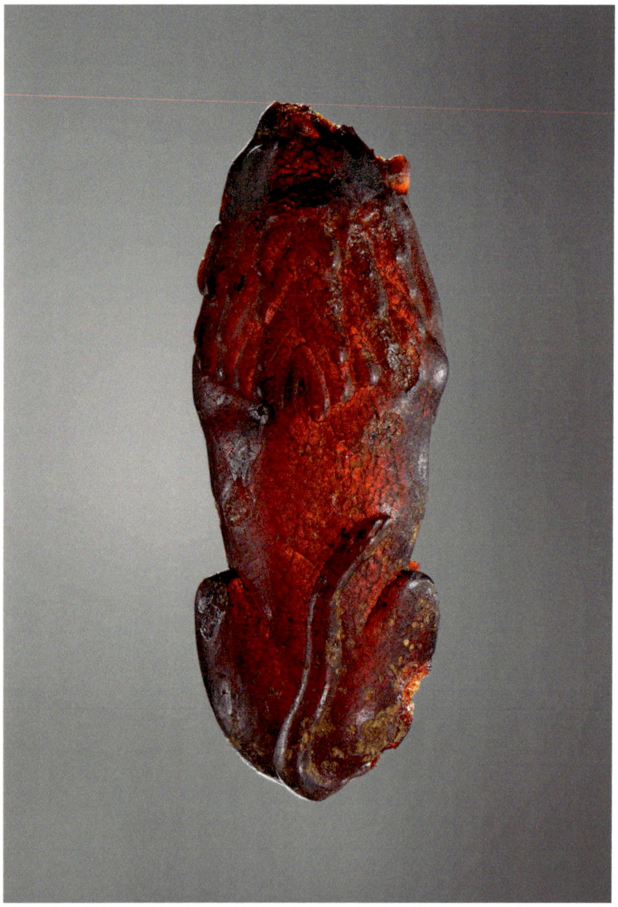

Accession Number	76.AO.78
Culture	Etruscan or Campanian
Date	525–480 B.C.
Dimensions	Length: 105 mm (as preserved; estimated original length 115 mm); width: 40 mm; depth: 18 mm (at chest); Weight: 35.5 g
Subjects	Egypt; Ionia, Greece (also Ionian, Greek); Lion; Transparency

Provenance

–1976, Gordon McLendon (Dallas, TX), donated to the J. Paul Getty Museum, 1976.

Condition

The pendant is intact, with a smooth, firm surface in most areas. The tip of the nose, the right side of the animal's muzzle, and the two distal digits of the right forepaw are broken off. There are also small breaks on the right haunch and foreleg. Small chips and pitting occur on both the dorsal and ventral surfaces and on the animal's sides. Some of the pitted areas contain yellowish degraded material. The amber is crazed all over. Near the right shoulder blade is a large inclusion, which is also visible on the ventral surface. This piece is translucent and dull reddish brown in ambient light and pale ruby-red in transmitted light. After entry into the museum, the pendant was mechanically cleaned and then treated with an amber-oil distillate that increased the translucency and darkened the color.

Description

The lion is dorsally convex and ventrally concave, full through the upper body, with the right shoulder area thickest in volume. There is a noticeable asymmetry of form, with the body and head curving slightly toward the left and the right lower leg protruding outward. Delicate sculptural transitions in the abdomen area suggest its softness. Subtle modulations in the carving indicate the subcutaneous structure of the shoulders and back. Between the root of the tail and the hocks on the underside is a long declivity, made when the amber was being prepared for carving. The thick, tubular tail begins its curve to the left, courses right, and curves again to the left, with the tip of the tail lying on the right flank. The tip of the tail has four lobes.

The lion is positioned with all four legs drawn up under the body, the right front paw placed under the chin and the left curled up under the chest. The left foreleg is represented in side view, the outer two toes in overlapped profile, while the right front foreleg and paw are straight and the footpad flat. The right foreleg is unusual in that its dewclaw is indicated, as a modeled zigzag indentation. The hind legs draw close together against the abdomen, the haunches and legs as though in profile, the feet flattened, with all of the toe pads in a line. Each of the four feet has four large hemispherical toe pads.

The lion's forehead has a deep medial indentation and two lateral indentations sweeping back to the mane. The almond eyes are set back, deep-set under the eyebrow ridges, which flow in a naturalistically undulating line to the wide cheekbones. Despite the losses to the top of the pendant, the nose can still be traced. Just above the break, there are two horizontal wrinkles. The muzzle, mouth, and chin are smoothly defined, with a groove for the mouth. The upper lip has two bulbous undulations on the surviving side and overlaps the lower one. The chin is short, and the throat is subtly carved to suggest the underlying anatomy. A shallow groove separates the face from the mane. The mane itself is bilaterally divided, with the median line of the forehead continued into the fur. Heavy tufts of the mane are separated into a somewhat symmetrical array. The mane hangs down onto the sides of the figure's head and continues onto the shoulders and chest. On the ventral surface, above the left foreleg, five hanks of hair are rendered. Large, softly modeled, folded ears commence at the front edge of the mane and are tipped upward. The antihelix is represented as a raised fillet.

The curve of the lion's body, the thinness of the pendant, the concavity of the ventral side and convexity of the dorsal, and features such as the declivity on the flank of the animal indicate the general outlines of the original shape of the amber blank. It was likely carved from an amber sheet (see introduction).

Evidence of manufacture includes the scrape marks located on the larger planes of the legs, back, and ventral surfaces. Around the eyes and in between the hanks of the mane are traces of a graver. Multidirectional abrasion marks are found near the transitions between body parts, such as the intersection of haunch and body.

The break in the muzzle offers a clear view of a triangular set of perforations for suspension: there is a cross-bore just below the top of the nose bridge and two perforations connected to an exit hole at the tip of the nose. Suspended from a filament in the cross-bore, the lion would have hung with its head uppermost, its body curving slightly downward to the right. The medial line of the lion's mane would have been roughly perpendicular to the ground.

Discussion

76.AO.78 belongs to a large group of Archaic and Sub-Archaic ambers carved fully in the round in the shape of resting animals and demonic creatures—lions, rams, bulls, goats, gazelles, hippocamps, human-headed bulls (Achelous), sirens, and sphinxes. The compositions emphasize the dorsal and ventral sides of the animals. Their faces are flush with the ventral plane, one or both paws are beneath the chin (or lower jaw), and both pairs of legs are drawn up close to the body. The more convex side is used for the animal's back. No two of these are alike, because in each one the animal is accommodated to the peculiarities of the amber nodule. When the pendants were new and the amber translucent, the features of one side would have been blurrily visible from the other, inviting handling so as to see the figures from every angle. The pendants would have needed to be turned over in order to see the entire animal, especially the sculptural tour de force of the curled-under paw.

Of this type of amber animal pendant, the couchant lion is the most numerous. The group includes, in addition to 76.AO.78, two pendants in Numana from a tomb at Sirolo; one from a recent find at Bologna (Tomb 12); two pendants in London, British Museum 64 and 65; one in Copenhagen; one each in private collections in Hamburg, New York, and London; and two on the art market, one in London and the other in Basel.[1] A lion nearly identical to BM 65 is part of a group of a lion attacking prey, the bow decoration of a bronze fibula from Tomb 72 at Belmonte Piceno. Related in format are a pendant in the form of a feline (77.AO.81.8, cat. no. 32) and two other fibula bow decorations from Tomb 72 at Belmonte Piceno, one of a lion attacking a deer and the other in the form of a pair of addorsed lions' heads.[2]

The pendant closest to 76.AO.78 is BM 64. The correlations include the overall conception of the design, the treatment of the volumes, the subtle transitions from one plane to another, and individual details such as the placement of body and limbs on the amber nodule, and the curve of the tail. Next in kin is BM 65, which is, however, less organically modeled and more simplified in key details, such as the mane and the number of toe pads (it has three each to the four of 76.AO.78), and has a flange, a bead-and-reel device, at the rump, which would have suspended the pendant head downward. These are enough alike in lion type and format that it might be hypothesized that they descend from a common sculptural invention, one close to 76.AO.78. The other lions' heads are carved in a more schematic manner, flatter and harsher, and perhaps likely descend from a variant model or prototype.

The style of 76.AO.78 (and its closest relations) is elucidated by comparison to objects in various scales and media, but especially to large-scale marble sculpture. It is similar to a number of Archaic marble lions from the Greek East and Lydia,[3] with one of the best parallels being the fragmentary marble lion's head from Ephesus in the British Museum.[4] 76.AO.78 shares with these marble lions an analogous formation of the head, a comparable fine-featured, slightly pointed face, similar modeling, especially around the eyes, and most noticeably, a parallel treatment of the long mane locks. Characteristically, the manes of this group are divided into coifed hanks rendered as separate, plastically modeled lobes. In the arrangement around the head and on the back, the design varies from a full petal-like blanketing to a bifurcated overlap patterning. 76.AO.78 has in common with a complete marble lion from Miletos in Berlin and a fragmentary lion from Miletos in Paris[5] an exposure of the ventral surface of the paws and the same rounded flame-shaped hanks of the mane.

The hypothetical prototype of 76.AO.78 is as likely one of amber as of ivory and was probably the work of a South Ionian artisan or someone trained in an ambient where the language and techniques of South Ionian sculpture were alive. This and the likely late-sixth-to-early-fifth-century date of 76.AO.78 argue for manufacture in the west, in Italy. So does the material: it is in these decades that large pieces of jewelry-grade amber were used for ornament-amulets found in Etruria and the areas bordering it—in the mid-Adriatic, Campania, and parts of inland South Italy. Manufacture on the west rather than the east coast of the peninsula is supported by similarity to some Campanian small bronze lions and to the lions incorporated into the top of the handles of several types of Etruscan *bucchero* vessels.[6] The lions' backs emerging from the surface are remarkably similar in conception.

The provenance of only the fibula lions from Tomb 72, Belmonte Piceno, is established. BM 64 and 65, from the Sir William Temple bequest, are said to have come from Armento—the same alleged provenance as that of three other stylistically related amber pendants from the Temple bequest, the "Man-headed bull-couchant (Achelous)" (BM 68), a couchant sphinx (BM 69), and a couchant gazelle (BM 70).[7] In turn, this ex-Temple group is stylistically related to the large sphinx pendant from Tomb 102 at Braida di Vaglio, a burial of circa 500 B.C.[8] While Belmonte Piceno, Armento (if the provenance is accurate), and Braida di Vaglio were the final resting places, they are not necessarily the places of manufacture of these ambers. In this author's view, the carved ambers under discussion are Etruscan or Campanian, made by an artisan from South Ionia or under the strong influence of a South Ionian invention.

The tradition of small sculptures of resting animals and other creatures carved fully in the round was age-old by the sixth century B.C. Amulets in this format are documented in third-millennium Sumerian and Egyptian graves. Related to such imagery are the contorted and whirling animals, such as trussed antelopes or geese, cut into Bronze Age Aegean seals or Egyptian containers, or the ritual utensils (spoons or containers) shaped as a swimming nude woman or a running dog. Parallels among objects closer in date to 76.AO.78 and the related amber animal subjects are the amber pendants that represent sirens and other flying anthromorphs. *Their* antecedents are perhaps the precious metal pendants of birds and flying figures, such as the seventh-century owls from Ephesus or the bee-goddesses from Kameiros.[9]

However, the contemporary precious objects most like 76.AO.78 and the other amber resting animals carved fully in the round are a number of extraordinary Greek gems (dating from the late sixth century onward) of *equidae* in unusual poses, among them rolling and falling horses and centaurs shown in various activities.[10] The horse-subject gemstones found in Cyprus, East Greece, and the Greek islands[11] and the seventh-century Ephesian and Rhodian gold pendants may point to the area of origin for the sculpted "invention" behind 76.AO.78 and its relatives. It is not only the pose, the sculptural accomplishment, and the nature of the subject that link these diverse classes of objects. They also have in common a similar sculptural approach, softness in the anatomical description, minimal use of line, and smoothing of the planar transitions, all of which underline the artistic connection.

The lion of 76.AO.78 conflates many aspects of a lion's behavior: the elongated body, lowered head, and crouching hind legs suggest that the animal is lying in wait, ready to pounce; its ears are laid back and the nose wrinkled, indications of the animal's ferocity. Yet the closed mouth and the curled right paw belie imminent action.

Such an image may have conjured up the power of the deities of the wild, and the heroes and gods who conquered the animal. When the object was worn, the wearer may have taken on by assimilation the power of such a vanquisher. On one level, a lion amulet might have brought to its wearer the celebrated power, bravery, and ferocity of the legendary animal (probably never seen by any of the amber lion carvers). By assimilation, it may have endowed its wearers with the same qualities. A lion amulet might also incorporate danger-averting and protective functions for the owner. The lion of 76.AO.78 is vigilant, a guardian and protector, and as an amulet, it would have been powerful in both life and death. However, might the lion amber pendants also have incorporated aspects of more ancient customs? In Mesopotamia, the lion was a favorite metaphor in literature for warlike kings and fierce deities. In the Neo-Assyrian period, the lion was considered "a generally magically protective type, known as *urgulû*."[12]

In Egypt, the lion was a symbol of the sun-god Ra, and by extension a symbol of the god Amun. As a desert dweller, the lion was believed to have regenerative capabilities and as such was an essential amulet for the dead. If the amber lion pendants incorporate Egyptian symbolism (as well as aspects of Egyptian style), what better material for them than amber, a solar substance? What better form than a tear?

At the time of their making or burial, these amber lions may have been described with the Etruscan or Italic equivalent of the Latin *fulvum* (English: tawny, like the hides of wild animals), which Pliny records was used to describe one kind of amber.

Two of the lion ambers indicate how they might have been worn in the grave: the Belmonte Picero lions formed the bows of bronze fibulae, and we might assume were worn on the upper part of the deceased's body. The smaller lion in the London private collection was one of three amber pendants hanging from a hip-encircling reticulated amber girdle. In both cases, the location (and the method of attachment) could have focused the amuletic powers of the ambers. The lion pins would guard the heart and soul; the girdle pendant would protect the pelvic region. For the latter, a Late Antique prescription used in "aggressive magic" is relevant: amulets with the head of a lion were believed to have the ability to calm the womb (the roaring of the lion counteracting the "roaring" of the womb) and were to be worn in its proximity.[13]

NOTES

1. The pair in Numana, Antiquarium Statale 1548a–b, come from the sixth-century B.C. "Tomb of the Regina Picena," Circolo 1, Fossa A, 1989 excavation, and were found with two lions' heads: see *Ambre* 2007, p. 174, nos. III.121–22; and M. Landolfi, "La tomba della Regina nella necropoli picena 'I Pini' di Sirolo-Numana," in *Eroi e regine* 2001, p. 358, n. 129. Landolfi compares them to a lion on the art market illustrated in Mastrocinque 1991, p. 77, fig. 22. Both of the Bologna lions come from female grave contexts of the late sixth to early fifth century. The lion from the necropolis of the ex–Manifattura Tabacchi in Bologna was one of five necklace pendants from Tomb 12/2004/2005, Soprintendenza per i Beni Archeologici dell'Emilia Romagna 247773: *Ambre* 2007, pp. 151–53, no. III.85. The other lion is one of eleven pendants from Tomb 11/1986 from the necropolis of the Giardini Margherita, now in Bologna (Museo Civico Archeologico SBAER240863): ibid., pp. 154–55, no. III.88. For the British Museum lions, see Strong 1966, pp. 75–76, 78, nos. 64–65, pl. XXV. H. Hoffman, AA (1969): 364, fig. 49a–b, discusses amber lions in a Hamburg private collection, in Copenhagen, and in the Norbert Schimmel Collection, New York. For the last, see also Hoffmann in *Norbert Schimmel Collection* 1974, no. 72. The lions on the art market in London and Basel are unpublished.
2. For the fibula decorations from Tomb 72 (Ancona, Museo Archeologico Nazionale 11014 [lion and prey (perhaps a bull)], 11015 [lion attacking a deer], 11016 [addorsed lions' heads]), see Rocco 1999, p. 75 (as mid-Adriatic production); Cristofani 1996, p. 138 (as probably Magna Graecian); Negroni Catacchio 1989, pp. 679–80 (as mid-Adriatic); Strong 1966, p. 76 (as imports from southern Italy); Brown 1960, p. 100 (as locally made by an immigrant craftsman from Etruria [perhaps from Orvieto or Chiusi], because they served a purpose to which amber was commonly put in Picenum); and P. Marconi and L. Serra, *Il Museo Nazionale delle Marche in Ancona* (Ancona, 1934), p. 29.
3. H. Gabelmann, *Studien zum Frühgriechischen Löwenbild* (Berlin, 1965); Tuchelt 1970; P. Müller, *Löwen und Mischwesen in der archaischen griechischen Kunst: Eine Untersuchung über ihre Bedeutung* (Zurich, 1978); C. Ratté, "Five Lydian Felines," AJA 93 (1989): 379–93; and B. Pfeiler, "Die Silberprägung von Milet im 6. Jahrhundert v. Chr.," *Schweizerische numismatische Rundschau* 45 (1966): 5–26.
4. British Museum B 140: Brown 1960, LXIII a.
5. Berlin, Antikensammlung Sk 2708; Louvre Ma 2790: Hamiaux 1992, p. 58, no. 50 (dated end of the sixth century to beginning of the fifth century).
6. Compare the lions of the handles of Etruscan *bucchero* such as those found at Poggio Civitate: see Berkin 2003.
7. The three British Museum examples are Strong 1966, p. 78, no. 68, pl. XXVII ("Man-headed bull-couchant [Achelous])," no. 69, pl. XXVIII ("Sphinx couchant"), and no. 70, pl. XXVIII ("Gazelle[?] couchant").
8. For Tomb 102 at Braida di Vaglio, see Bottini and Setari 2003.
9. Laffineur 1978.
10. See E. Zweirlein-Diehl, "A Centaur Playing Kottabos," in Tsetskhladze et al. 2000, pp. 397–401 (with list of examples and bibl.).
11. For a gray chalcedony cut horse of about 500 B.C. in a Swiss private collection, see Boardman 2001, p. 398, no. 1027. See also Zweirlein-Diehl 2000 (in n. 10, above), p. 399.
12. Black and Green 1992, p. 119.
13. Bonner 1950, p. 126.

32. Pendant: Female Animal (Lioness?)

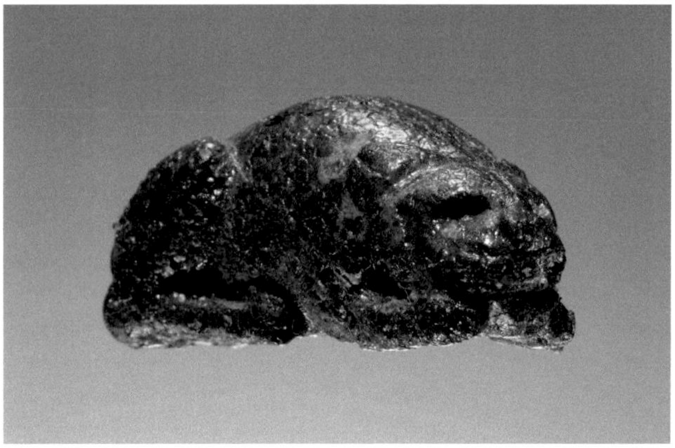

Accession Number	77.AO.81.8
Culture	Etruscan
Date	500–480 B.C.
Dimensions	Length: 55 mm; width: 23 mm; height: 23 mm; Weight: 7.2 g
Subjects	Animals; Fertility; Inclusions; Lion

Provenance

–1977, Gordon McLendon (Dallas, TX), donated to the J. Paul Getty Museum, 1977.

Condition

The pendant is intact except for small losses to the left lower jaw and center of the chest. The surface is crazed and cracking, with flakes missing from the cortex. Before entry into the donor's collection, the pendant was mechanically cleaned and treated with a thickly applied surface consolidant that added shine to the surface. In ambient light, the pendant is light brown, and in transmitted light, red-brown. The untouched surface inside the perforations may indicate the appearance of the amber before treatment.

Description

The original shape of the amber blank may be reflected in the compact, droplike shape of the pendant. The bottom is flat and the top convex, tapering at the nose end. The animal is stretched out, with her open-eyed head placed on her paws. The head and body curve slightly to the right, with the animal's left shoulder and leg extended farther on the left. The head is square, with a full forehead that gently slopes to the bridge of the nose, a flat muzzle, and a small mouth. The jaws are square. Dividing the head from the body at the thick neck is an engraved line. From it spring two triangular ears. The animal's back is full and rounded, arching all the way through the lumbar region. The front and back legs have a similar shape. The haunches are drawn forward, with the lower legs next to the abdomen and advanced past the point of the knee to midbody. The back lower legs are comparatively thin and angular, with long, hooklike feet. In comparison, the front feet are small. The left foreleg is farther forward under the head; an object may be held beneath it. The five pairs of large dugs lie in neat rows.

Because of the state of preservation and the chemical treatment of the amber, only the traces of engraved lines are witness to the pendant's manufacture. A perforation with holes each 2 mm in diameter passes directly under the neck in line with the root of the ears. If suspended by the boring through the head, the animal would have hung head upward; if a suspension device were attached to the holes in the rump, it would have hung nose downward. A pair of stopped bores, each 2 mm in diameter, are located at the sides of the tail, to a depth of about 3 mm. Below the left side of the lower lip is a similar bore, 2 mm in diameter and about 2.5 mm deep.

Discussion

Feline, dog, sow, or hare? The physiognomic characteristics of 77.AO.81.8 are not like those of any other animal, or of any other ancient representation of an animal, known to me. It has no parallels within the corpus of pre-Roman carved amber. Nevertheless, in its general format, 77.AO.81.8 is similar to many other types of amber in the form of dormant animals—for example, 76.AO.78 (cat. no. 31), to name one Getty pendant. The curve of the upper part of the animal's body, the short, fat body, and the positioning of the right paws are similar to the features of a group of Orientalizing amber dog pendants[1] and to many Archaic amber lions. The ancestral schema of the amber lions may be Mycenaean. 77.AO.81.8 is particularly close to an extant pendant, the gold couchant lion from Tomb 5 at Hagia Triada (circa 1500–1450 B.C.).[2] This said, the form of the head and ears, the manner in which the animal crouches, and the dugs invite comparison to a small group of Orientalizing representations of felines. (The legs and feet of 77.AO.81.8, however, are more lagomorphic than leonine.) The short, curled tail is the one big difference between 77.AO.81.8 and 76.AO.78 and the other related amber felines (as well as the comparable Bronze Age gold lions). Although a few Etruscan felines appear to have short tails,[3] the short curly tail of 77.AO.81.8 is more like that of a pig or some breeds of dog, including *Canis familiaris Studer*, the big, smooth-haired, heavy dog with small ears represented in Mesopotamian art.[4]

The stopped bores near the jaw area and to either side of the tail are equally without parallel. They were likely used for attachments, perhaps a collar at the neck if the animal is a dog, or for the attachment of pendants (nursing young?). The prominence of the milk-laden breasts of the amber animal emphasizes the fertility and regenerative aspects of the pendant amulet. If it represents a dog, there may have been an association with guardianship, protection, and healing. The ancient Near Eastern association of the dog with Gula, "the great physician," is allied to the later importance of the dog in the iconography and cult of the healers Apollo Asgelatas and Asclepius.[5] Some Egyptian Late Period amulet types of glazed-composition sows (the sky goddess Nut, or Isis?) were intended to endow their wearers with fecundity.[6] Comparable are Archaic Etruscan painted representations, such as the felines painted on vases (primarily found in tombs) by the Micali Painter, or the mammiferous feline painted on the walls of the Tomb of the Lionesses at Tarquinia, where the lioness acts not only as mediator but also as nurse.[7] Nigel Spivey suggests that the lactating felines of Etruscan funerary art allude to breastfeeding, to the feeding of children, to the infantile condition, and to rebirth, and that the passage to the afterworld is expressed simply by a return to the infantile state.[8] J. Bulté shows how in Egypt, images of a lactating lioness (or of a figure with a feline body and a human head) were not uncommon as the subject of faïence (glazed-composition) amulets, which she shows to be associated with happy maternity ("l'heureuse maternité").[9] A lactating feline carved from amber must have been a powerful amulet, one in which the fertility associations and regenerative aspects of the material were enhanced by the subject (especially if the attachments were tiny kittens).

NOTES

1. For the small amber dogs, see 82.AO.161.2 (cat. no. 27).
2. Heraklion Museum 140 (from Tomb 5 at Hagia Triada): Higgins 1980, p. 65, pl. 5B; and A. Marinatos and M. Hirmer, *Crete and Mycenae* (London, 1960), p. 48.
3. Brown 1960, pp. 176–77.
4. D. Bonatz, "Dog," in *Iconography of Deities and Demons: Electronic Pre-publication* (last revision November 4, 2008), www.religionswissenschaft.uzh.ch/idd/prepublication_2.php.
5. Ibid.
6. Andrews 1994, p. 35.
7. See also A. H. Ashmead, "Etruscan Domesticated Cats: Classical Conformists or Etruscan Originals?," in De Puma and Small 1994, pp. 144–64.
8. N. Spivey, "Il Pittore di Micali," in *Il Pittore di Micali*, exh. cat. (Rome, 1988), p. 19.
9. Bulté 1991, pp. 52–55, 84, chap. 7.

Lions' Heads

After rams', lions' are the most numerous of all pre-Roman animal heads in amber. The Getty collection reflects the relative popularity of these two animal subjects: there are four lions' and fifteen rams' heads. A feature found on almost all the amber animal heads is the collarlike finial section in imitation of a metal mount, which shows that the amber examples imitate pendants made entirely of gold (or another precious metal) or of another material such as ivory set in a metal mount.

The list of amber lions' heads now includes the four examples in the Getty Museum, 76.AO.80 (cat. no. 33), 76.AO.81 (cat. no. 34), 77.AO.81.9 (cat. no. 35), and 77.AO.81.10 (cat. no. 36); a pair now serving as the finials of two gold bracelets of Hellenistic date in the Louvre, findspot unknown;[1] a pair from an amber-rich find at Novi Pazar, St. Peter's Church (Etruscan? late sixth century B.C.), and a single lion's head from a tomb at Atenica of the same date;[2] a single(?) lion's head from Canosa(?) in London (British Museum 78);[3] and the dozen or more tiny lions' heads on three Etruscan necklaces, one in Paris from a controlled excavation and two others on the London and New York art markets said to be from Etruria.[4] On each necklace, the lions' heads are joined by an equal number of rams' heads and plain beads.

Parallels for the amber heads include complete amber lions, as Donald Strong noted about a lion's head in the British Museum,[5] and lion-foreparts pendants, one excavated with context from the girl's Tomb 102 at Braida di Vaglio, near Melfi,[6] and another from Armento, now in London.[7]

There is a difference between these Late Archaic lions and the earlier, sixth-century examples in the Getty and elsewhere (i.e., 77.AO.81.2, cat. no. 5; 77.AO.81.3, cat. no. 6; and two in Paris). The later Etruscan and other Italian-made lions of the Late Archaic are characteristically a mélange: the style is a complex blend not borrowed from any one source.[8] The earliest amber lions' heads in the Getty, and a related few amber examples, demonstrate visibly their Oriental antecedents—their Assyrian, Hittite, and East Greek connections. In the case of 76.AO.80 and 76.AO.81, the Assyrian elements are salient through comparison to the beasts of Ashurbanipal. The East Greek elements are brought out by comparison to marble lions such as those from Miletos and Ephesus,[9] to an East Greek terracotta vessel in the shape of a lion protome,[10] and to various carved gems and tiny ivories. Of the last, salient comparisons are the lions (and lions' heads) engraved on a number of Ionian Greek gemstones, notably a plasma scarab signed by Aristoteiches[11] and a pair of Ionian ivory lions' heads from Smêla, whose eyes are inset with amber.[12] The longevity of the Ionian lion types in adornment might be exemplified by the finials of a pair of late-fourth-century B.C. silver bracelets with gold lions' heads from Pantikopaion, now in St. Petersburg.[13]

Comparable Etruscan contemporary lions' heads in other media (which have known provenances) come from southern Etruria, mainly Vulci, but also from Cerveteri, Orvieto, and Tarquinia. These include the tiny gold finials of a blue glass bracelet from Vulci,[14] gold pendants in Berlin, Paris, and Edinburgh,[15] and the larger lions' heads of hammered bronze, especially an example from Orvieto, now in Boston.[16] The above-noted examples have in common a similar anatomical form, a similar modeling of the eye area, tipped-up noses, deeply carved mouths, and comparable ferocious expressions. The Orvieto bronze is remarkably like 76.AO.80 in the length, depth, and form of the head, and in the schema of the dagged mane ruff.

The only extant Etruscan painting of lions' heads is found on one of the many necklaces and garlands hanging from the branches of the sacred grove painted in the first chamber of the Tarquinian Tomb of Hunting and Fishing (see introduction, figure 40). The painted lions' heads correspond closely with extant contemporary gold, ivory, and amber lions' heads. This depiction suggests an amuletic or religious function for the objects.[17] Demonic forces are attracted and repelled; propitious forces are invited.

As noted earlier, a lion of amber combined a potent subject with a potent material, one where the magical aspects of amber and subject, the color of the material and the representation, were matched up. Lion and amber were from earliest times associated with the sun, as was much-prized carnelian, which is very like amber in appearance. Carnelian was traditionally employed in Egypt for pendants of lions' heads and lions' foreparts.[18] Most early Greek and Etruscan gemstones of lion subjects are also carnelian. A subject—such as the lion—that enhanced the inherent danger-averting, protective, and regenerative aspects of these solar materials might have been a straightforward choice.

Amber lions' heads likely served as permanent amulets—that is, as both ornament and magical object. For almost two thousand years before the series of amber and gold head-pendants were produced, lion, lion's-head, and lion-foreparts amulets similar to them in schema and materials had been popular in Egypt. As a symbol of the sun-god, Ra, the lion was, by extension, a symbol of the pharaoh. Worn in life, a lion-subject amulet could symbolize fierceness and bravery, endowing its owner with the same qualities; thus, in life or in death it could function as a protective, danger-averting amulet. The lion-foreparts amulet, unique to the Egyptian late Old Kingdom and First Intermediate periods, was set at the neck to protect the deceased from a second death and endowed the owner with the ability to come forth from the realm of the dead and "become an excellent spirit."[19] The Egyptian beliefs in the regenerative capabilities of the lion (characteristic of all desert dwellers) are assumed to underlie the symbolism of the lion's-head amulets that were current from the late Old Kingdom through the Late Period.[20] Both lion-foreparts and lion's-head types were almost exclusively carved from carnelian.

Because of the Assyrian typological and stylistic aspects of 76.AO.80, the important role played by the lion in adornment and architecture during the Neo-Assyrian period should be recalled: as noted earlier, the lion was a generally magically protective type (known as *urgulû*).[21]

There are demonstrable connections in Greek myth and the material culture of Greece among Apollo, the sun, and the lion. If the trees on the walls of the first chamber in the Tomb of Hunting and Fishing depict Apollo's sacred grove, as some scholars have posited, the necklace with (amber?) lion's-head pendants (and the necklace with ram's-head pendants hanging on another tree) may take on special import. If, however, the dancers of the Hunting and Fishing tomb are directly connected to Fufluns/Dionysos, the god of wine, as Sybille Haynes proposes, the tomb may illustrate Dionysian religiosity.[22] In any case, the dance is apotropaic and purificatory. In either interpretation, the lions' and rams' heads were appropriate to the painted events.[23]

Although the Getty amber lions' heads are generally similar in format and function to one another (as well as to amber heads in other collections and to protomes in other media), each is idiosyncratic. The Getty lions' heads demonstrate both a close relationship to existing types and models and the distinctive hand of individual carvers. 76.AO.80, 77.AO.81.9, and 77.AO.81.10 are pendants, bored laterally in the neck area. 76.AO.81 is perforated with a large rostrocaudal through-bore and has a beveled edge. This indicates a usage different from those of all other amber lions' heads. 76.AO.81 may have served as a finial, such as a finial bead on a necklace, the likely purpose of one of the amber ram pendants in the Getty collection (77.AO.81.12, cat. no. 52). However, the size of the hole and the delicacy of the carving of the lion's mouth suggest other functions; possibly it served as the added spout of a small vessel. In addition to the pendants, there are two other amber lions' heads in the Getty collection, the finials of 77.AO.83 (cat. no. 38), a plaque with a walking boar as its subject.

NOTES

1. For the lions' heads in the Louvre, see Metzger 1991 and the discussions under 77.AO.81.2 and 77.AO.81.3.
2. For the material from Novi Pazar, St. Peter's Church, and Atenica, see Palavestra and Krstić 2006.
3. Strong 1966, p. 81, no. 78, pl. XXX.
4. The amber necklace included in the 1992 Paris exhibition *Les Etrusques et l'Europe* was *hors catalogue*. One of the other necklaces was on the London art market in 1982 (it included nine small lions' heads and two rams' heads of very similar size [height: 12 mm; length: 15 mm; width: 13 mm], and one large lion's head [height: 21 mm; length: 30 mm; width: 24 mm]); the second, with even smaller pendants, is in a New York private collection. I thank B. Aitken for facilitating my study of the New York necklace.
5. Strong 1966, p. 81.
6. For the Braida di Vaglio necropolis, see Bottini and Setari 2003.
7. Strong 1966, p. 80, no. 75, pl. XXIX.
8. Brown 1960, p. 94.
9. See cat. no. 31, n. 3 and n. 4.
10. This early Corinthian example and other, East Greek, examples are generally accepted by scholars as Rhodian. As W. A. Biers, "A

Lion in Kansas City," in Clark and Gaunt 2002, p. 35, outlines, the lion vase has antecedents in metal drinking vessels of the Neo-Assyrian and Achaemenid periods. The secondary(?) use of plastic vases in Etruscan graves may be connected with an amuletic aspect of the containers and the goods. For the magical and medicinal aspects of oils, scents, and "perfumes," see Brunner-Traut 1970 (in "The Archaic and Afterward" in the introduction, n. 214); and L. Manniche, *Sacred Luxuries: Fragrance, Aromatherapy, and Cosmetics in Ancient Egypt* (Ithaca, NY, 1999). See also *Lost Scents: Investigations of Corinthian "Plastic" Vases by Gas Chromatography-Mass Spectrometry*, MASCA Research Papers in Science and Archaeology 11, ed. W. A. Biers, K. O. Gerhardt, and R. Braniff (Philadelphia, 1994).

11. Boardman 1968, p. 134, placed the plasma scarab in East Greece: "The inscription points to the Ionic islands of the Cyclades rather than Ionia. Beazley adduced telling parallels with Cypriot coins of the early fifth century from Amathus and Golgoi. Cf. also the terracottas of East Greece." Boardman 2001, p. 421, includes additional bibliography for the plasma.

12. Boardman 1980, p. 259, fig. 301; G. Minns, *Scythians and Greeks* (repr., London, 1971), pp. 78, 193 (fig. 85), 266.

13. Hermitage P. 1854.289: D. Williams and J. Ogden, *Greek Gold: Jewelry of the Classical World*, exh. cat. (New York, 1994), pp. 156–57, no. 96. Etruscan burials included numerous lion subjects on items of adornment or as singletons. The single gold lion's heads in Paris (Bibliothèque nationale, Cabinet des Médailles, Luynes 502: Brown 1960, p. 105) and Edinburgh (Scottish National Museum of Antiquities FF 34: ibid., p. 106) were likely single finds. The girl's grave Tomb 102 at Braida di Vaglio contained 290 worked ambers, 83 of them figured, but only one lion subject, a pendant in the form of a lion's foreparts. A pair of gold lion's-head pendants (provenance unknown; circa 500 B.C.) in Berlin (Antikenmuseum GI 416/417: Cristofani and Martelli 1983, p. 294, no. 157) and a fragmentary fifth-century gold necklace from Vulci in the Vatican (Museo Gregoriano Etrusco 13542: ibid., no. 156), with three lions' heads and four clasps that once held amuletic objects such as teeth, document the burial of multiple lions' heads.

14. Rome, Museo Nazionale Etrusco di Villa Giulia 59791 (from Vulci): Cristofani and Martelli 1983, pp. 173, 297, no. 174 (with significant comparisons).

15. For the pendants in Berlin, Paris, and Edinburgh, see n. 13, above.

16. Boston, Museum of Fine Arts 55.497. See also the hammered bronze head once on the Rome art market (provenance unknown): Brown 1960, p. 99, pl. XL b.

17. Ibid., p. 106, was the first to draw the connection between existing jewelry and the painted images. On the tomb, see Steingräber 2006, pp. 20ff., especially the illustrations on p. 96 (the nineteenth-century watercolors of the first chamber by G. Mariani) and the detail of one bush on p. 97; P. Romanelli, "Le pitture della Tomba della Caccia e della Pesca," *Monuments et Mémoires, Fondation E. Piot* 1, no. 2 (Rome, 1938); R. Holloway, "Conventions of Etruscan Painting in the Tomb of Hunting and Fishing at Tarquinia," AJA 69 (1965): 341–47, where he is the first to propose the apotropaic aspects of the subjects, including the objects hanging from the branches in the grove, and his more recent "The Tomb of the Diver," AJA 110 (2006): 374–75; L. Cerchai, "Sulle Tomba 'del Tuffatore' e 'delle Caccia e Pesca': Proposta di Lettura Iconologia," *Dialoghi di Archeologia* 5 (1987): 113–34; A. Rouveret, "La tombe tarquinienne de la Chasse et de la Pêche: Quelques remarques sur la peinture de paysage à l'époque archaïque," RA (1992): 170–71; Simon 1998; and Haynes 2000, pp. 228–30. All these sources provide significant comments on the grove, which is discussed later in the section introducing rams' heads.

18. Some archaeological ambers have been identified as carnelian and vice versa: see, for example, Todd 1985 (in "Where is Amber Found," n. 37), generally in reference to the Bronze Age material; and Waarsenburg 1995, p. 426, n. 1111, concerning a carnelian scaraboid from Satricum (Rome, Museo Nazionale Etrusco di Villa Guilia 10809), which was published as being of amber. The color and transparency ranges of carnelian and red jasper are comparable to those of amber. On carnelian in the ancient world, see Caubet 1999.

19. As Andrews 1994, p. 79nn., notes, this amulet type is specifically required by Coffin Text no. 83. The Egyptian lion-foreparts amulet is very close in form to the two extant ambers of the type, British Museum 75 and a sphinx pendant from Tomb 102 at Braida di Vaglio. For the latter, see Bottini and Setari 2003 (with earlier bibl.); and *Treasures* 1998, pp. 224–25.

20. Andrews 1994, p. 65.

21. Black and Green 1992, p. 119.

22. The Dionysian interpretation "is confirmed by the representation of satyrs with drinking horns reclining in the gable of the entrance wall" (Haynes 2000, p. 229). See I. Krausfopf, "The Grave and Beyond," in *The Religion of the Etruscans* (Austin, TX, 2006), pp. 77, 82–83, for a critical summary of arguments for Fufluns/Dionysos in the Etruscan tomb and interpretations of the Tomb of Hunting and Fishing, with bibl., including M. Torelli, *Il rango, il rito, e l'immagine: Alle origini della rappresentazione storica romana* (Milan, 1997); M. Cristofani, "Mystai kai bakchoi: Riti di passagio nei krateri volterrani," *Prospettiva* 80 (1995): 2–14; C. Weber-Lehman, "Spätarchaïsche Gelagebilder in Tarquinia," RM 92 (1985): 19–44; and E. Simon, "Die Tomba dei Tori und der etruschische Apollonkult," Jdi 88 (1973): 27–42. Simon believes the large plants in the tomb represent laurel trees.

23. For further discussion of the ram's-head necklace and the painted grove, see "Rams' Heads" introduction.

33. Pendant: Lion's Head

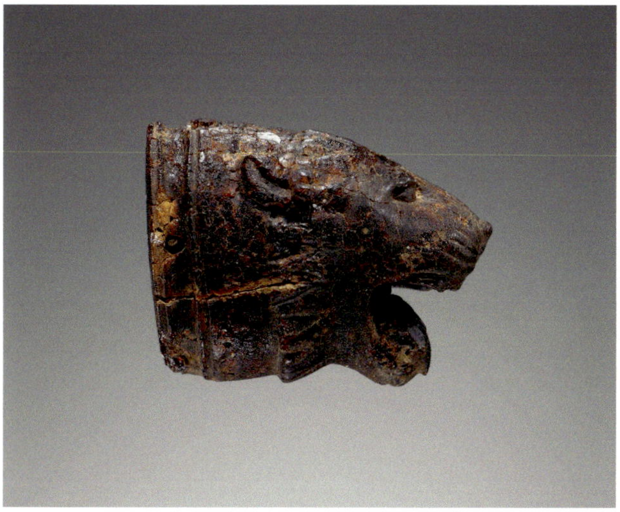

Accession Number	76.AO.80
Culture	Etruscan
Date	550–500 B.C.
Dimensions	Height: 28 mm; width: 22 mm; depth: 38 mm; Weight: 13.6 g
Subjects	Funerary use of amber (also Burial); Lion

Provenance

–1976, Gordon McLendon (Dallas, TX), donated to the J. Paul Getty Museum, 1976.

Condition

The pendant is in good condition, with a generally smooth surface. There is a large nonmodern chip behind the right side of the chin and four broken teeth; there are also many tiny modern chips and cracks and crazing over the entire surface. Inclusions are visible in one crack on the right side of the neck. All over the surface are small blotches of a pale yellow-ocher patina. Degraded amber fills many of the fine cracks. The piece is dull brownish red in ambient light and bright red in transmitted illumination.

Description

The pendant consists of the head and upper neck portion of a ferocious lion, mouth open wide, tongue extended, teeth bared, and ears flattened against the head. The forehead, medial indentation, areas around the eyes, and top of the snout are richly modeled, suggesting a subcutaneous musculature (even if it is not anatomically correct). The eyes are rendered plastically, in profile appearing narrow and hooded and in frontal view open and rectangular, in a gaze of focused concentration. Emphasizing the raised elliptical domes of the eyeballs, their upper lids are undercut.

The pert nose is delineated by two outward-curving nasal wrinkles. The tip of the nose is flat on the frontal plane and tilted upward in profile. A raised philtrum separates the halves of the muzzle; the wrinkled snarl lines on either side are engraved with four lines parallel to one another on the frontal plane and turned up on the sides. The edge of the mouth is smooth. The flews are not turned out. The lips are undercut to reveal the ridges of the gums and the teeth. Only the stumps remain of the upper and lower canines. The chin is narrower and shorter than the upper jaw. The extended position of the tongue (slightly asymmetrical to right) and the cavern of the deeply hollowed palate emphasize the lion's ferocious expression. From the front, the head is oval, with the mane fitting tightly around the head; in profile, the head is long, with the mane flowing from the edge of the face to the edge of the skull. The ruff is composed of a pattern of pinnate, plastically rendered hanks of hair on top; shorter, triangular sections of the mane are carved in the interstices. On the sides and under the neck, the mane is more schematically rendered and the forms flatter. The soft folded-over ears emerge from the mane (the right one is more legible than the left).

The left side of the head is indented along a broad, shallow groove, making the head slightly asymmetrical, and gives evidence of the original shape of the amber blank from which the object was carved. The pendant was perforated with two 2 mm holes, each initiated on one side of the finial (collar?) and exiting caudally through the base. The finial consists of a 3 mm flute flanked on each side by raised fillets. When suspended, the head would have hung downward, with the top of the head showing.

Discussion

The signs of use wear of 76.AO.80 are significant. The pulling troughs on the upper edges of the perforation demonstrate longtime use where gravity or human activity has caused the suspension cord to saw the amber. This is the only animal head in the Getty collection to demonstrate this use; the other ambers in the collection with visible wear by abrasion are female head-pendants. Unlike on the female head-pendants, however, there is little corresponding wear on the face. There is no internal evidence to suggest the manner of the pendant's use, whether as personal adornment (and whether it was talismanically rubbed or kissed) or as an ornament suspended from a branch, animal harness, or architectural element.

Counterparts for 76.AO.80 were found in female graves in controlled excavations, and it is possible that this piece came from a similar context. While representations of both ram's head–wearing women and men are documented in funerary imagery, there is no comparable image of a lion's-head wearer of either sex. Whatever may have been the pre-interment function of the amber, gold, and bronze lions' heads, their ultimate role was funerary, just like the necklaces with lion's heads hanging from the branches in the grove of the Etrusco-Ionian Tomb of Hunting and Fishing (see the "Lions' Heads" introduction).

34. Spout or Finial: Lion's Head

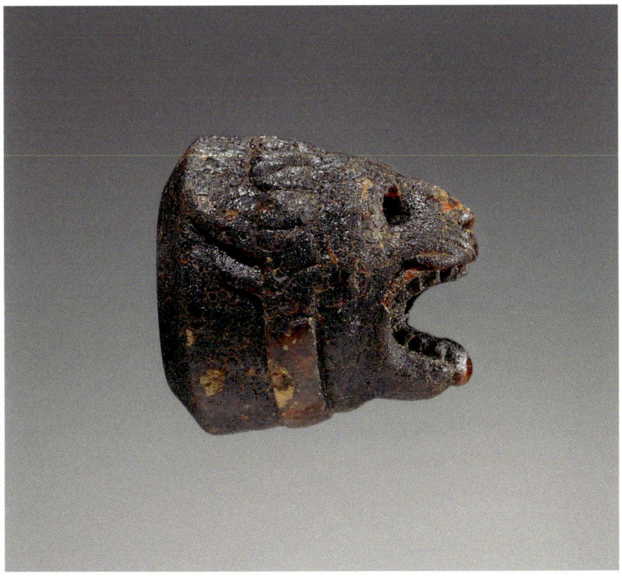

Accession Number	76.AO.81
Culture	Etruscan
Date	525–480 B.C.
Dimensions	Height: 19.5 mm; width: 17 mm; depth: 20 mm; Diameter of through-bore: 6.5 mm; Weight: 3.1 g
Subjects	Lion

Provenance

–1976, Gordon McLendon (Dallas, TX), donated to the J. Paul Getty Museum, 1976.

Condition

The object is intact, in very good condition, with minimal cracking and crazing overall. There are minute chips in the mane on the right side and behind the right eye, and more degraded small areas on the lower right side of the head, marked by a lighter yellow-ocher alteration area. In ambient light, the object is dull brown; illuminated by transmitted light, it is ruby colored. No inclusions are evident.

Description

This amber is worked in the shape of the head and neck of a ferocious lion. The head is almost perfectly square. In frontal view, the object is tubular. In profile, the face tapers toward the nose and chin. The supraorbital areas are plastically modeled, with fine ridges carved to represent the eyelids. The eye sockets are deep, hollow, and asymmetrical: the right one is higher and more circular than the left one, which is amygdaloidal in form. The sockets may originally have held inlaid eyes (ivory and amber?).

The snout slopes down to a step above the short, rounded nose. The nares and nostrils are detailed with fine horizontal grooves. Two additional horizontal engraved lines, wrinkles, cross the bridge of the snout just above the nose. Four engraved diagonal lines rise from each side of the upper lip, wrinkling the lion's muzzle. The philtrum is indented. The mouth of the lion is a large, hollow cavity. The lips are drawn back tightly and undercut to reveal the gums; a full set of incisors and matched small teeth rim the mouth. Below are flaps of the jaw, marked with two incised lines on each side at the front. The end of the tongue protrudes and is slightly transluscent.

The back of the head, which is beveled, is larger in circumference than the mane. The mane is a raised collar. On the upper part of the head, the hanks of hair are rendered as long pinnate lobes, with shorter triangular fillers. On the underside, the mane is not detailed. The lion's large ears commence at the back of the mane, lying flat to the head and pointing straight backward.

There are many tool marks. Incision lines mark the wrinkles of the muzzle and the nose and the details of the mouth. Abrasion is evident on the back of the head and around the lower edge of the collar. Faint drill marks remain inside the bore.

Discussion

The style (and form) of the lion of 76.AO.81 is not closely paralleled by any other lion's head. However, an amber lion pendant in London (British Museum 64), of unknown provenance, bears a familial resemblance. Both look as if they would mew rather than roar. The similarities are most obvious in profile or top view. The rendering of both animals is schematic, and they seem tame. They have simplified ruffs, flat muzzles, and wooden mandibles, and the eyes are crudely modeled bulges.

There is one other figured amber with a through-bore, a stylized head of a ram in the Getty collection (82.AO.161.4, cat. no. 53). Both lions' and rams' heads may have served similar purposes. Perhaps they served as the tiny inset spout of a small vessel, as a decoration on a small container, or as the finial bead of a necklace or other item of adornment. The faïence feline-head amulets on an Eighteenth Dynasty string of beads in New York are one prototype for the last.[1] An object in the Norbert Schimmel collection, a unique Egyptian blue (light) finial of a lion's head holding in its open jaws a Negro head, provides another idea. If reconstructed on this model, the Getty lion's mouth may have held a human leg or head, part of an animal, or small prey. Greek models for such a use are the incorporated animal protome spouts on terracotta and metal vessels, the earliest dating to the seventh century B.C. The lion's-head spouts on three gold rhyta from the fourth-to-third-century B.C. Panagyurishte Treasure (Bulgaria) are a later model for such a hypothetical use.[2]

The lions of 76.AO.81 and the London pendant (BM 64) fit in well with the group of Late Archaic–period Etruscan lions, small gold heads, and various objects of bronze brought together by W. L. Brown.[3] As he showed, these lions show strong stylistic relationships to earlier Etruscan lions and Greek models from both Ionia and Magna Graecia. The closest among them are the gold lions' heads and a bronze lion once on the Rome art market. They have a comparable form of the mane and the lower face, including the semicircular flaps, and deep-set eyes. A bronze head in Boston from Orvieto has a similar disposition of the teeth and lower jaw (including the lower flaps), and the lion's head of a *lacunaria* in Perugia from Castel San Mariano has a remarkably similar profile and the same short face, set-in eyes, and small tongue.[4] 76.AO.81 is also effectively compared in type to the heads of the lions and chimeras on a group of Late Archaic Etruscan gold fibulae.

The distinctive physiognomic elements of the bronze comparisons place 76.AO.81 near to Perugia. The style of the pendant also shows a connection to Cerveteri and Vulci, the locations where the gold pendants are thought to have been made.

Whether 76.AO.81 was an ornament finial or a tiny spout, the lion's head would have played its traditional roles in protection and the aversion of danger. If liquid poured from its mouth, such a use would relate it to the popular lion's-head waterspouts on sacred (and secular) buildings[5] and to fountain heads.

NOTES

1. New York, Metropolitan Museum of Art, 1926 26.7.1364. These are pictured in *Hatshepsut* 2005, p. 203, no. 122.
2. Plovdiv, Archaeological Museum 3200–3202: *Ancient Gold: The Wealth of the Thracians, Treasures from the Republic of Bulgaria*, exh. cat., ed. I. Marazov (New York, 1998), p. 142, nos. 68–70.
3. Brown 1960, chap. 5.
4. Perugia, Museo Archeologico Nazionale dell'Umbria 1390: Brown 1960, p. 101, pl. XLI.
5. For discussion of the historical and typological background of lion's-head waterspouts in Italy, with a succinct analysis of the two diverse traditions there, Etrusco-Italic and Magna Graecian Sicilian, and with a study of their relationships to earlier and contemporary examples, see P. Pensabene, *Terracotte del Museo Nazionale Romano I: Gocciolatoi e protomi da sime* (Rome, 1999), pp. 19–24.

35. Pendant: Lion's Head

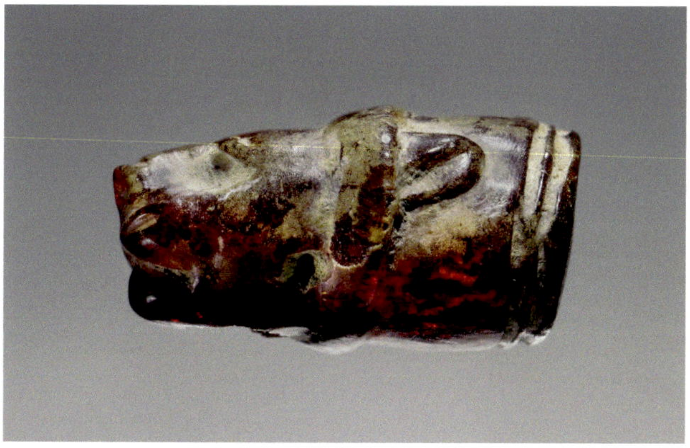

Accession Number	77.AO.81.9
Culture	Etruscan
Date	525–480 B.C.
Dimensions	Height: 9 mm; width: 17 mm; depth: 9 mm; Weight: 1.3 g
Subjects	Lion

Provenance

–1977, Gordon McLendon (Dallas, TX), donated to the J. Paul Getty Museum, 1977.

Condition

The lion's head is intact and in excellent condition; the surface is smooth and firm. There are only a few small chips, on the throat area and on the back of the mane. Pale yellow mottling covers much of the surface. Pale yellow-ocher patina (degraded amber) surrounds the eyes and lies in other crevices. The amber is dark brownish red in ambient light. In transmitted light, it is translucent and bright reddish orange. One long inclusion runs the length of the head.

Description

In form and profile views, the narrowness and flatness of this head are striking. The height is proportionally low in comparison to its width and length. From the dorsal view, presumably the main view when the pendant was suspended, the impression of a compressed composition disappears. In fact, it reads very much like that of the head of 77.AO.81.10 (cat. no. 36), which is three-dimensional.

From its general structure to the details of the eyes, mane, and chin, the head is a combination of flattened and plastically modeled forms. Engraved lines mark the mouth, the muzzle wrinkles, the circumscription of the eyes, the eyebrows, and the fillet at the back. The slanted-back eyes, of an almost rectangular form, are deeply carved, leaving the eyeballs protruding above the cheek plane. The folded-over ears, of triangular shape with soft points, arise from the back of the mane. The snout is smooth and softly curved, stepping down to the tip of the nose, which tilts upward. The nose-to-chin line is even and tapers in slightly. The nostrils are drilled.

The muzzle itself is flat and wrinkled, with five folds of flesh that are rendered by uneven diagonal grooves radiating from the mouth rather than from the philtrum. Three ruffled bulges break the line of the mouth on the sides. The underside of the jaw and throat area is flat. The mane is schematically rendered as a smooth collar. At the back termination are two parallel grooved lines.

Patterns of tiny scratches caused by abrasion are visible on the back of the piece. Two holes 1 mm in diameter are bored into the base, originating near the outer edge, sloping downward and inward toward the center. A pair of 1.5 mm–diameter perforations is drilled under the throat area: one bore originates below the right ear, and a second begins below the left corner of the jaw. These latter perforations might have been secondary to those on the base. When suspended by the base borings, the lion would have hung nose downward. If it had been suspended from the cross-bore through the head, it would have hung nose upward.

Discussion

The pose of the head and neck encodes a mixture of lion behaviors: repose is implied both by the closed mouth and by the relaxed eyes; the inclination of the head suggests motion; and the position of the ears, which lie flat against the head, signals a display of anger.

Although related in style to the following head (77.AO.81.10), especially in dorsal view, this lion differs in small physiognomic details from other representations in amber and other media. Distinguishing features of this head are the flatness of the pendant, the flat collar of mane devoid of fur markings, the muzzle wrinkles, and the ruffled upper lip. According to W. L. Brown, the last might be considered a regressive trait in lion representation of the period.[1] There appears to be no other example of an amber pendant perforated in this manner. The only parallel for such a system known to me is an Elamite lapis lazuli bull's head from Susa.[2] On that head, when the cross-bore in the head section is connected to the base set, the resulting holes make an anchored suspension loop.

NOTES

1. Brown 1960, passim.
2. P. O. Harper et al., *The Royal City of Susa: Ancient Near Eastern Treasures in the Louvre*, exh. cat. (New York, 1993), p. 152, no. 97 (with bibl.).

36. Pendant: Lion's Head

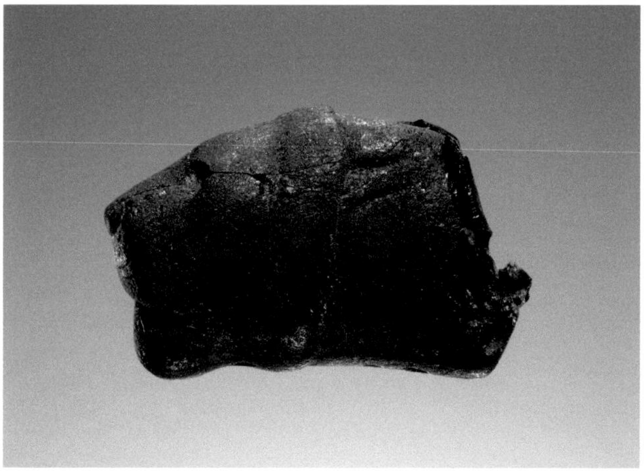

Accession Number	77.AO.81.10
Culture	Etruscan
Date	500–480 B.C.
Dimensions	Height: 15 mm; width: 16 mm; depth: 26 mm; Diameter of suspension hole: 2 mm; Weight: 3.6 g
Subjects	Etruscan culture; Jewelry; Lion

Provenance

–1977, Gordon McLendon (Dallas, TX), donated to the J. Paul Getty Museum, 1977.

Condition

The surface of the pendant is firm but rough because of the degradation of the amber. The figured part of the pendant is intact. On the reverse are an old loss and more recent breaks on the back of the neck and top of the head. On the obverse, the many fine cracks in the surface layers extend through the piece; there is a large crack through the left eye and left part of the head. There are no apparent inclusions. The surface is a dull matte yellow-brown to reddish brown on the old surfaces and a rich, glassy brownish red-orange on the newly broken areas. In transmitted light, the amber is transparent and bright red-orange.

Description

This pendant represents a calmly posed lion whose ears lie flat against its head. The neck portion is cylindrical. The modeling of the head is smooth. The forehead is gently rounded, and the snout slopes to the down-curving, rounded nose. The eye sockets are bored fairly deeply, which suggests that they might have been inlaid. The muzzle is relatively large and full, divided longitudinally by a shallow indentation; no philtrum is indicated. Fine lines are engraved on the muzzle to indicate creases. Similar lines detail the ears, the closure of the mouth, and the slits of the nostrils. The chin is distinctly bulbous.

Raised only slightly from the surface of the head, the collar of mane is carved with an undulating edge to suggest the texture of the ruff. There are finely engraved lines suggesting the hairs of the mane. Large ears, with the helixes flopped over, point downward (or backward?), and commence at the back of the ruff.

The very slight asymmetry of the piece (the lion's head cants slightly to its left) might indicate something of the original shape of the amber blank. There is a perforation, 2 mm in diameter, for suspension that passes laterally through the pendant about 2 mm from the surface of the base. It appears that a piece of the base was broken off in antiquity, leaving exposed a transverse section through the suspension perforation. The pendant would have hung nose downward.

Discussion

Within the corpus of lions' heads, this example stands out. It has no close parallel in amber or any other material. Although the face exudes serenity, the representation also includes more aspects of lion behavior. The flattened ears are in an attitude usually associated with roaring, leaping, and anger or fear.

The artistic style of this cat reveals its breeding: its artistic-morphological characteristics depend on earlier traditions of lion representation. The softness of modeling is reminiscent of East Greek carving, as a comparison to some Lydian felines makes clear.[1] Ultimately, the pointed, folded ear laid back in anger on the head is borrowed from Assyria. The narrow back of ruff has mainland Greek and Magna Graecian parallels, and the lack of a back mane is common in Italian creations.[2]

77.AO.81.10 is comparable to a series of Etruscan bronze objects dated by W. L. Brown to the Late Archaic period, comparisons that locate this head to north-central Etruria. Other bronzes might situate this amber pendant more precisely, to the early fifth century B.C., and to Orvieto specifically. Among these comparanda are the hammered bronze finial of a chariot pole from near Orvieto, now in Boston (dated by M. Comstock and C. Vermeule to the early fifth century),[3] and, from the same find, the four couchant lions of a wheeled brazier.[4]

The stepped taper of the pendant's base might be evidence that the head was set into a metal mount. This head then could have served as a finial on a bracelet or as a pendant in a necklace.

NOTES

1. See C. Ratté, "Five Lydian Felines," AJA 93 (1989): 379–93. The piece might also be compared to any number of Egyptian New Kingdom images of lioness divinities.
2. Brown 1960, chaps. 5–6, passim.
3. Boston, Museum of Fine Arts 55.497 (Gift of the Estate of Dr. Jacob Hirsch): Comstock and Vermeule 1971, p. 484, no. 712; and Brown 1960, pp. 99–100, pl. XLc.
4. Brown 1960, p. 91, pl. XXXVIIb (formerly in the collection of Dr. Jacob Hirsch, New York).

Boars

37. Pendant: Foreparts of a Recumbent Boar

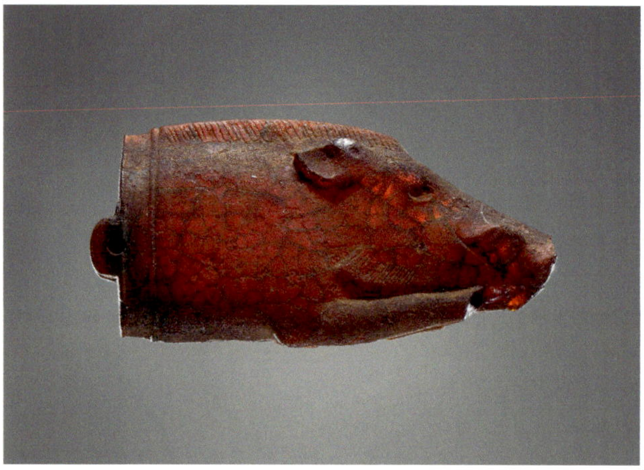

Accession Number	76.AO.84
Culture	Etruscan
Date	525–480 B.C.
Dimensions	Length (rostrocaudal): 50 mm; depth: 13 mm; height: 24.5 mm; Weight: 9.3 g
Subjects	Amulets; Boar; Ionia, Greece (also Ionian, Greek)

Provenance

–1976, Gordon McLendon (Dallas, TX), donated to the J. Paul Getty Museum, 1976.

Condition

The pendant is in good condition, although it has suffered some surface degradation and losses. The surface is grainy, with cracks and crazing overall. The tip of the snout, a section of the right orbital area, and the tips of the hooves, especially on the right, are missing. There are small chips on the tips of both ears, the right side of the mandible, and the underside of the left leg. Except at the new breaks, the piece is opaque, the surface brownish tan with a pale yellow patina; in transmitted light, the reddish brown translucency of the inner structure is visible at the breaks. There are no visible inclusions.

Description

The pendant is carved fully in the round, including the head, forelegs, and upper trunk of a wild boar. The body is truncated about midway, at a point immediately behind the caudal end of the mane. The animal is in the pheonotype of repose, with its forelegs drawn up under its head. In side view, the narrow body portion of the pendant is rectangular and the head long and triangular. In frontal view, it is somewhat flat.

The anatomy of the boar is sleekly modeled, with the details of musculature revealed by subtle surface modulations in the region between the shoulder and the base of the skull. The surviving eye cavity is shallow and unfinished at the bottom edge. The lip is sharply upturned where it meets the protruding tusks and other teeth, which must have numbered three on each side, although none remain on the damaged side. The sensitively carved, leaf-shaped ears lie close to the head and point straight back. Incised marks representing bristles mark the jowls. The ridge of mane is smooth and rounded and projects only slightly above the backbone; it is incised with fine diagonal striations that point outward. The shoulder is minimized in comparison to the head and plump forelegs.

The narrowness of the animal in frontal view suggests that the artist was constrained by the thinness of the original amber blank from which this piece was worked. When suspended, the animal would have been seen in profile and hung head downward.

Discussion

There exist a number of boar-subject amber objects, but no close parallel for this pendant. The boar is an uncommon subject in pre-Roman amber and an infrequent subject generally in Greek and Etruscan jewelry; it is far more common in other media, as a figurative subject on arms and armor or ceramic vases or stamped on coinage. Except for 77.AO.83 (cat. no. 38) and a fragmentary plaque in New York,[1] the other wild-boar subjects in amber are pendants or the decorative bows of fibulae. In addition to this example, there are three other pendants in the form of boars, all in the recumbent position: a pendant in New York, a more schematic example in London (British Museum 77), and a softer type, one of the numerous pendants from Tomb 102 at Braida di Vaglio.[2] The boars-as-bows of fibulae are standing: an example in Cleveland arches its back (the animal is similar in form and style to 76.AO.84);[3] the fibula bow in a London private collection is stiff and simplified in modeling and unrelated to any other Greek or Etruscan representation.[4] Also probably meant to represent a recumbent boar is a small amber in London (BM 76), titled by Donald Strong "pendant in the form of the foreparts of a pig."[5]

In style, these wild boars invite comparison with contemporary Archaic representations in other media—small bronzes, coins, gems, and paintings on vases. The boars of 76.AO.84, 77.AO.83, the New York plaque, and the Cleveland fibula bow decoration compare especially well with the boar represented on an amygdaloidal serpentine gem in Berlin, which has been connected to Melos.[6] The ambers and the gem share common forms of eyes, ears, whiskers, and mane, to cite but four key traits. They also share many sculptural similarities with a number of small sculpted bronze boars: a winged boar at bay in the Norbert Schimmel collection;[7] a close parallel in the Hispanic Society, New York;[8] a relative of theirs, although later and without wings, in the Walters Museum in Baltimore;[9] and two early-fifth-century bronzes, a Campanian (possibly) wild boar at bay from a large, circular vessel in Boston[10] and a standing boar, one of the Etruscan votive bronzes from the votive deposit of Fonte Veneziana, Arezzo.[11] The two winged boars in New York present both dating and location problems, as Hans Hoffman outlined. The stylistic conventions for (the few extant) sculpted boars are virtually identical with those of the boars painted on Early and Middle Corinthian pottery. The absence of a break in the bronze boars' manes, their taut modeling, and the rich use of incision argue against an Ionian origin for the bronzes. Hoffman concluded that the two are Greek and to be dated to the second half of the sixth century, a general attribution and date appropriate also for 76.AO.84.

76.AO.84 may have come from the same original context as the kore (76.AO.77, cat. no. 8) and three of the ram's-head pendants in the Getty collection: 76.AO.82 (cat. no. 39), 76.AO.83 (cat. no. 40), and 77.AO.81.7 (cat. no. 41). This is plausible because of the similar condition and the similar style of the five carvings. It is suggested in the entry for the kore that it is a work of a South Ionian artisan. The resemblance of 76.AO.84 to the Melian-attributed Greek serpentine gemstone might support an East Greek connection. Further comparisons with early East Greek coinage of the late sixth to mid-fifth century B.C. with boar devices, such as that of Clazomenae and Samos, should bear fruitful results.

The wild boar plays an important role in European myth, for it was perhaps the most ferocious wild animal in Europe once lions were extinct. In Greek myth, Peleus was chased up a tree by a lion and a boar; Atys (the son of Croisus), Attis, and Adonis were slain by boars. Ultimately, it was the boar sent by Artemis (furious that King Oeneus of Calydonia had forgotten to include the goddess in his annual sacrifices to the gods) that led to the death of Meleager.

As a symbol, device, or amulet the boar could work amuletically by assimilation: the wearer, object, or building would take on the characteristics of the animal. In direct magic, a boar could frighten off danger and by extension was protective. The boar is a common blazon on Archaic armor; it might be woven into the decoration of a divine robe or mounted as protection on a building. Boar waterspouts on the corners of a *sima* might play an even more important role than lions' heads flanking the sides of a temple.

Although many ancient representations of the foreparts of an active boar, including winged boars, exist—on gems, on coins, and as the blazons of shields painted on Archaic-period Greek vases—much rarer is the recumbent boar, whether complete or excerpted, as in the case of the foreparts. In general, a recumbent animal is a tamed or sleeping animal, not an active one, and this may be key to its magical function. The 76.AO.84 boar may be latently powerful; it rests in readiness, its aggressiveness to be called upon when needed to viciously protect its charge and attack danger.

NOTES

1. Fragmentary boar plaque: Metropolitan Museum of Art 1992.11.17, Purchase, Renée and Robert A. Belfer Philanthropic Fund, Patti Cadby Birch, and The Joseph Rosen Foundation Inc. Gifts, and Harris Brisbane Dick Fund, 1992.
2. Boar pendant (Etruscan or Italic) in New York: Metropolitan Museum of Art 1992.11.16, Purchase, Renée and Robert A. Belfer Philanthropic Fund, Patti Cadby Birch, and The Joseph Rosen Foundation Inc. Gifts, and Harris Brisbane Dick Fund, 1992 (*Art of the Classical World* 2007, pp. 295, 473, no. 341). Boar-foreparts pendant in London: British Museum 77 (Strong 1966, p. 81, no. 77, pl. XXIX). Pendant from Tomb 102, Braida di Vaglio: Bottini and Setari 2003, p. 40, pl. XLV, no. 133.
3. Cleveland Museum of Art 1978.124.
4. Unpublished.

5. Strong 1966, p. 81, no. 76, pl. XXIX.

6. Berlin, Antikensammlung: A. Fürtwangler, *Die antiken Gemmen* (Leipzig, 1900), no. 92; J. Boardman, *Island Gems* (London, 1963), p. 23, no. 2, pl. 1; E. Zwierlein-Diehl, *Antike Gemmen in deutschen Sammlungen 2. Staatliche Museen Preussischer Kulturbesitz, Antikenabteilung, Berlin* (Munich, 1969), no. 115, pl. 272; and Hampe and Simon 1981, p. 196, figs. 306–7.

7. *Norbert Schimmel Collection* 1974, no. 24 (H. Hoffman).

8. Ibid. cites as parallel the bronze boar in the Hispanic Society, New York; for which, see A. Garcia y Bellido, *Hispanica Graeca* 2 (1948): 28, 95–96, no. 13.

9. D. K. Hill, *Catalogue of the Classical Bronze Sculpture in the Walters Art Gallery* (Baltimore, 1949), no. 275, also cited by Hoffman (see n. 7, above).

10. Boston, Museum of Fine Arts 10.163 (from Sirolo): Comstock and Vermeule 1971, p. 309, no. 435.

11. Florence, Museo Archeologico Nazionale 294: Colonna 1985, pp. 174–79, no. 10.2; P. Bocci Paccini, "Alcuni bronzetti arcaici della 'Fonte Veneziana,'" in *Studi di Antichità in Onore di Guglielmo Maetzke* (Florence, 1984), pp. 119–23; and P. Bocci Paccini, "La stipe delle Fonte Veneziana ad Arezzo," StEtr 48 (1980): 73–91 (with additional bibl.); also illustrated in *Civiltà degli Etruschi* 1985, p. 251, fig. 3.25. Like many of the other bronzes in the deposit, the boar is strongly Ionian in style.

38. Plaque: Addorsed Lions' Heads with Boar in Relief

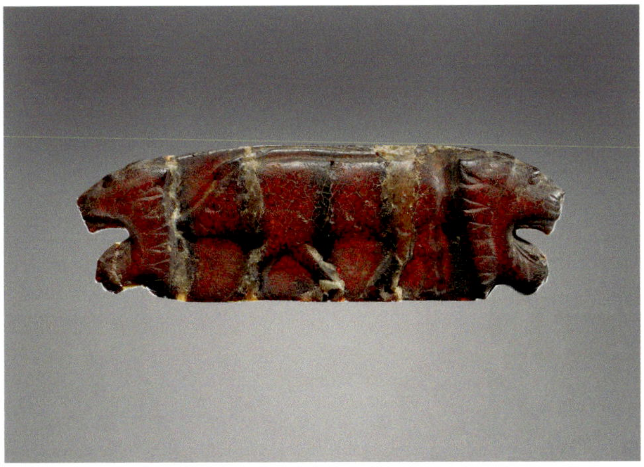

Accession Number	77.AO.83
Culture	Etruscan
Date	500–480 B.C.
Dimensions	Height: 36 mm; width: 82 mm; depth: 12 mm; Length of boar: 46 mm; Height of left lion's head: 23 mm; Height of right lion's head: 26 mm; Weight: 19 g
Subjects	Artemis; Boar; Etruscan culture; Lion

Provenance

–1977, Gordon McLendon (Dallas, TX), donated to the J. Paul Getty Museum, 1977.

Condition

Before its entry into the museum, the piece was broken into five pieces and repaired. The breaks, along or through four of the five vertical perforations, were filled with a tan resinlike material, creating a banded appearance. The surface is stable, its upper surface layers mottled yellow-brown, with sporadic pitting, small chips at the edges of the breaks, and fine cracks overall. Yellow residue of degraded amber appears in cracks around the outline of the boar and behind the lions' heads, around the lions' eyes, and in their mouths and manes. After it arrived at the museum, the object was treated with amber-oil distillate, which probably darkened its color. The background area is more matte in appearance than the (shinier) boar. The piece is darkish tan to brownish red in ambient light and is a dark ruby-red in transmitted light. There are no visible inclusions.

Description

In the central obverse panel of the flat, rectangular piece, a left-facing boar is carved in shallow relief (on the reverse, the central area is plain). The boar fills the panel, its forelegs extended widely, with its right foreleg back and its left forward. The rear legs are closer to each other, with the right hind leg at the back and the left hind leg forward.

The boar's heavy head is in the relaxed position that is associated with walking; its chin is on the same level as its abdomen. The top of the head dishes in slightly above the snout, which is long and pointed. The large, squarish almond-shaped eye is set low in the face and is framed above by a heavy bulge. Only the lower part of the ear remains, the top obscured by one of the repairs.

A ridge of bristles extends from the forehead and appears to continue all the way to the lower back. Swellings at the shoulder and above the elbow suggest the animal's powerful musculature. The full-tipped tail hangs down behind the animal's hocks.

The heads of two open-mouthed lions, facing outward, flank the boar. While the lions' heads are similar, they are not identical. The two obverse profiles are more carefully detailed than are those on the reverse. The head composed of the obverse right and reverse left is larger and more fully modeled; the ruff is more plastically rendered, and it has fully defined ears on both sides. The head composed of the obverse left and reverse right is smaller and more sketchily treated, and its volumes are simplified.

Despite all this, the two lions' heads may be considered typologically to be of one style. The face is short and the muzzle is flat and slightly inclined. A bulge above the protrusion of the eye and the depression of the cheek below the eye socket are rendered naturalistically. The snout continues the incline of the head, but the nose tilts up very slightly at the tip. The muzzle is divided by an indented philtrum, and the folds are separated by four wrinkles. The lips are drawn back, revealing a set of small teeth, both sets of canines, and the gum ridges. The tongue is outstretched. The lower lip is boldly scalloped, and the chin is short and globular. A sharp indentation sets off the angle of the mane from the jaw. The collar of mane is wider at the top, tapering and curving inward at the neck. The individual hanks of hair, formed into unevenly shaped triangles, are set off from the face by shallow indentations but still give the impression of growing from the head; at the back, they are squared off. On the edges are the compacted frontal faces of the lions.

Engraved tooling lines are found on the lions' manes, teeth, eyes, noses, nostrils, and muzzle wrinkles. Such marks are also found on the boar's back and above its hooves. Linear abrasions (which appear to be file marks) are seen on the lions' jaw flaps and manes, and scraping or abrasion marks appear around the contour of the boar.

The piece is perforated vertically with five holes. Each hole was drilled from top and bottom: the holes meet midway. From left to right on the obverse, the holes in set a (above left lion's ear) are 2 mm top and bottom; set b (behind boar's ear), 2 mm top, 2.3 mm bottom; set c (at boar's mid-body), 2 mm top, 1.75 mm bottom; set d (at rear termination of boar's bristles), 2 mm top, 1.8 mm bottom; and set e (behind lion's mane), 2 mm top and bottom. Neither the top nor the bottom edge of the piece is truly flat. The upper edge of the pendant is slightly convex, while the lower has soft, concave undulations.

Discussion

There is no parallel for this plaque in amber or in any other material. It is one of three pre-Roman-period examples of figured amber plaques with low-relief carving. The two others are the Getty *Addorsed Sphinxes* (78.AO.286.2, cat. no. 28) and a hemispherical plaque with a recumbent boar in the Metropolitan Museum of Art.[1] The Getty plaque is very like ivory carving and can be compared to a wide number of Etruscan ivory and bone plaques.[2] In the modeling of the figures and the depth of the carving, the plaque looks back to the ninth-century ivory plaques, from Assyria and North Syria especially, made for insertion into furniture.

The lions of 77.AO.83 are generally comparable to a number of Etruscan lions of the Late Archaic, a large and varied group gathered together by W. L. Brown, with the closest parallels dating to the fifth century.[3] In style and scale, the plaque finials are related to two small, possibly Campanian bronzes found at sites on the mid-Adriatic: a lion in Boston (from the marshes of Sirolo),[4] which was once the rim or shoulder decoration of a *lebes* or another large, circular vessel, and a lion from a *dinos* rim from Amendolara (in Ancona).[5] The bronzes and amber lions' heads have in common facial shape and modeling, especially around the eyes, and have similarly scalloped flews, angular breaks at the jaws and mane, and flamelike locks.

From the same vessel as the Boston lion is a bronze boar at bay. It is related in type and style to the boar of 77.AO.83, as is the bronze boar votive from the extraurban deposit of Fonte Veneziana, Arezzo, in Florence.[6] Each is sculpted with a sympathetic simplification of the features and softness and subtlety in the modeling. The boar of the amber plaque and the two bronze boars are the slim creatures of the sixth century, not the heavier-set creatures of fifth-century Greek gemstones.

The boar and the lions of 77.AO.83 are tame beasts. The boar gives little indication of its legendary strength or its viciousness when at bay; instead, the most dangerous beast to roam the ancient countryside and the fiercest adversary of the most skilled hunter trots meekly to stage left. The lions of 77.AO.83 make the motion of roaring but instead only drop their nutcrackerlike jaws, tamely opening their mouths.

Based on the form of the amber plaque, the horizontal orientation of the subjects, and the five large vertical perforations of equal size and placement that transverse it, it might be supposed that the plaque was part of a complex object. Strands must have run through the piece, with those above connecting to another section of this object, a necklace, or a fibula. Those below may have attached another section, or perhaps terminated in small pendants. Alternatively, the plaque may have been the main portion of a comb. If the body of a comb, it may have resembled the earlier ivory comb with sphinxes from the Tomb of Ivories, Marsiliana d'Albegna.[7]

Why the combination of boars' and lions' heads on an amber plaque? In northern Europe, in the Greek-speaking world, and in Italy, but not in Egypt, the boar was an especially important subject, in particular for protective purposes.[8] It is common as a device on armor and figures as the subject of waterspouts (but only on the corners of *simas*). Lions could serve as protective and danger-averting symbols and represented the sun and regeneration. The boar was the most ferocious animal in the pantheon of the hunt. The most famous boars recounted in Greek tales were the ravaging boars of the countryside, best known from the important hunts organized to destroy them, which include Herakles' killing of the Erymanthean boar and the hunt for the Calydonian boar. And as Ivan Mazarov notes,

> *Like the wild youth, the boar is a largely undifferentiated creature: it is both herbivorous and predatory, and although it lives on land, it prefers marshes and swamps. Since ancient times, the boar has been a recurring metaphor for the ferocious warrior. It has also signified death.... The boar could test the hero's virtue. If he won, the powers of the defeated antagonist passed on to him. In Thrace, the principal adversary of a pretender to the Thracian kingship was the wild boar.*[9]

The imagery of this amber plaque might have been suitable for both male and female owners. If it was found in a woman's or child's tomb, the association of the boar in the feminine realm would be of importance. The pendant may refer to Artemis in her aspect as Agrotera (or to the Etruscan Artumes). As the goddess of the hunt, the divinity roamed mountain forests and uncultivated land hunting for wild animals, including especially lions but also panthers, hinds, stags, hares, and boars. As a wild and fierce animal, the boar was regarded as a symbol of one side of the goddess's nature, capable of unleashing sudden, violent destruction on humans and property, and her sacred buildings were often decorated with images of boars' heads.[10]

The joining of lion and walking boar on 77.AO.83 is significant. The two animals are featured in combat or paired in calmer modes throughout ancient art. The subjects are common shield devices in Greek and Etruscan art, both individually and in combat. The two animals are also linked in one of the early stories of Greek myth: Adrastus, the king of Argos, had learned from an oracle that he must yoke his daughters to a boar and a lion. He then saw Polynices and Tydeus, an exile from Calydon, fighting. One of them had a boar painted on his shield, the other a lion. Adrastus immediately recognized the true meaning of the oracle, stopped the fight, and married his daughters to the combatants. In the *Iliad*, the combat of two great warriors is likened to that of a boar and a lion.[11] Artemis, alternatively, may be the key to this representation in amber. She is associated with both wild animals and was a protector of women and girls, one of whom was likely the recipient of this amber. The use of boars and the pairing of boars with other animals of the hunt might have been the most powerful kind of "fighting fire with fire" imagery.

NOTES

1. Metropolitan Museum of Art 1992.11.17, Purchase, Renée and Robert A. Belfer Philanthropic Fund, Patti Cadby Birch, and The Joseph Rosen Foundation Inc. Gifts, and Harris Brisbane Dick Fund, 1992.
2. For Etruscan ivory carving generally, see Y. Huls, *Ivoires d'Étrurie* (Brussels, 1957); and M. Martelli,"Gli avori tardo-arcaici: Botteghe e aree di diffusione," in *Il commercio etrusco arcaico: Atti dell'Incontro di Studi, 5–7 dicembre 1983* (Rome, 1985), pp. 207–48.
3. Brown 1960, chaps. 6, 9.
4. Boston, Museum of Fine Arts 10.162 (circa 480 B.C.): Comstock and Vermeule 1971, p. 309, no. 435.
5. C. Albizzatti, *Dedalo* 1 (1920): 153–61, pl. 157; and H. Jucker, AA (1967): 628–29, fig 17f.
6. Florence, Museo Archeologico Nazionale 294: see cat. no. 37, n. 11.
7. Grosseto, Museo Archeologico e d'Arte della Maremma 93437: Bartoloni et al. 2000, p. 133, no. 89; and *Civiltà degli Etruschi* 1985, no. 3.14.23.
8. The boar is rarely represented in Egyptian art, but in at least two wall paintings the animal is connected with Seth, the archenemy of Osiris. The animal was almost never the subject of body adornment, but some small amulets are known: figurines with a sow nursing her litter were intended to endow their owners with fecundity, fertility, and good luck, a subject carried over into Greek engraved hard stones (see Andrews 1994, p. 26; and P. F. Houlihan, *The Animal World of the Pharaohs* [New York and London, 1996], pp. 25–28).
9. I. Mazarov, ed., *Ancient Gold: The Wealth of the Thracians, Treasures from the Republic of Bulgaria*, exh. cat. (New York, 1998), pp. 59–60.
10. See Bevan 1986.
11. *Iliad* 16.823. On the imagery of boars and lions in Archaic Greek art, see, for example, F. Hölscher, *Die Bedeutung archaischer Tierkampfbilder* (Würzburg, 1972); and for the hunting of animals, J. K. Anderson, *Hunting in the Ancient World* (Berkeley, 1985).

Rams' Heads

The subject of a ram or ram's head for adornment, amulets, and amuletic jewelry is age-old, and the animal is one of the earliest to appear in these forms. Rams' heads are among the most numerous of pre-Roman amber subjects and range considerably in style.[1] The earliest ones can be dated to the third quarter of the sixth century B.C., but the first documented examples date to the last decade of the sixth. There are a number from fifth-century contexts and a group from the fourth century. Some, such as 76.AO.82 (cat. no. 39), are minor masterpieces of Archaic art, while others are vaguely blocked out and defined with schematic scratchings. The differences among the rams' heads throughout the period of their production suggest that they were made by any number of carvers, some highly skilled, others not, for acquirers from all parts of the Italian peninsula and areas within close sailing distance. The findspots include central and south Italy, two sites in ancient Illyria, and Alalia (Corsica). Many others that can be dated by style to the sixth and fifth centuries have come to light without secure documentation. Of the examples in the Getty collection, three are dated here by style to the sixth century, and twelve examples to the fifth.

Amber rams' heads are usually found in pairs or larger numbers, along with other amber objects—figured and plain pendants, beads, and fibulae. In the sixth century, rams' heads were the most numerous of all figured subjects in amber. They were joined with korai; female head-pendants; birds; heads of lions, boars, gazelles, and horses; and floral and shell subjects. Ram subjects in amber are not documented in the company of human or humanoid male subjects—satyrs, Dionysos, Herakles, or any of the other unnamed bearded or unbearded males represented in amber. Rare, too, is the interment of ram subjects with demonic subjects. Two exceptions are the rams' heads from Tomb 102 at Braida di Vaglio, which also included a sphinx, and the rams' heads in the Petit Palais, Paris, from the Sala Consilina burial, which also included a range of flying figures, perhaps harpies or sirens.[2]

The parallels for the amber heads in other media have led to the conclusion that the earliest amber rams' heads were made in the second half of the sixth century, in an ambient where Greek specialists (Laconians, Ionians, Islanders, and other Eastern Greeks) were working, but the locations are elusive. Comparanda for the sixth-century amber rams' heads include the rams on the handles of Greek bronze vessels, especially Laconian ware, and on other kinds of Greek bronzes—statuettes, mirrors, and other utensils; the sheep subjects engraved on Ionian Greek gems; a singular silver ram pendant in New York; the pairs of rams on a silver East Greek oinochoe found in Lydia; and the rams' heads struck on coinage from Cyzicus, Cyprus, Lesbos, Delphi, and Melos.[3] Plastic vases in the form of recumbent rams and ram protomes, the small plastic ram protomes on Greek and Etruscan ceramics, and the drawings of rams on black- and red-figured vases of various fabrics also provide a mine of further comparanda.

Most of the amber rams' heads appear to date to the fifth century. Parallels include gold works dated by specialists to the late sixth century, but which may be later. These include a Greek gold ram's-head pendant from Eretria in Berlin, and a gold fibula with the tip in the form of a ram's head from Ruvo in London.[4] The amber rams' heads dated to the second half of the fifth century have several good comparisons in precious metals, among them a gold necklace with nine rams' heads excavated from a tomb of circa 450–425 at Roccanova and the spiral earrings ornamented with rams' heads worn by the nymph Arethusa on Syracusan tetradrachms of 405 B.C.[5]

The best evidence for early-fourth-century amber rams' heads comes from a sporadic find at Cumae and three graves from the Andriuolo necropolis at Paestum, the latter datable to circa 380–370 B.C.[6] In each case, rams' heads were all part of necklaces that also included female heads. Both types are schematic.

An important antecedent for the joining of rams' heads and other figured elements in jewelry is the "Island Greek" or Lydian (perhaps) crescent-shaped gold pendant from Aydın (Tralles) of about 600 B.C.[7] In addition to rams' heads, this pectoral pendant includes the figure of *Potnia Theron* with snakes, rosettes, and sun disks, and griffins', lions', and bulls' heads. *Potnia Theron* is the divinity who oversees plant life, the creatures and demons of the earth, the chthonic realm, and the sky. The Aydın pendant thus might be read as the ontological statement of a divinity of great fecundity and protectiveness. Despite the difference in scale, the role of the ram here is similar to that of the ram protomes of the large stone *perirrhanterion*, the ritual purification basin, from the Isthmian sanctuary of Poseidon: the bowl is held up by women who stand on lions, holding them by leash and tail, interspersed with large rams' heads.[8] A ritual object such as the seventh-century terracotta lamp from a Gela sanctuary (the rams alternate with female protomes) may be an excerpt of such an ontological declaration.[9]

Rams' heads are also a popular subject of "Phoenician" glass pendants[10] and the subject of one uncommon type of Egyptian scaraboid, a blue frit amulet of the earlier sixth century made at Naucratis.[11] The blue frit scaraboids are unusual in their treatment as seals with representational motifs or inscriptions on the flat sides. Their distribution is significant: Greece, the Black Sea, Magna Graecia (Taranto), and one example from Cerveteri.[12] Related Egyptian amulet types, representing the king of the gods, Amun Ra, are the flat-backed amulets of a ram's head with a disk and uraeus, or uraeus alone, in hollow gold and lapis lazuli as well as glazed-composition and frit, a feature of burials from the Third Intermediate period onward.[13] Such is the role of the rams' heads flanking the necks on the pair of glazed-composition flasks found in the "Isis Tomb" at the Polledrara cemetery at Vulci (British Museum GR 1850.2-27.57): the pair of rams' heads invoke Amun, and the hieroglyphic inscription expresses greetings for the New Year, a potentially dangerous time of transition, just as was death.

Archaic representations of ram's-head pendants in use include sculpted and painted examples, worn by both females and males, both deities and mortals. Some examples are telling: the terracotta statuette of an enthroned chthonic(?) female deity from Agrigento wears three superimposed pectoral strings, with bull, ram, and satyr heads.[14] The subject of a terracotta urn lid (from the Monte Abatone necropolis, Cerveteri) is a reclining woman who wears a necklace with seven pendants, two of them rams' heads.[15] The male banqueter painted in the pediment of the main wall of the Tomb of Hunting and Fishing at Tarquinia wears three ram pendants on a carrier. In addition, several necklaces (one with three lions' heads) hang from branches in the grove of the first chamber of the same tomb.[16] The rams' heads (amber) worn by the Monte Abatone figure are painted reddish in color, the same color of the two tiny beads at the top of the necklace, in contrast with the yellow-orange (perhaps gold) central pendants. The three rams' heads worn by the male banqueter in the Tomb of Hunting and Fishing are the same reddish color as his skin and contrast in color with the yellow- and white-limned vessels (perhaps of shiny bronze or silver) held by the banquet's participants. The ram's-head necklace in the first chamber is also painted a reddish color, in contrast with the yellow-orange palmette and lion pendants (also gold, perhaps) in the murals of the first chamber.[17]

Erika Simon interpreted the trees of the first chamber of the Tomb of Hunting and Fishing as laurels and a sacred grove of Apollo as further evidence supporting her larger argument concerning the importance of Apollo in the funerary art and customs of Etruria (in contrast to his Greek nature).[18] If the grove is indeed Apollo's and the animal-head necklaces hanging from the trees are of amber, this could be taken as further evidence of the connections between amber, Apollo, and life in the tomb as well as the funereal realm. Although the rams' heads depicted in southern Etruria predate by several decades the Delphic coin types with confronted pairs of mounted rams' heads (perhaps pendants rather than rhyta), the coins may substantiate the Apollonian connection.[19] Alternatively, it is possible that the tomb is Dionysian, as Sybille Haynes argues.[20] Stephan Steingräber points out its Dionysian aspects.[21]

The ram has always enjoyed favor as a potent symbol because of the animal's legendary strength and virility (and hence its creative powers), and its characteristics as a leader and protector of his flock. In the ancient Near East, and from earliest times throughout the Mediterranean, the ram was associated with powerful divinities and heroic figures, wealth and the elite, and sacrifice. In South Italy, ivory figurines of recumbent animals, including the ram (decorations of fibulae, for the most part), were excavated at Motte delle Timpone (Francavilla Marittima).[22] At Argos and Perachora in Greece, ivory ram figurines have appeared in sanctuaries devoted to Hera, and at Ephesus, in that of Artemis. The single greatest number of ivory rams has come from the sanctuary of Ortheia (Artemis Orthia) at Sparta. In each case, the ram is linked with a powerful female divinity and a potent and high-value material. The series of early-sixth-century bronze Laconian *hydriai* (whose shape might be directly connected with women) with vertical handles animated by female heads, recumbent rams, and lion could be other instances of this affiliation.[23] The fourth-century necklaces with rams' and women's heads from Paestum, the last gasp of pre-Roman amber carving, may reflect the same pairing.

The traditional importance of the ram's head as a subject in the ancient Near East is exemplified by two lapis lazuli pendants dating to the third millennium, possibly from Iran, and by a calcite Jemdet-Nasr-period amulet-seal of circa 3000 B.C.[24] In Egypt, the ram was connected to several key deities.[25] The ram with downturned horns was a symbol of the god Amun, and when he wore the solar disk between his horns or incorporated other solar iconography, the ram's head was one of two guises of Amun Ra (the other was a goose). A ram's head in amber, the subject enhanced and focused by the material from which the amulet was made, would put its wearer under the protection of the deity represented and would by assimilation offer the wearer access to its particular powers.[26]

The ram, the most highly valued and sexually potent of domestic animals, was from earliest times the most prestigious sacrificial victim. Greek drinking vessels (rhyta) in the form of rams' heads are the most numerous by far and had an ancient ancestry. In Greece, the ram's-head rhyton, as Hans Hoffman first argued, is associated with tragic heroes (he who must die, i.e., be sacrificed himself).[27]

A ram's head might have been worn to show the patronage of, or devotion to, a deity. For a Greek or an Etruscan, a ram's-head pendant may have been an exotic, "Oriental" magical amulet, a talisman of protection, one that symbolized the power or knowledge of Egypt, the Punic world, or the Near East.

In the Greek-speaking world, the most famous stories of the ram's apotropaic powers concern acts by Hermes, the Olympian responsible for the increase and protection of flocks. At Tanagra, Hermes averted a pestilence from the city by carrying a ram around its walls. A series of ram-bearer statuettes found at Medma (Calabria) attest to the widespread influence of the cult of Hermes in the West.[28] It was Hermes, too, who sent the golden ram that flew Phrixos to safety in Colchis. The magic of the volant ram did not cease at its sacrifice: the Golden Fleece displayed in the grove of Ares was believed to be magical.

Throughout Greek culture, the ram figures prominently as a metaphor of strength and courage (thus the association with Ares). Accordingly, Homeric heroes are likened to thick-fleeced lambs (*Iliad* 3.197). In Attic vase painting, rams are sometimes represented in an explicitly sacrificial context. More commonly, the context is heroic, with the ram's sacrificial role implicit only. Such is the case in the story of Phrixos, or of Odysseus. Both the ram that carried Odysseus from the Cyclops's cave (*Odyssey* 9.436ff.) and Phrixos's mount are sacrificed as soon as they have finished their tasks. Their sacrifice is part of the story. The emblematic power of the Golden Fleece recalls the story of Atreus and Thyestes: the kingdom belonged to him who owned the golden lamb.

Almost all of the small-scale individual rams' heads have been found in graves or, in the case of their representation in art, in funerary settings. This is critical to a better understanding of the subject in adornment. Although ram's-head adornments might be hung from trees in a painting, and worn by both male and female reclining figures, the amber heads from documented contexts have come exclusively from female tombs. It is likely that in each burial, the rams' heads functioned as ornament and amulet, the subject and material combining to create an elite object, a potent ornament, one with a battery of allusions—religious, divine, heroic, mythic, magical. It may also have worked in aggressive magic or medicine. Many Late Antique gems are engraved with a ram-headed god, one wearing the symbol of the sun (based on Amun Ra), and are specifically connected with the uterus. Such amulets were thought to check any morbid condition, to prevent conception, or to favor and facilitate parturition.[29]

The solar aspects of amber may well have underscored the connection of the pendant subjects with regeneration, with the Egyptian *ba*, with solar divinities, with heroes (Odysseus, Phrixos, or Jason), or with a magical figure such as Medea. The Aydın (Tralles) pendant suggests the place of the ram in the universe of a powerful female nature divinity. Of the Olympian gods, if Apollo (the solar divinity) were brought to mind (and to work) by the ram's head, then perhaps there was an association with his son Phaethon and the Heliades—whose shining tears shed in mourning for their brother Phaethon were hardened by the sun and turned into amber. If Hermes was evoked, it might allude to not only his legendary magical act at Tanagra, but also his role—and the ram's—as *psychopompos*. As this survey reflects, the subject of ram imagery in ancient art deserves continued study.[30]

NOTES

1. Recumbent and couchant rams of amber are much rarer than rams' heads. This author knows of only one complete ram from a controlled excavation dating to the fifth century B.C.: a necklace with eighty-four plain beads and a ram pendant from Tomb 21, a collective grave in the cemetery of Valle Oscura (Marianopoli) of circa 530–470 (Marianopoli, Museo Archeologico 925: R. Panvini in Pugliese Carratelli 1996, p. 694, no. 44). Ram pendants are among the earliest subjects for amulets and for ornamentation in the ancient Near East. In Greece, bronze ram pendants are current in the Late Geometric: see Langdon 1993, p. 148; I. Kilian-Dirlmeier, *Anhänger in Griechenland von der mykenischen bis zür spätgeometrischen Zeit* (Munich, 1979), pp. 186–88; and C. Rolley, *Les statuettes de bronze*, vol. 5 of *Fouilles de Delphes* (Paris, 1969), p. 81, no. 120, pl. 21.

2. For Tomb 102 at Braida di Vaglio, see Bottini and Setari 2003. For the Sala Consilina ambers in the Petit Palais, see introduction, n. 219.

3. Stibbe 2006 (esp. chap. 3) has looked closely at the rams of the handles of bronze vessels. Nevertheless, much remains to be done in the analysis of the animals.

4. For the Greek gold ram's head from a necklace (Berlin, Antikensammlung GI 15), see B. Deppert-Lippitz, *Griechische Goldschmuck* (Mainz, 1985), p. 121, no. 69; for the London gold fibula, Marshall 1911, no. 1408; and Higgins 1980, pl. 30A.

5. For the necklace from Roccanova (Taranto, Museo Archeologico Nazionale 6452–59, 6461–63), Guzzo 1993, p. 230, VC6. For the Syracusan tetradrachm (a comparison first made by Higgins 1980, p. 128), see, for example, E. Boehringer, *Die Münzen von Syrakus* (Berlin, 1929), no. 423.

6. The now-lost tomb contents of a sporadic find from the Cumaean necropolis were recorded in 1913 by E. Gabrici, "Cuma," MonAnt 22 (1913): col. 91, fig. 37. At the Paestum necropolis, two rams' heads were found in the early-fourth-century Tomb 19 (Museo Archeologico Nazionale 24904: Pontrandolfo Greco 1977, p. 51–52, figs. 18, 1, and 22, 6), three dating to the beginning of the second quarter of the fourth century come from Tomb 22 (21330: ibid., p. 36, figs. 2, 4 and 2, 8), and three from the early third quarter of the fourth century were found in Tomb 20 (24962: ibid., p. 37, figs. 3, 2 and 3, 6).

7. Paris, Louvre: BCH 3 (1879): pls. 4–5; and Higgins 1980, p. 115.

8. Corinth Museum, numerous fragments: M. C. Sturgeon, *Isthmia: Excavations by the University of Chicago under the Auspices of the American School of Classical Studies at Athens*, vol. 4, *Sculpture I: 1952–1967* (Princeton, 1987), no. 1. The sculpted basin, which stood guard over the entrance to Poseidon's shrine, was for ritual purification. Sturgeon believes that the caryatid female figures represent the Mistress of the Animals and calls her Artemis, considering the lions' and rams' heads to reinforce the iconography. (She notes the rarity of rams' heads in Greek plastic art.) She sees Artemis's role in this unique object (the only *perirrhanterion* to have rams' heads) as that of a protective guardian, or goddess outside the doors.

9. Museo Archeologico Regionale di Gela 7711 (from the extraurban sanctuary of Predio Sola, Gela): P. Orlandini, "Gela: La stipe votiva del Predio Sola," MonAnt 46, no. 1 (1963): 33–41, figs. 14–16, pls. 8 ac, 9 ab.

10. Very few rod-formed glass pendants in the form of a ram's head are preserved. The type is first found in the seventh century, in a small, not very carefully executed version, and survives down to the first century B.C. The later examples are larger and are rendered more naturalistically. The heads are of white or dark glass, the horns in the opposite color. The animal's eyes, ears, and horns are various colors. The various kinds of "Phoenician" glass pendants (bearded male heads, demonic heads, rams' heads, birds, bells, grape bunches, and *phalloi*) were made the centerpieces of precious metal necklaces throughout the Mediterranean. The few examples from controlled excavations in Italy have come from graves. The rams are attributed to Carthaginian workshops. See Uberti 1988, p. 482, no. 758; E. M. Stern and B. Schlick-Nolte, *Early Glass of the Ancient World, 1600 BC–AD 50: The Ernesto Wolf Collection* (Ostfilden, 1994), pp. 180, 190–91; D. F. Grose, ed., *Early Ancient Glass* (New York, 1989), pp. 82–83; V. Tatton-Brown, "Rod-Formed Glass Pendants and Beads of the 1st Millennium," in *Greek and Roman Glass*, vol. 1, ed. D. B. Harden (London, 1981), pp. 152–53; and M. Seefried, "Glass Core Pendants Found in the Mediterranean Area," *Journal of Glass Studies* 21 (1979): 17–26. One seventh-century example from Narce was strung with gold repoussé pendants in the form of winged Hathoric figures: see Marshall 1911, no. 1453, pl. 23. The late Catherine Lees Causey was of essential aid with the glass literature.

11. A. F. Gorton, "Lions' and Rams' Heads," in Tsetskhladze et al. 2000, pp. 110–14 (with previous bibl.), believes "the frit rams are undoubtedly the inspiration for the later Greek gold rams' heads pendant seals, such as the example in London from Kourion."

12. For the Taranto example, see Hölbl 1979, vol. 2, p. 214, pl. 63.3; and for the Cerveteri seal, ibid., p. 29, no. 98. References are from Gorton 2000 (see n. 11, above).

13. Andrews 1994, p. 30. See also Waarsenburg 1995, p. 445, n. 1219.

14. Agrigento, Museo Archeologico Regionale "P. Orso" AG 1145 (from the 1953–55 excavations, sector to the southwest of the sanctuary of the chthonic deities): G. Castellana in Pugliese Carratelli 1996, p. 683, no. 96; and E. De Miro, *Le Valle dei Templi* (Palermo, 1994), p. 59, fig. 61. Strings of *boukrania* (as several superimposed pectoral ornaments held in place by attachments at the shoulders) are worn by some Archaic female divinities from Magna Graecia. A votive mask phenotype from the extraurban sanctuary of Predio Sola at Gela wears two necklaces, one of taurine heads, the other possibly of acorns:

see R. Panvini in Pugliese Carratelli 1996, p. 680, no. 93. See also the terracotta seated deities from Athana Lindia.

15. Rome, Museo Nazionale Etrusco di Villa Giulia 167: illustrated in Haynes 2000, fig. 177.

16. See Steingräber 2006 (with earlier bibl.).

17. In the banqueting scene of the second chamber, the jewelry worn by the female banqueter and many of the vessels of metallic shapes are painted white or yellow-orange.

18. Simon 1973 (see "Lions' Heads" introduction, n. 22); and E. Simon, "Apollo in Eturia," *Annali della Fondazione per il Museo "Claudio Faina"* 5 (1998): 119–48.

19. For the Delphi coins with paired rams' heads (or rhyta, as first identified by C. Seltman, supported by C. M. Kraay), see Kraay 1976, p. 121, with reference to C. Seltman, *A Book of Greek Coins* (London, 1952), p. 14.

20. See "Lions' Heads" introduction, n. 22.

21. Steingräber 2006, p. 95.

22. These are among the oldest excavated parallels for recumbent rams on fibulae.

23. Ram attachments are found on the upper and lower ends of bronze vessel handles in company with various other subjects. Among them are nude male figures that hold lions by the tail and stand on rams, suggesting a wider interpretation for the subjects.

24. One of the lapis lazuli rams'-head pendants is in the Metropolitan Museum of Art (55.65.8), and the other is in the Norbert Schimmel collection, New York; for the latter, see P. O. Harper in *Norbert Schimmel Collection* 1974, no. 102. For the calcite amulet-seal of a ram's head, see *Norbert Schimmel Collection* 1974, no. 103.

25. Andrews 1994, pp. 15, 30.

26. Ny Carlsberg Glyptotek 3571: Johansen 1994, pp. 86–88, no. 37 (from which this summary is taken).

27. H. Hoffman, *Sotades: Symbols of Immortality on Greek Vases* (Oxford, 1997), p. 12.

28. For the cult of Hermes in Medma, see M. Paoletti, "I culti di Medma," in *Santuari della Magna Grecia in Calabria*, exh. cat., ed. E. Lattanzi et al. (Naples, 1996), pp. 95–97. For Medma, see Paoletti and Settis 1981. For a recent collation of bibliography on the *kriophoroi* of the region and the subject generally, see M. T. Iannelli in *Magna Graecia* 2002, p. 189, no. 30.

29. Bonner 1950, p. 85.

30. Unlike the rich literature on the horse and the lion in ancient art, to my knowledge there is no corresponding analysis of ram imagery. The topic is a rich one, and the in-depth studies of some classes of material, Greek bronze vessel attachments (e.g., Stibbe 2006) and Greek vase rhyta (Hoffman 1997, in n. 27, above), lead the way. Research should reveal not only something about artistic approaches to the subject and the transmission and modulation of imagery but also some understanding about ram *realia*.

39. Pendant: Ram's Head

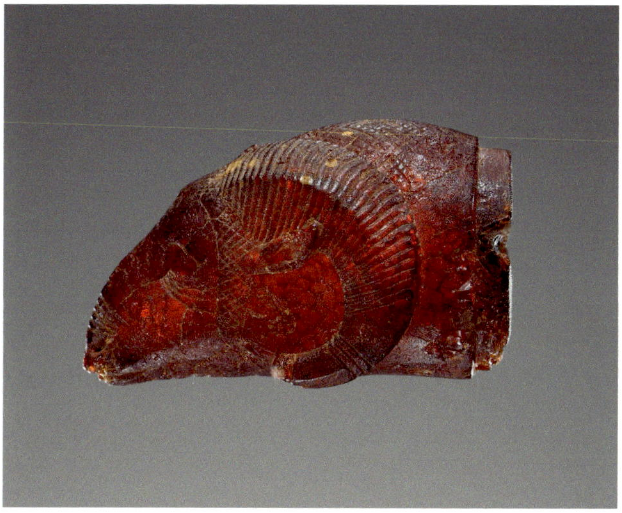

Accession Number	76.AO.82
Culture	Etruscan
Date	525–480 B.C.
Dimensions	Length: 36 mm; width: 20 mm; depth: 18 mm; Weight: 9 g
Subjects	Ionia, Greece (also Ionian, Greek); Jewelry; Ram

Provenance

–1976, Gordon McLendon (Dallas, TX), donated to the J. Paul Getty Museum, 1976.

Condition

The pendant is in good condition, with only minor modern chipping at the animal's right horn (at the base) and on the lower lip and right jaw. The surface is sound, but there is considerable crazing and many areas where small flakes have chipped off. The caudal edge at the base plate has been rabbeted, most likely to fit into a now-missing metal mount. Part of the back has been broken off, revealing two plugged bores embedded with the remains of metal (silver?) wire or pins. Traces of metal corrosion product surround the attachments. The surface is red-orange, and under strong light the core of the amber appears transparent and bright red-orange. No inclusions are evident. There are some tiny patches of yellowish residue on the surface.

Description

The ram is extraordinary for its morphological specificity. The zygomatic process arches upward in a dramatic, swelling curve, and the eye sockets are deeply drilled. It is possible that inlaid eyes were originally inserted in these cavities. At the inner corner of the eye, three lines indicate folds of flesh below the eye. The ears are sharply raised from the surface of the horn. Five lines are incised across the lower nose and slope down toward the mouth. The nostrils are carved in shallow relief.

Finely spaced diagonal cross-hatching defines the fleece on both poll and cheeks. The horns are relieved from the cap of fleece. Regular, evenly spaced ridges are incised for two-thirds of the length of the horn, three more lines are indicated at the midpoint of the remaining third, and one last ring circles the tip of the horn. The horn tips splay away from the face. The pattern of ridges is meticulously rendered, with a gentle undulation in the line; the spacing between the ridges narrows from the base toward the tip.

The anatomical detail of this pendant, combined with the interplay in the surface pattern and the precision in artisanship, marks this as the work of a master artisan. The counterpoint created by the placement of the curved, striated pattern of the ridges of the horns adjacent to the reticulated cross-hatching of the fleece offers but one example of the maker's careful attention to detail.

Unpolished tool marks lie deep in the curve of the horns; this was perhaps done to create shading. In other areas that would have been more difficult to finish, such as the sharp angle of intersection between the rings of the horns and the fleece, the tool marks are polished out.

Discussion

This head of a ram may have come from the same original context as four other pendants in the Getty collection. The five—76.AO.82 (this ram's head), the rams' heads 76.AO.83

(cat. no. 40) and 77.AO.81.7 (cat. no. 41), the foreparts of a boar pendant 76.AO.84 (cat. no. 37), and the kore pendant 76.AO.77 (cat. no. 8)—share a similar state of conservation, technique, and style. All have very similar hollowed-out eyes and show a similar use of the graver and of polishers. The kore pendant is attributed here to a South Ionian artisan, or to one trained with an artisan from the area, on the basis of comparison to terracottas and marbles. There is no such corresponding body of material for the other four pedants.

The three rams' heads are very much alike, and although generally similar to many other amber rams or rams' heads, they have no close counterparts. These three are comparable in the deep V-shape of the horns on the top of the head, the deep relief of the fleece from the neck, and the wavy, closely spaced ridges of the horns. The ridges of 76.AO.83 and 77.AO.81.7 have the same pattern of spacing. The representation of the fleece is identical on 76.AO.82 and 76.AO.83 (it may once have been the same on the more worn example 77.AO.81.7). A filament threaded through a perforation in the collar area suspended the ram's head 76.AO.83. The decorative suspension device is very like that of the boar and kore pendants. The other two rams' heads, 76.AO.82 and 77.AO.81.7, must have been set into metal mounts, for they both have broken bores that once held metal pins (the breaks may have been caused by the expansion of metal corrosion products over the long period of the burial). It is possible that 76.AO.82 and 77.AO.81.7 were originally carved with incorporated devices and that the metal mounts are later additions.

40. Pendant: Ram's Head

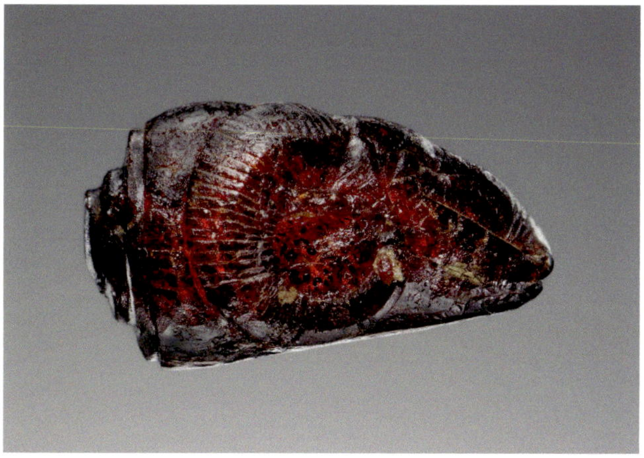

Accession Number	76.AO.83
Culture	Etruscan
Date	525–480 B.C.
Dimensions	Length: 24 mm; width: 19.5 mm; depth: 15.5 mm; Weight: 4.0 g
Subjects	Inclusions; Ram

Provenance

–1976, Gordon McLendon (Dallas, TX), donated to the J. Paul Getty Museum, 1976.

Condition

The head is in good condition, intact with small chips at the tips of the horns and overall fine surface cracking. A portion of the base is broken away. Wear, degradation, and a varnishlike coating (applied before entry into the Getty Museum collection) have obscured some finer details. One inclusion is visible at the nose. In ambient light, the head is dark brown with a reddish tint; in transmitted illumination, it is bright red-brown where translucent.

Description

The pendant form and the animal are very similar to the rams 76.AO.82 (cat. no. 39) and 77.AO.81.7 (cat. no. 41). Four lines cross the nose and slope down toward the corners of the mouth. Beneath the eyes are three small incised folds that curve toward the ear. The chin slopes smoothly upward under the overhanging upper lip; the underside of the chin is flat. The horns are relieved from the fleece. The missing horn tips would have turned outward. The poll, the cheeks, and the back of the head are cross-hatched. The horns are well elevated above the skull, and their ridges are closely spaced straight grooves extending two-thirds of the way from the root to the tip. Two additional rings occur at the midpoint of the last third of the horns. In comparison to 76.AO.82, the ears are upright, the muzzle is longer, and there is a more prominent rise or bump on the bridge of the muzzle.

Two 1.5 mm perforations at opposite edges of the back do not meet, suggesting that they were drilled after a metal mount was placed on the amber. A 3 mm stopped bore of indeterminate purpose is drilled into the back. The collar is bordered by a molded detail consisting of a small fillet surmounted by an ovolo topped by another fillet. There are clear abrasion marks in the molding.

Discussion

See the entry for 76.AO.82.

41. Pendant: Ram's Head

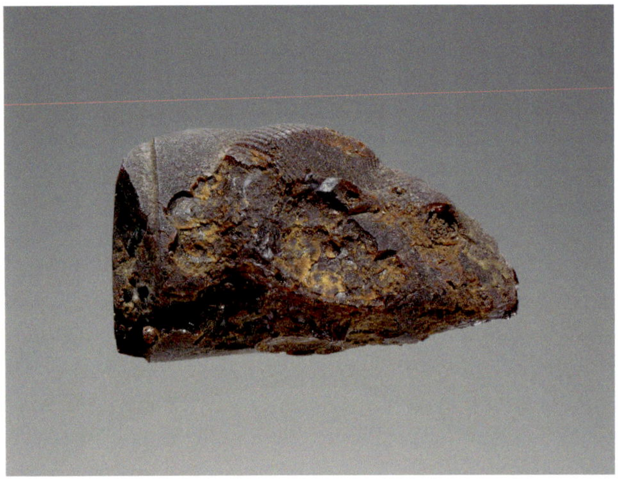

Accession Number	77.AO.81.7
Culture	Etruscan
Date	525–480 B.C.
Dimensions	Length: 40 mm; width: 24 mm; depth: 22 mm; Weight: 10.8 g
Subjects	Ram

Provenance

–1977, Gordon McLendon (Dallas, TX), donated to the J. Paul Getty Museum, 1977.

Condition

The pendant shows old breaks and chips as well as modern breaks, including at the tip of the ram's nose. There is extensive surface loss and chipping, at times deep, along the throat, neck, and horns. A large fissure is near the point where the back meets the surface at the lower neck. The amber is dark red-brown when viewed in ambient light, but transparency is evident in the areas of the modern chips. In transmitted light, the amber is dark red.

Description

This pendant is very like both 76.AO.82 (cat. no. 39) and 76.AO.83 (cat. no. 40), but it is closer in details and conception to 76.AO.82. It may be by the same hand. The surviving parts of the eyes are almost identical in treatment to those of 76.AO.82. There is fine, closely spaced cross-hatch engraving on the nape of the neck and top of the head. Wavy parallel lines define the horn surfaces and contrast with this cross-hatching.

Abrasion marks remain on the back. The base is sharply recessed, which suggests that this flange is functional rather than ornamental. A metal cap or mount probably once covered the caudal end. Two 2 mm lateral bores are drilled from opposite sides of the pendant, but they do not meet. A fragment of metal remains in the right socket.

Discussion

See the entry for 76.AO.82.

42. Pendant: Ram's Head

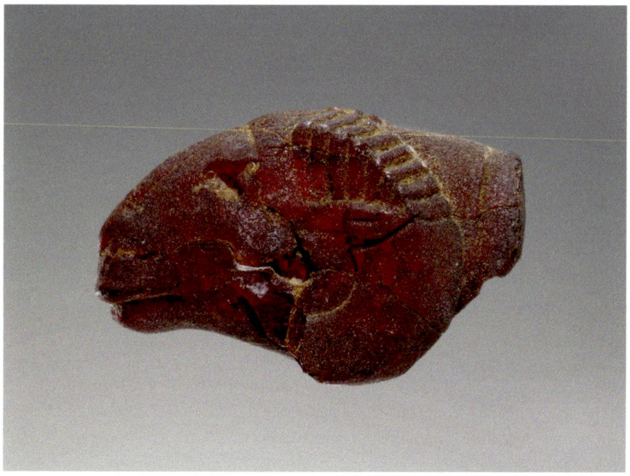

Accession Number	77.AO.81.11
Culture	Italic or Etruscan
Date	500–400 B.C.
Dimensions	Length: 21 mm; width: 18 mm; depth: 15 mm; Weight: 33.3 g
Subjects	Ram

Provenance

–1977, Gordon McLendon (Dallas, TX), donated to the J. Paul Getty Museum, 1977.

Condition

The pendant is largely intact, except for a chip missing from the tip of the left horn and a large, wedge-shaped chip missing from the base, which leaves the suspension hole exposed. The amber is laced with fine, widely spaced cracks. An inclusion is visible at the throat. The amber is dark brown in ambient light, perhaps due to consolidation with amber oil. There is some yellow-ocher degradation residue in the carved interstices. In transmitted light, the pendant appears dark red. Some transparency is noted in the areas of modern chips. A 2 mm horizontal perforation for suspension runs through the collar area of the back, about 1 mm below the surface. Probable metallic residue remains in the stopped bore on the proper right side of the head.

Description

The head is finely rendered. Incised wavy ridges on the horns begin at the forehead and extend to the intersection with the tip of the ear; the remaining involution of the horn is smooth. The tips of the horn flare outward. The eyelid is distinct, with a sharp outer edge, and the eye sockets are deeply bored. The cap of fleece on the poll rises above the plane of the face; the separation is delineated by a lateral incision. Shallow cross-hatching indicates the fleece on the poll and over the neck area. The depression of the throat is indicated by a faintly incised triangle under the mandible. The upper lip overhangs the lower lip, and the chin is rendered distinctly. The nostrils are incised. The ears are long, straight, and narrow and lie flat on the horns, with an incised line visible in the middle of the ear.

Discussion

See the entry for 76.AO.82 (cat. no. 39).

43. Pendant: Ram's Head

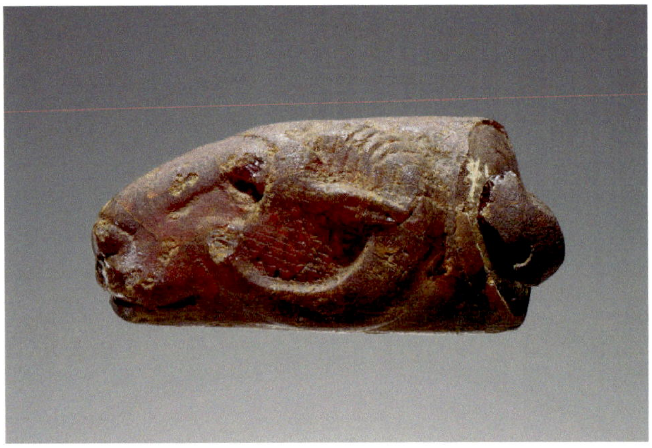

Accession Number	77.AO.81.13
Culture	Italic or Etruscan
Date	500–450 B.C.
Dimensions	Length: 32.5 mm; width: 19 mm; depth: 15 mm; Weight: 5.3 g
Subjects	Ram

Provenance

–1977, Gordon McLendon (Dallas, TX), donated to the J. Paul Getty Museum, 1977.

Condition

The pendant is largely intact, although there are several small old chips at the throat, on the back, and on the rear and bottom of the left horn; there is a modern chip on the right horn. Each upper eyelid has small chip losses. A section on the right side of the suspension device was broken off and reglued before the pendant's entry into the Getty Museum. There is some loss associated with this area. The surface was apparently treated with a consolidant, such as amber oil. The surface has ocher-colored degradation material in interstices, particularly on the throat. The amber is a dull red-brown except for the shiny areas of the exposed fractured inner surface, which show the material's transparency. In transmitted light, the pendant is dark red. The only visible inclusion is in the fissure that rises to the surface at the back.

Description

The pendant is almost tubular. The horns do not rise higher than the poll, and they curve out timidly. The muzzle is somewhat squared off. The eye sockets are hollowed. The broad, flat nose has a ridgeline across its tip, much in the way that the nostrils of some pre-Roman amber lions are indicated (see, for example, 77.AO.81.10, cat. no. 36). The line extending from each inner canthus along the length of the nose is clearly indicated. The underside of the chin is flat. The upper lip puffs out slightly on both sides of the mouth.

The poll rises between the base of the horns from a shallow but broad V-shaped incision centered on the forehead; it then spreads outward, curling around and behind the eyes. A pattern of minute, shallow cross-hatched incisions defines the fleece across the forehead and on the cheeks. Shallow, regular incisions circle each horn, whose surface is then smooth from the point of the ear to the tip of the horn.

A 1.5 mm perforation runs through the suspension device, which is decorated with a bead-and-reel design.

Discussion

This pendant has no close parallel.

44. Pendant: Ram's Head

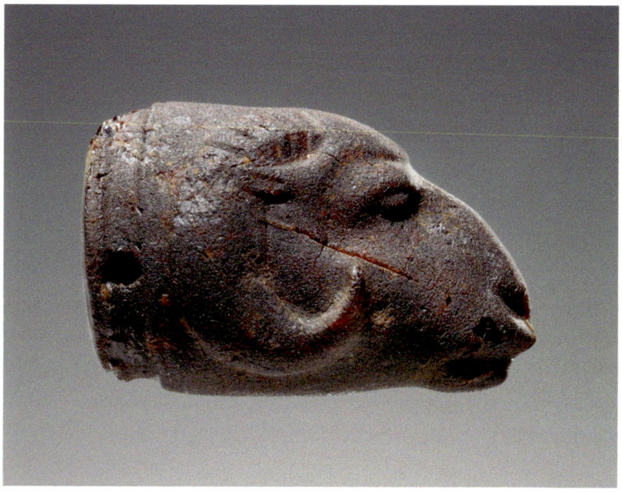

Accession Number	77.AO.81.14
Culture	Italic
Date	500–400 B.C.
Dimensions	Length: 30 mm; width: 20 mm; depth: 19 mm; Weight: 6 g
Subjects	Ram

Provenance

–1977, Gordon McLendon (Dallas, TX), donated to the J. Paul Getty Museum, 1977.

Condition

The pendant is intact except for large chips at the back, behind the left ear, at the right side of the upper lip, and at the inner tip of the left horn. There is a shallow crack on the right cheek. There is also a small vertical crack extending from the tip of the nose to the upper lip. The surface, firm with a slight granular appearance, may have been treated with a consolidant. In ambient light, the pendant is dull brown. In transmitted light, the piece is dark red and opaque.

Description

This pendant is almost cylindrical. The eyes and eye areas are plastically modeled: the eyes are almond-shaped, the inner canthi recessed, the lids indicated, and the eyebrow arches prominent. The small ears lie flat against the horns. The muzzle is broad and somewhat truncated; the nares are carved in a V-shape. The upper lip is naturalistically full and hangs over the lower one. The throat is almost flat, except for a slight concavity behind the chin. The horns originate from a domed poll, which rises sharply from the plane of the face. The base of each horn has six engraved lines; the rest of the horn's surface is smooth. The tips of the horns project outward. At the base of the pendant is a collar, which is made up of an indented ring and a flaring terminus.

The pendant was hung from a triangular system of holes: a 2 mm perforation passes horizontally within the indented ring section of the collar and joins to another perforation 2 mm in diameter in the center of the back.

Discussion

Despite its small size, this pendant reveals many idiosyncratic elements of the animal: the plastic eye area, the form of the mouth, and the shape of the head.

45. Pendant: Ram's Head

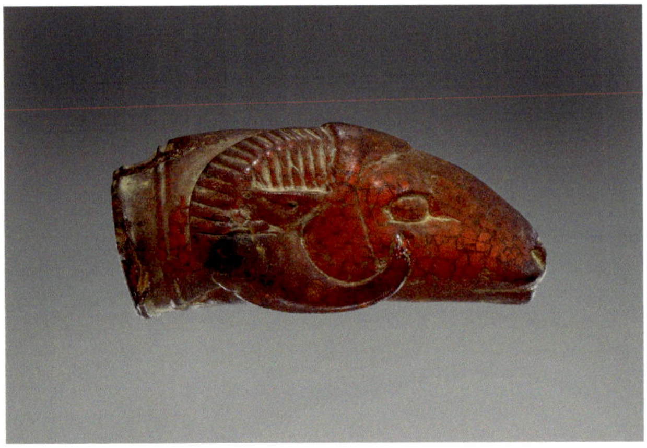

Accession Number	77.AO.81.15
Culture	Italic
Date	500–400 B.C.
Dimensions	Length: 36 mm; width: 19 mm; depth: 15 mm; Weight: 5.8 g
Subjects	Ram

Provenance

–1977, Gordon McLendon (Dallas, TX), donated to the J. Paul Getty Museum, 1977.

Condition

The pendant is intact except for a chip on the proper right edge of the pendant back and minute chips at the tip of each horn. The surface shows crazing. The pendant has a reddish brown patina, with some yellow mottling at the forehead, at the left naris, and under the chin. The pendant may have been consolidated with amber oil. It is transparent in ambient light. In transmitted light, the pendant is bright orange, and fine subsurface cracking is visible. No inclusions are evident.

Description

The ram of this pendant has a long and sleek head, the tapering muzzle tipping down as in nature. The ocular cavity is simplified, the salient features indicated. The plastic, almond-shaped eyes are defined by surrounding incised grooves. The area below the eye is flat and the arch raised. A short, shallow groove extends downward from the inner corner of each eye toward a deeply incised naris. The division between the face and the fleece is sharp, with the line of the wool sweeping in a curve behind the eyes. The upper edge of the fleece is cut away sharply over each ear, resulting in a ledgelike ridge. Each horn is encircled with fourteen or sixteen closely spaced engraved ridges that extend from the base of the horn to a point just past the intersection with the tip of the ear. The remainder of the horn is smooth. The horns have sharp ends and flare outward. The collar area is narrower in circumference than the neck; it is an inverted flute or a flute with narrow fillets on either side.

A deep groove, 1.5 mm wide, begins at the upper edge of the back and extends 8 mm into the area between the horns, probably the result of removing a fault in the material. Two other grooves, 8 and 10 mm long, also likely resulting from fault removal, are located diagonally across the underside of the neck. Two 1.5 mm perforations were drilled 4 mm apart in the center of the back. The perforations are obstructed with dirt. It is possible that a filament was threaded through the holes. The pendant would have hung head downward.

Discussion

This head and the next four—77.AO.81.16 (cat. no. 46), 77.AO.81.17 (cat. no. 47), 77.AO.81.18 (cat. no. 48), and 77.AO.81.19 (cat. no. 49)—look to be by the same hand. 77.AO.81.16 has a similar diamond-shaped poll and similar widely spaced ridges on the horns. The eyes are plastically modeled, which is especially apparent when the pendant is viewed on the dorsal side. The upper edge is sharply cut away over each ear, resulting in a ledgelike ridge. In profile, the animals show the same morphological features of the lower part of the head: they both have a long, narrow muzzle with a flat or slightly concave chin and neck. 77.AO.81.17 is very much like this pair, but it has a different decorative finial. One of the rams' heads in London (British Museum 82) is especially close in form to 77.AO.81.15.[1]

NOTES

1. Strong 1966, p. 83, no. 82, pl. XXX.

46. Pendant: Ram's Head

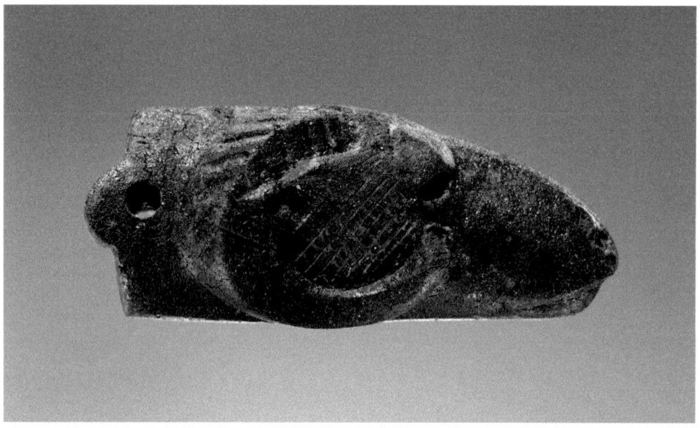

Accession Number	77.AO.81.16
Culture	Italic
Date	500–400 B.C.
Dimensions	Length: 32 mm; width: 16 mm; depth: 13 mm; Weight: 3.5 g
Subjects	Ram

Provenance

–1977, Gordon McLendon (Dallas, TX), donated to the J. Paul Getty Museum, 1977.

Condition

The pendant is intact, with only minute chips at the tip of the left ear and at the tip of the right horn. There is crazing overall. Several cracks are visible, most notably under the left side of the chin. There is a small vertical crack extending across the left side of the mouth. The piece has been treated with amber oil. The surface is a dark, dull red. In transmitted light, the pendant is dark red but does not itself transmit very much light. No inclusions are evident.

Description

This head and the following three—77.AO.81.17 (cat. no. 47), 77.AO.81.18 (cat. no. 48), and 77.AO.81.19 (cat. no. 49)—are attributed here to the same hand and are thus described as a group in this entry. (The previous head 77.AO.81.15, cat. no. 45 is also likely by the same hand.) Three—77.AO.81.16, 77.AO.81.17, and 77.AO.81.19—are nearly identical in weight, mass, type, form, and style. All four have nearly identical plastic suspension devices incorporated into the pendant design: in each, a hole is drilled through the neck, and not through the device. 77.AO.81.18 is slightly different, showing a greater affinity to live sheep, the head being higher and deeper, and the angle of the lower skull more natural. The other three heads are in comparison longer and narrower in profile and more slender in top or bottom view. Identical features of all four include the shape of the hollow eyes—hemispheric concavities with heavy overhanging lids. All four have fleece indicated on the poll, the back of the neck, and the cheeks, with fine, irregularly spaced cross-hatching. The horns of each are carinated from base to tip in a chevron pattern, and in each small ears lie directly on the horns. On the upper side, the ears are cut away, leaving small, shelflike plateaus. On each, the rendering of the ears and the horns is especially careful. The head of 77.AO.81.16 is longer and more slender than the heads of the other three. On all four, tiny V-shaped incisions indicate the nares.

Traces of tools remain on all the pendants, indicating the use of a sharp tool (or sharp edge) for the cross-hatching, a graver, and abrasive materials for the polishing. Inside one of the sets of holes, the marks of a drill are apparent. On all four of the pendants in this group, the hole for suspension is located in the neck, very near to the plane of the back. The holes range in diameter from 1.62 mm to 2.2 mm. All four pendants would have hung nose downward.

Discussion

The differences among the pendants of this group are probably owing to the form of the original drops of amber from which they were made. In the case of this pendant, the blank must have been more rectangular than those of the other three.

47. Pendant: Ram's Head

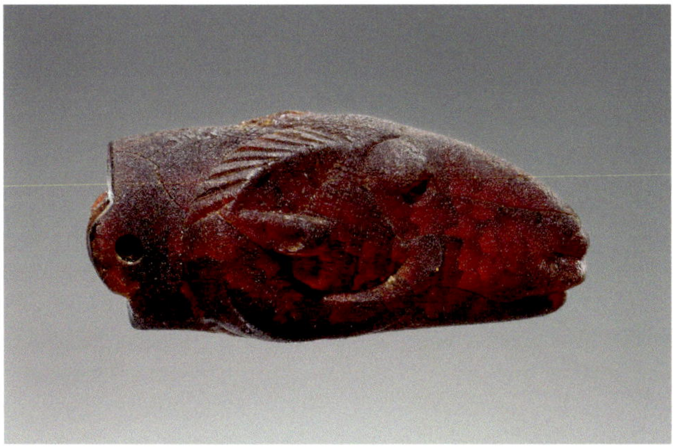

Accession Number	77.AO.81.17
Culture	Italic
Date	500–400 B.C.
Dimensions	Length: 30.5 mm; width: 18.5 mm; depth: 13 mm; Weight: 3.9 g
Subjects	Ram

Provenance

–1977, Gordon McLendon (Dallas, TX), donated to the J. Paul Getty Museum, 1977.

Condition

The pendant is intact, with only small chips missing from the tip of the left ear and the tips of both horns. The amber is laced with cracks. The piece has been treated with amber oil. The pendant surface is a dull, blackish red. In transmitted light, subsurface cracking is visible and the pendant is deep red.

Description

This pendant is nearly identical to the previous example (77.AO.81.16, cat. no. 46), with similar line engraving. However, the underside of this head is subtly marked by a slight depression along the line of the throat.

The 1 mm perforation for suspension passes through the neck portion of the pendant; fragments of metal possibly remain in the holes.

Discussion

The differences between this pendant and 77.AO.81.16 are slight: this pendant weighs slightly more and is minutely shorter in length and wider. This pendant reveals the removal of a fault in the amber in the underside of the head. It also retains fragments of metal in the perforation.

48. Pendant: Ram's Head

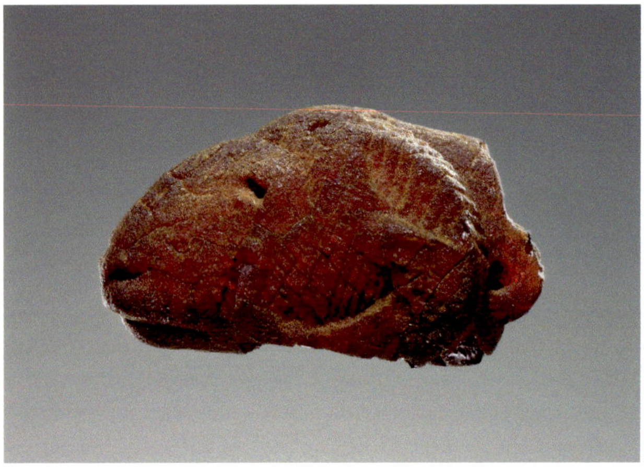

Accession Number	77.AO.81.18
Culture	Italic
Date	500–400 B.C.
Dimensions	Length: 28 mm; width: 20.5 mm; depth: 15 mm; Weight: 4.4 g
Subjects	Ram

Provenance

–1977, Gordon McLendon (Dallas, TX), donated to the J. Paul Getty Museum, 1977.

Condition

The pendant is intact except for a weathered pit loss in the midsection of the right horn and a large area of breakage in the throat area. Small chip losses are found at the tips of the horns, on the right side of the upper lip, and on the suspension device (bead-and-reel) base. Details are generally soft and the surface abraded. The surface is crazed overall; there are no visible inclusions. The piece is a light red-brown, with dusty yellow degradation in areas with incised detail and crevices. In transmitted light, the amber is orange.

Description

The head of the ram is broad, full, and relatively short from nose to base. In profile, the nose is arched; the nares are slightly incised. A U-shaped incision marks the mouth. The ears are cut away on the upper side, leaving small, shelflike plateaus. A ridge indicating the fleece rises sharply from the center of the poll and sweeps around each eye to the cheek. The fleece on the cheeks is rendered by irregularly spaced, shallow cross-hatching. The eyes are unevenly carved, with the right eye shallower and smaller than the left. The outer edges of the horns are carinated and ringed with ridges that extend from the poll to the tips of the ears in a chevron pattern. The horns flare widely on the forehead, and the poll rises slightly at the joining of the horns. The area between chin and neck is flat.

The short neck terminates in a base plate with a raised bead-and-reel molding, through which a 2 mm perforation for suspension has been drilled. The pendant would have hung nose downward.

Discussion

See the entry for 77.AO.81.16 (cat. no. 46). In comparison to the other three pendants in this group—77.AO.81.16, 77.AO.81.17 (cat. no. 47), and 77.AO.81.19 (cat. no. 49)—the nose is more arched, the neck shorter, the head tilted downward, and the ears more worn. This pendant is also the heaviest of the four.

49. Pendant: Ram's Head

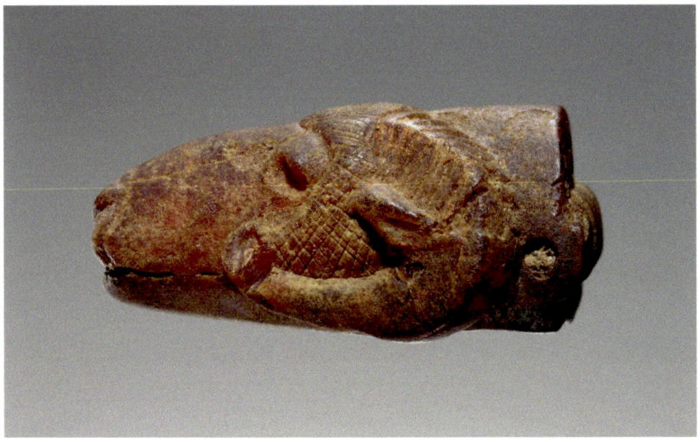

Accession Number	77.AO.81.19
Culture	Italic
Date	500–400 B.C.
Dimensions	Length: 27.5 mm; width: 20.5 mm; depth: 12.5 mm; Weight: 3.6 g
Subjects	Ram

Provenance

–1977, Gordon McLendon (Dallas, TX), donated to the J. Paul Getty Museum, 1977.

Condition

The pendant is in a very good state of preservation, except for tiny chips missing from the right lip, the back of the neck, and the tip of the left horn. Only the chip from the horn appears to be recent. A web of cracks covers the surface of the piece, several of which have yellowish residue in them. The pendant is a dull red-brown, with yellowish surface alteration layers on the left cheek and especially on the underside of the pendant. The surface is opaque. In transmitted light, the amber is translucent and red. There are no visible inclusions.

Description

In comparison to the other three rams placed in this group—77.AO.81.16 (cat. no. 46), 77.AO.81.17 (cat. no. 47), and 77.AO.81.18 (cat. no. 48)—the nose is broader and its tip blunter and the horns more flattened against the head. The ram's neck is cylindrical on top and flatter on the throat. All of these distinguishing features may indicate the limitations of the original amber nodule.

Discussion

See the entry for 77.AO.81.16. The 1.6 mm suspension perforation has remains of metal.

50. Pendant: Ram's Head

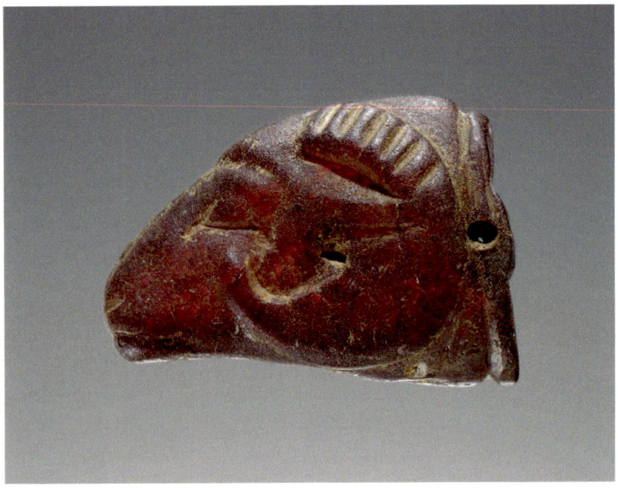

Accession Number	77.AO.81.21
Culture	Italic
Date	500–400 B.C.
Dimensions	Length: 25.5 mm; width: 14 mm; depth: 17 mm; Weight: 3.0 g
Subjects	Ram

Provenance

–1977, Gordon McLendon (Dallas, TX), donated to the J. Paul Getty Museum, 1977.

Condition

The pendant is intact except for a modern chip on the proper right side at the base. A fine fracture line extends from the bottom of the proper left horn to the center of the lower jaw. There is a deep gouge in the throat area and a small deterioration hole in the left cheek. The surface may have been consolidated with amber oil, resulting in a semiglossy, light brown surface color in ambient light. In the incisions are found dusty yellow-brown encrustations. In transmitted light, the amber is translucent and red and an extensive crack network underlying the surface is seen. There are no visible inclusions.

Description

Viewed dorsally, the ram's head is narrow and triangular, flaring only slightly from the tip of the nose to the back of the head. In profile, the head is more triangular in shape than most rams' heads in the Getty group. The plane of the face is delineated from the fleece by a shallow incision. The eye is amygdaloidal in shape, outlined with an engraved line, and a line indicates the tear ducts. The temple area swells slightly above the plane of the face. In profile, the nose curves smoothly downward. The right nostril is carved more deeply and is wider than the left. The ears are plastically modeled and lie flat on the horns. Seven ridges are carved between the root of the horn on the forehead and the horn's intersection with the tip of the ear. In profile, the horns are broad and flat. The chin and throat area has a slight swelling at the jugular notch. The collar area is set off from the animal by a crudely engraved line. On the base is a low, rectangular device that is schematically carved. Two horizontal lines cross the base, and vertical lines are incised at both sides.

Natural holes in the amber, perhaps resulting from original, now-missing inclusions, are evident under the neck and the left ear. The surface is worn, but traces of a sawlike edge tool are visible on the horns and on the pseudo-suspension device; traces of a graver appear around the eyes and the poll. A 2 mm perforation for suspension has been bored laterally about 1 mm through the collar section. The pendant would have hung head downward.

Discussion

This head is very similar in shape to 77.AO.81.18 (cat. no. 48), but the execution is much more schematic in most details, with the exception of the tear duct extension, which is anatomically precise.

51. Pendant: Ram's Head

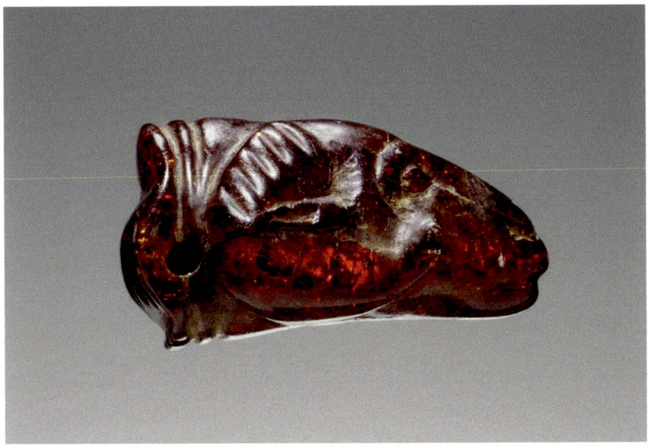

Accession Number	77.AO.81.22
Culture	Italic
Date	500–400 B.C.
Dimensions	Length: 23 mm; width: 16 mm; depth: 12 mm; Weight: 2.2 g
Subjects	Ram

Provenance

–1977, Gordon McLendon (Dallas, TX), donated to the J. Paul Getty Museum, 1977.

Condition

The pendant is intact except for a tiny chip on the tip of the right horn. The surface retains much of its polish, even though there are many fine cracks and crazing over its entirety. Some dusty yellowish residue is evident in the incisions, and dirt or encrustation remains in the suspension perforation. In ambient light, the pendant is red, shiny, and somewhat translucent. In transmitted light, the amber is orange-red. There is an inclusion in the partially cleaned cavity at the throat.

Description

In profile view, the head is a rectangle, broken only by the downward slope of the nose and the downward tilt of the chin. In top or bottom view, the head is triangular. The neck section is short. The pendant terminates in a pseudo-mount decoration that is larger in circumference than the neck. The somewhat cursory portrayal of the anatomy includes a more specific description of some features. The eyes have a slight bulge above for the arch, are plastic, and display a realistically long tear duct line; the ears show the swelling of the antitragus and the curve of the helix; and the chin is swelled and the throat subtly concave. The line of the mouth slants backward and downward. The ridges of the horns are suggested by five broad grooves between the poll and the ear (the rest of the horns are plain). The horn tips flare outward from the face. The cap of fleece is distinguished from the plane of the face by a shallow incision line.

The limits of the original shape of the amber nodule might be indicated by the flatness of the horns. Abrasion marks remain along the periphery of the cavity at the throat area. The collar of the pseudo-mount is made up of two engraved lines. On the back is a bead-and-reel–like device consisting of four broad vertical grooves. A 1.5 mm perforation for suspension passes laterally through the two engraved lines at the collar.

Discussion

This ram's head is by a hand different from that of any other in the Getty collection, and it has no related parallel.

52. Finial(?): Ram's Head

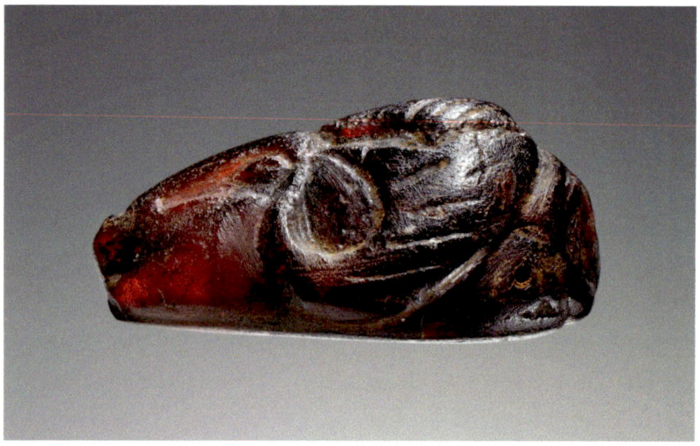

Accession Number	77.AO.81.12
Culture	Italic
Date	500–400 B.C.
Dimensions	Length: 41 mm; width: 27 mm; depth: 29 mm; Weight: 11.1 g
Subjects	Jewelry; Ram

Provenance

–1977, Gordon McLendon (Dallas, TX), donated to the J. Paul Getty Museum, 1977.

Condition

The pendant is intact and in good condition, although the surface is worn, as if from handling. The surface of the amber appears to be crazed, and fissures are visible on the underside of the pendant. There is a small chip loss to the lower edge of the nose, and a shallow fissure runs partially across the back. Numerous small scratches mar the surface and may be the result of heavy polishing. The pendant appears to have been treated with amber oil. In ambient light, the amber is red-brown. Some slight subsurface cracking is visible and the piece is bright red in transmitted light. Many inclusions are present.

Description

This ram is schematically rendered in comparison to most of the other ram pendants in the Getty collection. It has a broad, semicircular nose and a very low forehead. The muzzle is long and rounded. The horns divide into broad, flat volutes; the ridges on the horns are indicated by five deep and widely spaced horizontal incisions. The ears are represented by sunken amygdaloidal cavities beneath the horns near the top of the head rather than at the sides. A deep incision over the eye and an equally deep incision over the cheek are devices that serve to raise the eye away from the plane of the face. On the back is an oval raised section of amber delineated with five grooves (perhaps a pseudo-suspension finial).

There is wearing on the prominent surfaces, but the marks of a graver remain on the cheeks, throat, and chin. The grooves on the back appear to have been made by a sawlike rasping with a straight edge. There are two perforations through the pendant, either of which may have been used for suspension. A small bore, 1 mm in diameter, passes laterally through the pendant at the base of the head; the nostrils are perforated with a 2 mm bore.

Discussion

This ram has no parallel known to me for the morphology of the animal or the style. I also know of no other pseudo-suspension finial. The pendant is of considerable size, one of the largest known, and is comparable to the two largest pendants from the sixth-century B.C. Braida di Vaglio Tomb 102.[1] Two unusual aspects of this pendant are its two sets of holes and the wear on its prominent areas. It is also the only example that has a perforation through the nose. This pendant may have been used at some time as a finial for a necklace, since it is not uncommon for finial heads to meet nose to nose in ancient gold jewelry.

NOTES

1. For the amber from Tomb 102 at Braida di Vaglio, see introduction, n. 276. For the ram's head, see Bottini and Setari 2003, p. 40, no. 134, pl. XLVI.

53. Spout or Finial: Ram's Head

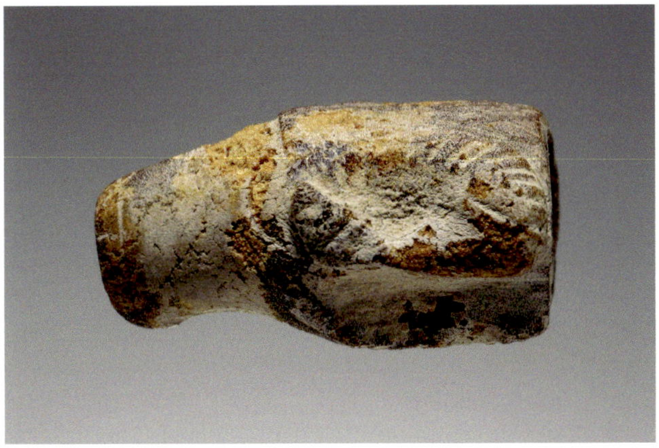

Accession Number	82.AO.161.4
Culture	Italic
Date	500–400 B.C.
Dimensions	Length: 26 mm; width: 17 mm; depth: 13 mm; Weight: 1.5 g
Subjects	Jewelry; Ram

Provenance

–1982, Jiří Frel, 1923–2006, and Faya Frel (Los Angeles, CA), donated to the J. Paul Getty Museum, 1982.

Condition

The piece is largely intact, suffering only a large break and loss to the right base of the right horn. A depression on the left horn may be ancient, as the rest of the horn curves around the flaw. The surface is covered by a thin, light tan alteration crust; freshly exposed areas have deteriorated further, to a yellow-ocher. In some areas, the surface appears slightly granular. Additionally, the surface is crazed and has yellowish residue in the cracks. The piece is yellowish gray in ambient light. At the break, the piece is orange. In transmitted light, it is opaque.

Description

The ram's head is long and slender, with a sloping muzzle. A slight ridge sets the fleece of the poll and cheeks off from the muzzle and horns. The surviving, left horn spirals in an oval, with irregular grooves indicating ridges. The eyes are closed. A shallow groove surrounds the muzzle; no other features are indicated.

A large perforation, 10 mm at the base and narrowing to 3 mm at the tip of the nose, runs through the pendant, likely for suspension. This form of suspension perforation is rare for amber pendants.

Discussion

The longitudinal suspension perforation and the schematic carving suggest that this head served as the spout of a small vessel, a design element of a ring, or the small finial of a necklace. The last is the most likely, since the opening is tapered toward the mouth, and the bottom of the ram's head is not flat. There are few Greek, Etruscan, or Italic examples of longitudinally perforated figured objects in amber or any other material (see also the Getty *Lion's Head* 76.AO.81, cat. no. 34).

Other Animal Heads

54. Pendant: Bovine Head

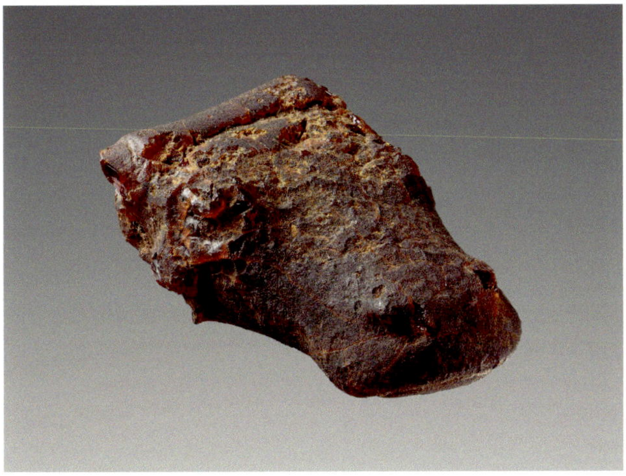

Accession Number	77.AO.81.20
Culture	Italic
Date	500–400 B.C.
Dimensions	Length: 35 mm; width: 24 mm; depth: 13 mm; Weight: 5.8 g
Subjects	Animals; Amulets; Egypt

Provenance

–1977, Gordon McLendon (Dallas, TX), donated to the J. Paul Getty Museum, 1977.

Condition

The horns of the pendant are broken off, with only stubs of exposed, unweathered amber remaining. The breaks on the horns appear to be modern. A large chip on the right ear and small chips on the fronts of the ears and nares also appear to be recent. The older degradation of the surface has resulted in overall pocking, flaking (especially on the forehead and the reverse of the pendant), and yellow-ocher material that is thickest in the larger craters and crevices, such as the interiors of the ears and the line of the mouth. Cracks are found under the chin and along the forehead, brow, and eyes. The cortex varies from brown to dark red-brown in ambient light. In transmitted light, the amber is translucent and a deep reddish orange, and extensive shallow cracking is apparent. There are no visible inclusions.

Description

Viewed in profile, this piece is slablike. Viewed frontally, the head's rectangularity is emphasized by the width of the muzzle and the flatness of the mouth's lower edge relative to the breadth of the head. The face, nose, and muzzle are smooth and almost level in plane, with the edge of the nares protruding just above this surface. The upper lip overhangs the lower, a groove separating the lips. The top of the pendant is flat; in front is the suspension spool and behind is the back of the animal's head. Judging from the remains of the breaks, the horns appear to have been about 1.5 mm in diameter at the base, likely curving outward and then upward. The eyes are high on the head, smallish, and plastically rendered, with the right eye in higher relief than the left. The ears, drooping downward, are shaped like short, broad leaves. The left ear is turned backward and is slightly more almond-shaped. The helixes are articulated by raised ridges; the ear openings are recessed. The decorative suspension spool may have replaced the forelock. The reverse side of the pendant, which includes the chin and throat, is nearly flat.

The face's slight asymmetry suggests that the pendant may be close in form to the original shape of the amber lump from which it was carved. A 1.5 mm perforation passes laterally through a bead-and-reel device even with the poll of the horns. The suspension device is slightly concave from end to end and is divided into approximately three beads by two grooves. A horizontal groove separates the device from the head. With the device carved into the forward part of the head, the pendant, when suspended, would lie flat, with the chin flush against the surface upon which it lay.

Discussion

The physiognomy of 77.AO.81.20 is described as much by subtle changes of surface modeling and plastic form as by linear definition. This pendant compares favorably with two other schematic bovine-subject ambers, one in the

British Museum (BM 79)[1] and another on the London art market. All three, despite their differences in style and morphology, present as hanging, detached heads seen from above. The poll and horns are at the top of the pendant and the muzzle at the bottom. They are flat, and their appearance suggests that they were worked from thin amber nodules. The undersides of all three are plain.

The sculptural description of 77.AO.81.20 relies on smooth planar transitions and modeling by abrasion; the others are more harshly worked, with greater use of the graver.

Bovine subjects, not counting two couchant man-headed, bull-bodied pendants in London and Paris,[2] are uncommon subjects in the corpus of pre-Roman figured amber. In addition to the three taurine head-pendants, six other amber carvings with bovine subjects are extant. They represent standing or recumbent bulls, cows, and calves. In every case, the head is reverted. The composition of these animals and animal groups follows a time-hallowed couchant type, one of the earliest animal composition types in the Near East.[3] The earliest example is the calf from the mid-seventh-century B.C. Tomb VI grave at Satricum.[4] Dating to the end of the sixth century is the recumbent bull (or cow) from Tomb 102 at Braida di Vaglio.[5] Three fifth-century examples include a double-subject amber (a recumbent cow or calf and a scallop shell) in Bologna,[6] a pendant of a recumbent cow with a milking calf (art market, New York), and a recumbent cow (once in the Gavin McKinley collection).[7] The bow decoration of a fibula from Belmonte Piceno (a calf or cow attacked by a lion) is related.[8]

To confirm the sex and age of bovine-subject ambers is essential for understanding why the subject was carved in amber and how such ambers were used, whether during the wearer's lifetime or for funerary purposes. The symbolism of the bull is age-old and is connected to hunting and conquering the animal, to its fertility, and to its guardian role in herd protection. Bull images might be clanic or mythic and symbolize a divinity, hero, king, pharaoh, or other ruler. Bull amulets can be classified as amulets of assimilation, conferring directly to the wearer the strength and virility of the animal, or of protection (especially if the subject was thought to symbolize a divinity).[9] Bull subjects were appropriate not only for a divinity or royalty; they also may have been specifically appropriate for an infant. As D. Plantzos, citing Menander, wrote, "A gold-plated iron ring with a device of 'bull or goat' helps to identify a baby."[10]

Alternatively, if the ambers depict a cow rather than a bull, the representation may incorporate other symbolic aspects and divine allusions. From earliest times in Egypt, the cow was considered to embody "all the most admired aspects of motherhood: she was fertile, protective, and provided sustenance for her young," and "from an equally early period, she was associated with Hathor, and later with Isis and the sky goddess Mehweret."[11] Amulets of a thin frontal bovine head with strongly curving lyre-shaped horns, first found in pre-Dynastic graves, are usually identified as cows' heads; they are associated with Bat or Hathor, who by the Middle Kingdom had completely assimilated the former and all her attributes.[12] From the Eighteenth Dynasty until the end of dynastic history, these amulets were used to depict Hathor.[13] As Carol Andrews notes, "Hathor-head amulets made of gold may be a punning reference to her epithet, 'the golden one.'"[14]

In the ancient Near East, the cow-and-calf motif is common from the Old Babylonian to the Neo-Assyrian period and has antecedents in earlier Near Eastern art, significantly in Sumerian art. It often appears to be a divine symbol and has been interpreted as an emblem of Ishtar or, perhaps more probably, of Ninhursaga, and was represented in apotropaic monumental sculpture at least in Urartu.[15] A rare Egyptian amulet type of the couchant calf, made of red-glazed faïence or cornelian, is possibly an amulet of rejuvenation.[16] A newborn calf in Egypt is the symbol of the infant sun.[17] A male *kriophoros*, or calf-carrier, in Greek art may represent two solar subjects, Apollo and the dawn.

In the opinion of this author, the softly modeled Getty pendant represents a cow rather than a bull. If this golden, sun-bright pendant carried with it a Hathoric association, the maternal and protective aspects of the amber object would have been emphasized, for Hathor was the celestial mother of the sun calf, protectress of the necropolis, goddess of love and music, nurse of the pharaoh, and consort of Horus.[18]

NOTES

1. Strong 1966, pp. 81–82, no. 79, pl. XXX (said to come from Armento). In style, this object is very like the pectoral ornament heads from Roscigno: see Losi et al. 1993; Holloway and Nabers 1982; and La Genière 1967.

2. The man-headed, bull-bodied carved ambers may represent the same kind of creature known in the Near East, a sun and fertility deity. Since amber had solar, fertility, and water-origin aspects, the bull-human anthromorphs would have been doubly powerful if they incorporated ancient Near Eastern aspects. For the pendant in London, see Strong 1966, p. 77, no. 68, pl. XXVII; for the very similar but smaller amber in Paris, A. de Ridder, *Musée Nationale du Louvre: Catalogue sommaire des bijoux antiques* (Paris, 1924), no. Bj 2123.

Generally speaking, bovine subjects are not especially popular in Archaic and Early Classical–period Greek, Etruscan, and Italic jewelry. When they do appear, bulls walking, running, savaging other animals, or represented as prey are more common than cows or calves in the repertoire of gems, finger rings, and other precious metal objects.

3. The oldest recumbent animal composition known to me is the Natufian stone young ungulate of ninth- to eighth-millennium date found at Umm ez-Zuweitina (Jerusalem, Rockefeller Archaeological Museum). See R. Neuville et al., *Le paléolithique et le mésolithique du désert de Judée* (Paris, 1951), pl. 14; and A. Spycket, *La statuaire du Proche-Orient ancien* (Leiden and Cologne, 1981), p. 26, pl. 17. See also Waarsenburg 1995, p. 436, n. 1169.

4. Waarsenburg 1995.

5. Bottini and Setari 2003, p. 40, no. 131, pl. XLV.

6. Bologna, Museo Civico Archeologico (unnumbered). The find is from Tomb 144, from the 1878 find in the area of the Giardini Margherita. See *Ambre* 2007, pp. 126–27; and Negroni Catacchio 1989, p. 662, figs. 48–56. On the back is a representation of a cockleshell (*Cardium edule*). One other carved amber pendant in the form of a shell (another cockleshell imitation?) is documented from the large pectoral ornament of Tomb 102 at Braida di Vaglio: see Bottini and Setari 2003, p. 40, no. 128, pl. XLV. That Tomb 102 has both bovine- and shell-subject pendants in the long necklace is significant in light of the proposed subject of the Bologna amber. *Cardium edule* is an early amuletic form in Egypt, and one that has a long life in the circum-Mediterranean area.

7. I am indebted to the late Gavin McKinley, London and Capetown, who generously encouraged my study of the pendant.

8. The lion-calf bow decoration was found in the same tomb with two other equally unique fibulae decorations, one representing a pair of addorsed lion foreparts and the other a lion and lioness or a lion attacking a deer. See Negroni Catacchio 1989, pp. 675–76, figs. 484–86, 489, pl. 8 AC.

9. Andrews 1994, p. 61.

10. Plantzos 1999, p. 19. See Menander, *Epitrepontes* 388–90.

11. Andrews 1994, p. 61.

12. Ibid., p. 62.

13. Ibid., p. 20.

14. Ibid.

15. Black and Green 1992, p. 53. A unique Old Kingdom glazed steatite amulet of a cow with its head turned back to its milking calf may be directly related to ancient Near Eastern typology. Andrews 1994, p. 62, writes that it was found in a male burial and that it may have been intended to provide the deceased with a supply of milk to drink in the afterworld.

16. Andrews 1994, p. 61.

17. Desroches-Noblecourt 2006, pp. 22–24.

18. Andrews 1994, p. 20.

55. Pendant: Horse's Head in Profile

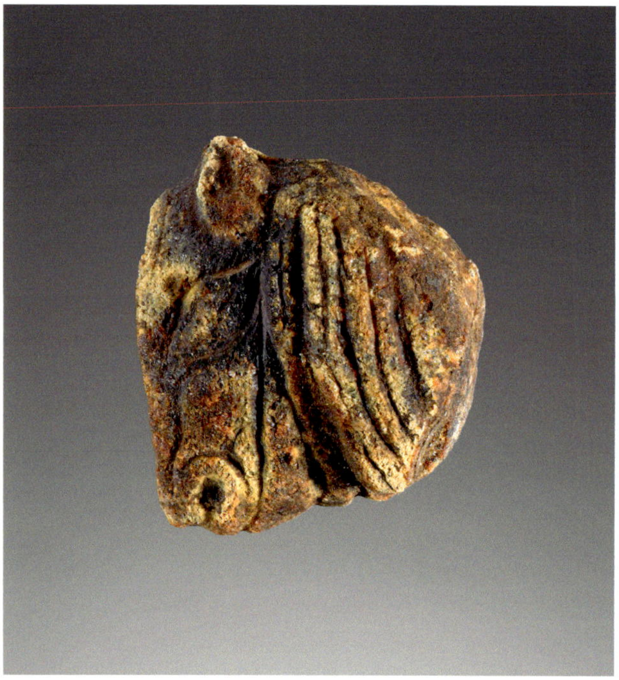

Accession Number	77.AO.81.6
Culture	Italic
Date	500–400 B.C.
Dimensions	Height: 37 mm; width: 36 mm; depth: 18.5 mm; Weight: 10.8 g
Subjects	Animals; Amulets; Dionysos, cult of (also Satyr); Etruscan culture; Inclusions; Magic

Provenance

–1977, Gordon McLendon (Dallas, TX), donated to the J. Paul Getty Museum, 1977.

Condition

The amber is intact, but the surface is in poor condition, degraded and grainy, with some old and weathered chipping and loss of surface detail at the tips of the ears and the base of the neck. On the reverse are small recent and larger old chip losses. The surface is blotchy red-brown, overcast by patches of light yellow and a rusty orange alteration surface layer. In transmitted light, the interior is a pale ruby color. There are no visible inclusions.

Description

The pendant, depicting the head and upper neck of a horse facing to the left, is plain on the reverse and worked on the obverse, with the design flowing onto the contiguous surfaces. The ventral termination occurs at the point just above the larynx, and the dorsal termination is at the base of the mane.

The ocular orbit bulges from the continuous raised line of the eyelids and is fullest at the center and depressed at the canthi. The leaflike left ear points straight upward and overlaps the partially represented right ear. The helix of the left ear is indicated by a raised line, while the opening is recessed. At the forehead, a small protrusion represents the forelock; one strand of hair, marked by a shallow groove, is indicated below and to the right of the suspension hole.

From the lower edge of the forelock to the tip of the rounded nose, the line of the face is almost straight. The muzzle and nares sweep up around the large circular indentation of the nostril. Below the eye is the protrusion of the cheek, gently undercut to emphasize the bulge of the nares. The line from the jaw to the mouth bows gently outward and is fullest through the cheek.

Behind the head is the fall of the mane, the hair rendered by eight unevenly spaced vertical grooves set at a slight diagonal angle parallel to the slant of the head. The edge of the mane is rounded but uneven and is set off from the flank by a groove. Between the mane and the head is a triangular section of amber representing the chest of the horse. The head is turned slightly toward the front.

The oblong contour, concave reverse, and convex obverse imply the shape of the original amber nodule. At the lower edge of the mane is a grooved indentation probably created by the removal of a fault. Rare evidence of the use of a pushed or driven tool is seen in the channel under the chin, which retains a succession of rippled cuts along the bottom, typical of a gouging tool. A 4.5 mm perforation for suspension passes through the front of the poll and exits behind the ears. Between them, cleverly worked into the design, is a stopped bore, likely a mortise for some kind of addition. When the pendant is suspended from the perforation, it is diamond-shaped, with the muzzle angled downward and the large, almond-shaped eye tilted upward.

Discussion

This head represents a horse as if in movement. Although it differs somewhat in style and type, it is of the same general form as three of the five other known amber pendants of horse subjects, a pendant in New York,[1] a pendant in a London private collection, and another on the Swiss art market.[2] Each one includes the head and neck of the animal in profile, with the head brought close to the body, the inside profile adjacent to the neck, and the neck arched. The animal is couped just above the jugular notch. Each example is perforated so that the horse's head is in a natural position of movement.

77.AO.81.6 is discussed here as the head of a horse, but on the analogy of the representation of some hippocamps in Etruscan art particularly, it may in fact be the head of a hippocamp (see the entry for 78.AO.286.1, cat. no. 29).

Two other horse heads are each of a different type. One is from a controlled excavation, Tomb 955 at Lavello-Casino, a "princely" tomb that provides important information about the context of figured ambers in the Basilicata. The woman buried in the fifth-century B.C. tomb was adorned with rich hair and body ornaments, including a girdle with five amber pendants. The largest is a pendant in the form of the foreparts of a rearing bridled horse, three others are illegible, and the fifth is a large female profile head-pendant.[3] The other horse subject is a pendant in London (British Museum 62, identified by Donald Strong as a grotesque head).[4] The Getty, New York, and London private collection pendants all have a double incised line at the bottom of the head, which emphasizes their bustlike format, and perhaps their meaning.[5]

The Getty horse's head is an artistic combination of a patterned representation and subtle modeling, suggesting that the carver integrated firsthand knowledge of the animal into an established prototype. The amber can be compared to Greek and Italian sculpture and vase painting of the Archaic period. The shape and proportion of the almond-shaped eye (like all amber carvings, it is characteristically without an incised pupil), the large fleshy nose with round nostril, and the form of the mane and especially the poll may reveal the artistic heritage of the Getty amber. The horse is of the type first seen in Orientalizing Greek vase painting of the seventh century, for instance the horses in the lion-hunt scene on the Chigi vase (a proto-Corinthian *olpe* from Veii) or on the Melian-ware amphora in Athens decorated with a representation of the Wedding of Herakles.[6] Its lineage can be traced to horses on Etrusco-Corinthian vases, and it is not far from horses and hippocamps painted on Etruscan black-figure wares. Sculptural comparisons from Etruria and South Italy especially demonstrate its Italian heritage, among them the bronze horse and rider from Grumento, of the mid-sixth century;[7] a bronze horse of the late sixth century, said to be from Locri, in New York;[8] and a hand-modeled terracotta of a horse (likely a patrice, or model for a mold) in Basel from around 600 B.C.[9]

The horse's head is a popular subject in sixth-century Greek and Etruscan vase painting and in Etruscan *bucchero*; it is found on coins, gems, bronze work, and funerary reliefs. Horses' heads are one of the many shapes of East Greece *aryballoi.*[10] Horses' heads are often paired on bronze and *bucchero* objects.[11] (Some seem to be of hippocamps rather than horses.) Some scholars believe that the large, bridled horses of one class of Attic black-figure amphorae (peaked from the end of the seventh century to the middle of the sixth) are the predecessors of the Panthenaic prize amphorae for equine events,[12] but they may have played roles in other ceremonies or in funerary ritual.

The most likely explanation for the horses' heads that encircle *bucchero* oinochoae and other shapes is that they had a funerary meaning. The number of vases, including plastic vases in the form of horses' heads, that have come from tombs suggests a direct connection between the subject, the tomb, and afterworld concerns, a possibility that deserves further attention. (This is not to deny the importance of the horse as a status symbol, its class and clanic associations, on the importance of horses in the elite culture of South Italy particularly.) The meaning of the horse's head in other contexts may shed light on the subject's "activity" as an ornament or amulet. On an Early Corinthian alabastron from Rhodes, what is the role of the curiously inserted large horse's head behind a centaur who grasps the arm of a woman? Does the scene represent Cheiron and Chariklo?[13] Horses' heads appear on some large early-fourth-century Metapontine terracotta reliefs of Dionysos-Hades reclining with a *kantharos* in hand (he is also joined by Kore and Iacchos): here the context of the horse's head is directly connected to the cult of Dionysos-Hades.[14]

In the ancient Near East, the horse's head, as Jeremy Black and Anthony Green point out, occurs as a divine symbol on a seal of second-millennium date and on Neo-Assyrian seals, as well as on a *kudurru* of the Babylonian king Nebuchadnezzar I (r. 1125–1104 B.C.), where a horse's head may represent a constellation. In the Neo-Assyrian period, the horse is the animal of the sun-god Šamaš (Utu), based on the associated winged disk.[15]

Why an ornament or amulet in the form of a horse's head? The subject is ancient: among the oldest known amber amulets from the Baltic is an equine (or elk's) head amulet of the Neolithic period.[16] The single horse's head is a rare subject in Egypt, the Aegean Islands, and Greece, which makes the subject's appearance in the Phoenician world stand out. A significant find from Orientalizing Italy is a Phoenician ivory protome in the form of a bridled, teeth-baring horse, which was included in the furnishings of the Barberini Tomb at Praeneste.[17]

Worn in life, the amber horse's head as pendant may have functioned as a sign of status: the horse, horse ownership, and the cavalier in the ancient world were markers of the political-religious elite. The horse might have acted as a metonym, naming the wearer as beautiful, or it might have alluded to the power of the horse-tamers of religio-mythological realms, or more directly to a heroic ancestor. As a permanent amulet, used in direct or aggressive magic, a horse's head might have conferred on its wearer the qualities affiliated with the horse or horse ownership, or the qualities of a deity or hero whose attribute was equine. A shining golden horse carved from solar amber could recall the great steeds that drew the chariot of the sun across the sky. Through magical assimilation, the wearer would be linked to Apollo, Eos, or Phaethon, for example. The link between the chariot of the sun, the new sun, and new life would follow on and link it to the age-old beliefs about the solar aspects of amber. Phaethon, Apollo's son, never suffered old age, instead became immortal, and was mourned in perpetuity by his sisters, who wept tears of amber. Eos, or the Etruscan Thesan, is the manifestation of the morning sun and aggressively abducts and pursues young men, who, too, will deny death. Thesan is an important solar and *kourotrophic* deity in Etruscan religion, and as A. Carpino reminds us, Thesan's love "could result in the attainment of immortality—the triumph over death."[18] In death, as a badge, a horse's head might also have brought to believers a lasting tie with the cult of Dionysos-Hades, securing the powers of the divinity for an individual's salvation after death.

A horse's head made from amber may have had added powers in warding away pain, for the horse was one of the many amulet types prescribed in Late Antiquity for abdominal pain[19] (which could include womb pain). As such, it would work in a "like banishes like" manner. An amber horse's head might ward off particular demons and dangers not only in life, but also after death, in the grave and on the voyage to the afterworld. If amber by itself could bring light into the tomb and symbolize the sun's regenerative power, a horse might bring it with the speed of the gods. At the very least, a horse's head could continue its protective and danger-averting functions, perpetually guarding the tomb.

NOTES

1. Metropolitan Museum of Art 24.97.117: Richter 1940, p. 32, fig. 103.
2. The other two heads are unpublished.
3. In Tomb 955, Lavello-Casino, the horse pendant was found on the front of the skeleton's pelvic area (she was placed on her back, with her legs drawn up, and turned to her right, as if in a seated position. The other large carved amber found in the Lavello-Casino burial, an unparalleled type of female head-pendant, was strung to the right of the horse pendant. For the female head-pendant, see the superb study by D'Ercole 1995. For the grave, see Bottini 1993; and *Magie d'ambra* 2005, p. 82.
4. Strong 1966, pp. 74–75, no. 62, pl. XXIV.
5. The same line (common on coins) is found on a few female head-pendants, including two of the heads from Tomb 164 at Banzi: see *Magie d'ambra* 2005, pp. 122–23.
6. The *olpe* is in Rome (Museo Nazionale Etrusco di Villa Giulia VG 22679). The Melian amphora from Melos is in Athens (National Archaeological Museum 354).
7. British Museum GR 1904,0703.1: Pugliese Carratelli 1996, p. 686, no. 113; and C. Rolley, *La sculpture grecque, I: Des origines au milieu du Vᵉ siècle* (Paris, 1994), pp. 123–24.
8. Metropolitan Museum of Art 58.180.1.
9. Antikenmuseum, Collection Ludwig BO 153: A. Bignasca in *Orient und frühes Griechenland: Kunstwerke der Sammlung H. und T. Bosshard*, ed. P. Blome (Basel, 1990), pp. 115–16, no. 172.
10. See Jean Ducat, *Les vases plastiques rhodiens archaïques en terre cuite*, fasc. 209 (Paris, 1966), pp. 107–12, pl. XV.
11. R. De Puma, CVA, *United States of America*, fasc. 31, *The J. Paul Getty Museum, Malibu*, fasc. 6 (Malibu, 1996), pl. 304 (with extensive bibl.).
12. For recent discussion of horse's-head amphorae, see *Centaur's Smile* 2003, p. 45, n. 4 (with references).

13. Rhodes, Archeological Museum 11150 (from Tomb 277 at Ialysos): CVA, *Italy*, fasc. 9, *Rhodes*, pl. 2.9; see also discussion and illustration in *Centaur's Smile* 2003, pp. 16–17, n. 97, fig. 13.

14. Metaponto, Museo Archeologico Nazionale 135679 (from the Metaponto theater votive deposit, first half of the fourth century B.C.): *The Wine of Dionysos: Banquets of Gods and Men in Basilicata*, exh. cat. (Rome, 2000), fig. 94.

15. Black and Green 1992, pp. 103–4.

16. Now lost. From Juodkrante, formerly in the University Museum, Königsberg: illustrated first in R. Klebs, *Stone Age Ornaments* (Königsberg, 1882), figs. 120–25.

17. Rome, Museo Nazionale Etrusco di Villa Giulia 13428: *I Fenici* 1988, p. 744, no. 940; and C. D. Curtis, *Sardis XIII: Jewelry and Gold Work* (Rome, 1925), p. 34, n. 56, pl. 15. 79.

18. A. Carpino, *Discs of Splendor: The Relief Mirrors of the Etruscans* (Madison, WI, 2003), p. 21, n. 109 (with references).

19. Bonner 1950, pp. 85–86.

56. Pendant: Asinine Head in Profile

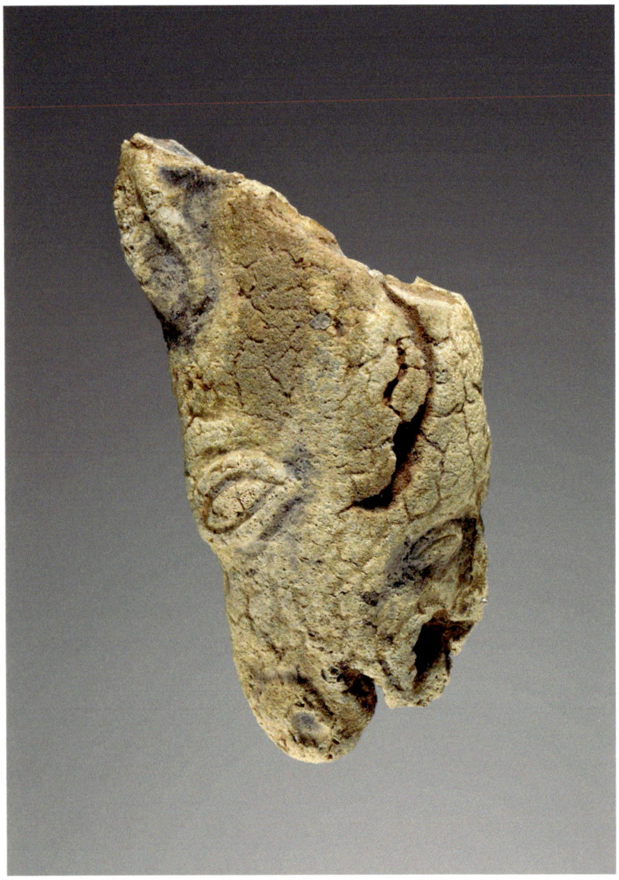

Accession Number	77.AO.81.24
Culture	Italic
Date	500–400 B.C.
Dimensions	Height: 48 mm; width: 59 mm; depth: 19 mm; Weight: 16.6 g
Subjects	Animals; Inclusions

Provenance

–1977, Gordon McLendon (Dallas, TX), donated to the J. Paul Getty Museum, 1977.

Condition

The piece is broken off at the neck area at a fissure; the tip of the left ear is missing. There are overall surface cracks. Large fissures are found at the jaw, at the mouth, on the cheek, and on much of the surface of the reverse. The surface has a thick, pale yellow deterioration crust, but little flaking has occurred; there is a cloudy area at the jaw. In ambient light, the piece is entirely opaque, predominantly yellowish tan, with patches of reddish brown on the nose and the temple area. In other areas, such as below the eye, the ear, and at the back (neck area), the amber is gray. In transmitted light, the piece is light orange. There are numerous inclusions.

Description

The obverse is convex and figured with the animal's head and neck facing left. The head is long and elliptical; the neck terminates at the right side in a curved irregular line that runs from the withers to the throatlatch. There is no indication of a mane. The eye is almond-shaped and has thick eyelids. The outside canthus turns downward. The bridge of the nose is rounded, and the nose has a large tip and rounded nostril. The mouth is open as if in laughter. Only the left ear is represented. It is upright and leaf-shaped, with the helix indicated by a raised line; the ear opening is recessed.

The form of the head seems to take advantage of the natural protrusions and undulations of the amber piece from which it was carved. The amber's shape may have been very like the finished product and may even have directed the subject and its disposition: the mouth seems to have been worked from a cleaned-out fissure and the jaw and eye from natural protuberances, the long ear appears to incorporate a depression, and the eye may have been formed from a small dome raised above the surface of the face. On the back of the head, near the break, is a section of the suspension perforation. The pendant is drilled with three large stopped bores, all about 5 mm in diameter. One bore enters just under the chin, proceeding upward for about 11 mm; a second, about 4 mm deep, enters the rear of the head and passes horizontally toward the front of the pendant; the third bore is on the reverse of the figure and proceeds to a depth of 13 mm.

Discussion

The terms *ass* and *donkey* are often interchanged today, and there appears also to have been some confusion in antiquity concerning representations of the nonhorse species of *Equus equidae*. Despite the schematic nature of the carving, the maker of the amber has emphasized the subject's asinine character: the animal is maneless, the ears are long, the nose is cupped at midlength, the muzzle is rounded, and the mouth is open as if braying. It seems even to be a specific breed, the wild ass, which, originally found in Africa, was domesticated by the third millennium B.C. (Variants of the wild ass have been bred for thousands of years and include donkeys.)

One other possible asinine-form amber is a head in the Vatican collections (findspot unrecorded), which, until now, has been considered a horse.[1] It is said to belong to a fibula (although it looks like a pendant) and was acquired at the same time as a second (much degraded) figured amber, which represents a bearded male in half-figure, who is carrying a *pithos* on his back (his arms reach backward). The head shape, ear position, ruff of hair on the jaw, and sparse mane, as well as the toothy grin, suggest to me an asinine rather than an equine subject. The style of the Vatican amber is entirely different from that of the Getty head. They differ also in manufacture.

Why an asinine amber? The association of asinine beasts with Dionysos, whose link with amber is well established, may be one reason. The ass served as transportation for Dionysos, and its presence might indicate the "hidden god" by association. An Attic rhyton in the form of a braying ass from Tomb 43 at Melfi-Pisciolo (Basilicata), a male tomb from the second half of the fifth century, is a tangible recognition of the Dionysian presence.[2] As Sarah Iles Johnston points out, the ass, the bird of prey, the horse, and the wolf were "the four animals whose traits the child-killer [demon] borrows in extant sources."[3] The ass is not usually a demonic animal in ancient Greece or anywhere else: the two exceptions are the association of the ass with the Egyptian Seth or Typhon and with the Near Eastern Lamashtu.[4] Johnston explains,

> *Lamashtu is often shown with ass's teeth and ass's ears on amuletic plaques and these teeth and ears are mentioned in ritual texts. Once, she is said to have an ass's form. Once, she is adjured to go away, like a savage ass! Body parts of asses can be used in amulets against Lamashtu, which may be a case of similia similibus.*[5]

Like the hyena, the ass was believed to have several obstetrical and pediatric uses, judging by Pliny's remarks in Book 28 of the *Natural History*,[6] where an ass's liver, worn as an amulet, was said to protect babies from epileptic fits. In Johnston's view, most recommended uses for ass parts in the magical papyri refer to the asinine form of Seth-Typhon and the "significance of the ass eventually became more broadly applied, too, so that it became a sort of all-purpose demonic animal, and its body parts became all-purpose amulets."[7] Campbell Bonner links the subject of the ass with aggressive amulets for women's pain in the abdomen (which in antiquity included the womb).[8] The roughly carved Getty pendant might even be said to be ugly, which might have aided its efficacy.[9] At the very least, an asinine amber was a potent amulet of healing. To be buried with a clamoring ass would be to be interred with an alert animal, ready to bray and avert danger.

NOTES

1. Vatican Museums, Museo Gregoriano Etrusco, inv. 13410.
2. Melfi, Museo Archeologico Nazionale del Melfese "Massimo Pallottino" 51488: *Magie d'ambra* 2005, p. 70; and *Popoli anellenici* 1971, p. 123, pl. LI.
3. Johnston 1995, p. 375.
4. Ibid., p. 377.
5. Ibid., nn. 40–41 (with important references).
6. Pliny, *Natural History* 28.77.
7. Johnston 1995, p. 385.
8. Bonner 1950, p. 121.
9. As Johnston 1995, p. 372, n. 3, points out, "Like demons throughout the world, child-killing demons generally are described as ugly."

Forgery

57. Statuette: Seated Divinity

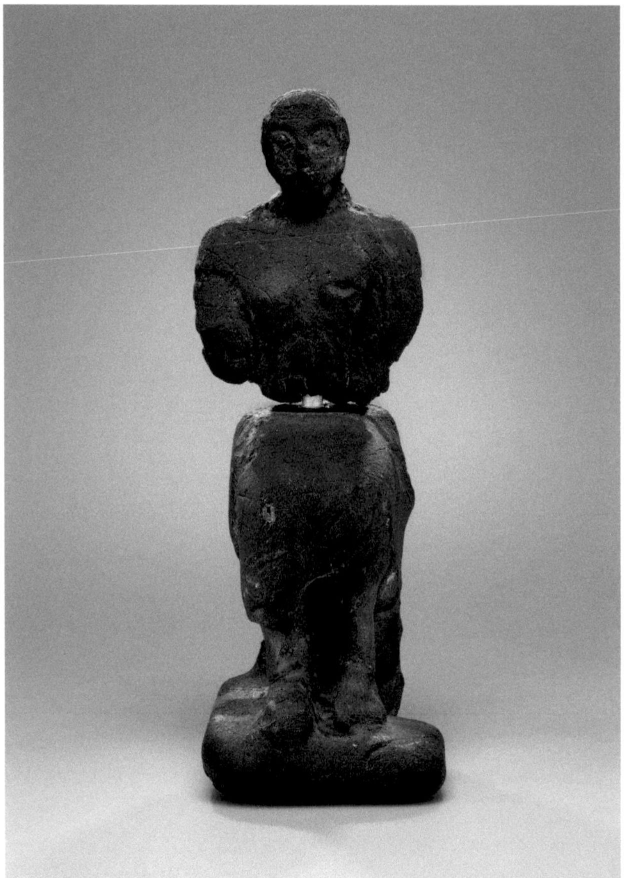

Accession Number	82.AO.51
Date	Modern
Dimensions	Height: 280 mm; width of base: 135 mm; Weight: 434 g

Provenance

–1982, Vasek Polak, 1914–97 (Hermosa Beach, CA), donated to the J. Paul Getty Museum, 1982.

Published

A. M. Shedrinsky, D. A. Grimaldi, J. J. Boon, and N. S. Baer, "Application of Pyrolysis Gas Chromatography and Pyrolysis Gas Chromatography/Mass Spectrometry to the Unmasking of Amber Forgeries," *Journal of Analytical and Applied Pyrolysis* 25 (1993): 77–95.

Discussion

After a preliminary assessment that this amber sculpture was a genuine pre-Roman work, further study of the sculpture and a chemical analysis of the amber showed it to be a modern object.[1] Stylistic similarities with the work of the forger who made the so-called Apollo of Fiumicino and an amber kouros, both once in the Grüneisen Collection, suggest that it is a work by the same hand.[2]

NOTES

1. For the chemical analysis, see A. M. Shedrinsky et al. (above, this cat. entry).
2. W. de Grüneisen, *Art Classique: Sculpture grecque, romaine, étrusque*, exh. cat. (Paris, 1925), pp. 1–3, pl. 1; and W. de Grüneisen, *Tableaux et esquisses de l'histoire de l'art: Apollon d'ambre trouvé à Fiumicino* (Paris, 1924). C. Albizzatti, "Analecta Gruenesiana," *Historia* [Milan], n.s., 1 (1927): 39–41, fig. 9, published the "Apollon" as a fake. I owe these references to R. D. De Puma (pers. comm., 1999).

Technical Essay: Analysis of Selected Ambers from the Collections of the J. Paul Getty Museum

Jeff Maish
Herant Khanjian
Michael R. Schilling

Introduction

Amber has been appreciated since antiquity for its unique aesthetic qualities in the production of small decorative objects. It has been a source of both mystery and curiosity, as it bridges the divide between the living and organic and the mineral and inorganic. It was initially selected for qualities such as color and hardness, with an eye toward an end market in jewelry production, and the Baltic Sea coastline has been, and continues to be, the largest source of the material.

The focus of amber studies over the past two hundred years has paralleled scientific developments in instrumentation and methods. Some of the earliest investigators used microscopy to view a hidden world of natural history and provide insights into past geological ages. More recent studies have analyzed the material itself in an attempt to better understand its chemistry, origins, and deterioration processes. This has included the identification of imitation ambers composed of natural and human-made compounds.[1]

Amber Characteristics

Although amber types have been classified generally, some ambiguities remain. Visual characteristics of amber such as color and translucency do not clearly relate to differences in chemical composition,[2] and some differences may relate more closely to inclusions, entrapment of air, and states of oxidation. Amber may also be defined by grade, color, or even geographic origin, such as Romanian or Sicilian. Ambers such as Baltic may be further subdivided into the categories allingite, beckerite, gedanite, or glessite, based in part on opacity, color, and friability.[3] Some subdivisions are also morphological. For example, amber with many tiny bubbles may be termed "bone" amber, whereas "foamy" amber has slightly larger bubbles. Amber typing can, therefore, be viewed from different perspectives ranging from morphological to chemical.

Amber Deterioration and Conservation

Although amber may have lain relatively dormant in geological deposits for thousands of years, its relatively recent collection, shaping, use, and reburial have often resulted in continued—and in some instances severe—deterioration. In general, deterioration manifests itself as a thick "corrosion" crust that not only obscures the translucent quality of amber but may also lead to flaking and loss of the carved surface. In the worst-case scenarios, the carved surface completely flakes off, leaving an ambiguously shaped amber core. Deterioration may continue in a collection's environment and be aggravated by pollutants, oxidation processes, and inappropriate environmental controls.[4] Recently, the degradation mechanisms and conservation treatments of archaeological amber have been studied using a variety of analytical instrumentation.[5]

Over the years, restorers and, more recently, conservators have attempted to reinforce fragile amber surfaces by applying a range of consolidative organic materials. Examples of past amber consolidants include dammar resin and "amber oil," a product of amber distillation.[6] A variety of waxes and natural and synthetic resins have also been applied. While preserving the morphological characteristics of carved amber, organic consolidants may interfere with future attempts to analyze or classify the amber. Therefore, the consolidation process should be carefully considered and, if carried out, fully documented.[7]

Scientific Analysis of Amber

The study of amber has kept pace over the past two centuries with the developments in scientific analysis. Microscopic studies beginning in the eighteenth century focused on the morphological characteristics of amber and the recognition of amber's botanical origins.[8] As methods for chemical analysis developed, so did the understanding of amber's complex chemical structure.[9] Considering the archaeological context of many amber finds, its characterization is further complicated by material degradation and possible interference from past stabilization treatments.[10] Beginning in the 1960s, analytical studies of amber relied heavily on infrared spectroscopy (IR)[11] and nuclear magnetic resonance (NMR).[12]

IR spectroscopy in particular was the first technique capable of readily identifying Baltic amber through the presence of a distinct succinic acid peak or "shoulder" in its infrared spectrum. However, the limits of this method were reached when it proved less successful in distinguishing among non-Baltic ambers. More-recent analytical studies have employed Raman spectroscopy,[13] capillary gas chromatography / mass spectrometry (GC/MS),[14] and pyrolysis–gas chromatography / mass spectrometry (Py-GC/MS),[15] which are capable of isolating a broad range of compounds that compose amber.[16] Combined with other analysis, this has led to proposals for the botanical origins of some ambers as well as common sourcing for previously distinct ambers.[17]

Current Research

The primary goal of the scientific investigation of a group of amber objects from the collection of the J. Paul Getty Museum was to verify that the ambers were indeed of Baltic origin. A secondary aim was to ascertain whether treatment with amber oil or other organic materials might interfere with the identification process. Samples were removed from the cores of twenty-six amber objects for analysis at the Getty Conservation Institute using Fourier-transform infrared spectroscopy (FTIR) and pyrolysis–gas chromatography / mass spectrometry with tetramethylammonium hydroxide for thermally-assisted hydrolysis and methylation (THM-Py-GC/MS). Surface samples were also removed from seven amber objects, in order to better understand the composition of weathered amber surfaces. For comparative purposes, tests were carried out on a number of reference materials, including Baltic amber, Dominican amber, copal resin, pine resin, sandarac resin, dried residue from amber-oil distillate, and amber varnish.

Fourier-Transform Infrared Spectrometry Procedure

The samples were analyzed on a Nic-plan infrared microscope equipped with a nitrogen-cooled MCT/A detector. Selected amber particles were placed on an infrared diamond window, flattened with a metal roller, and analyzed using a transmitted infrared beam apertured to 100 x 100 microns. The spectra are the sum of 100 scans at a resolution of 4 cm^{-1}. Infrared analysis of the samples produced spectra containing bands that correspond to amber. For example, a characteristic peak attributed to the carbon-oxygen bond at 1158 cm^{-1} distinguishes Baltic amber. Additional bands at 1737 and 1715 cm^{-1} are assigned to the ester and carboxylic acid groups, whereas peaks located at 1643 and 888 cm^{-1} are attributed to the exocyclic methylene group. Other components may be present in the samples at concentrations below the detection limit (5%).

THM–Pyrolysis–Gas Chromatography / Mass Spectrometry Procedure

Samples were tested on an HP 5972 gas chromatograph / mass spectrometer using a CDS Pyroprobe 2000, fitted with a valved interface at 330°C and purged with helium at 25 ml/minute. The split injector was at 340°C (30:1 ratio), and the MS transfer line was set to 300°C. A DB-5MS capillary column (30 M x 0.25 mm x 0.25 µm) was used, with helium at 44 cm/sec. The GC oven temperature program was 2 minutes at 40°C, then rising 6°C/minute to 310°C, and 13-minute isothermal. The solvent delay was 2.5 minutes. The mass spectrometer was scanned from *m/z* 35–700. Samples were placed into quartz tubes fitted with quartz wool, and three microliters of 25% tetramethyl ammonium hydroxide (TMAH) in methanol were introduced for derivatization. After 3 minutes, the tube was placed into a coiled filament probe, which was inserted into the valved interface. After purging for 3 seconds before pyrolysis, samples were pyrolyzed using the following temperature program: 200°C for 1 second, then ramped at 10°C/millisecond to 700°C, and held isothermally for 10 seconds.

Figure 1, an overlay of the FTIR spectra for Baltic amber and an amber object (76.AO.84, cat. no. 37), reveals characteristic spectral differences that make it possible to positively identify Baltic amber. The infrared spectrum of Baltic amber shows characteristic intense absorption bands at 2926, 2868, and 2849 cm^{-1}, attributed to C-H stretching modes of the CH_2 and CH_3 groups. A doublet for carbonyl C=O stretching peaks at 1738 and 1715 cm^{-1} is characteristic of ester and acid groups. Additional bands at 1259 and 1158 cm^{-1} are assigned to CO-O- modes of the succinate group, whereas the C-H bending modes for the terminal olefins are located at 888 cm^{-1}. Finally, the peak located at 1643 cm^{-1} is attributed to the exocyclic methylene group.

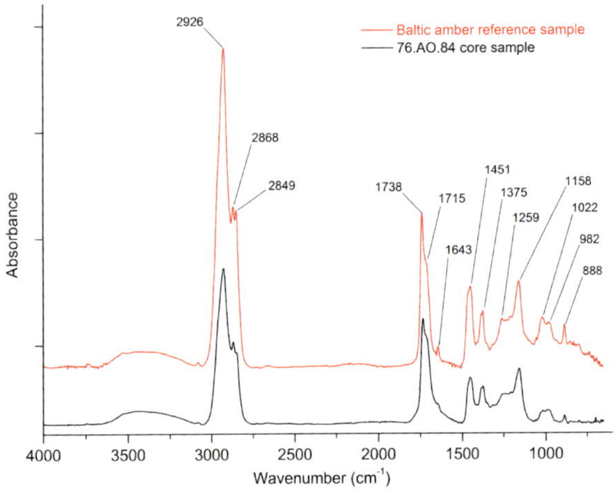

Figure 1 FTIR results for Baltic amber and amber object 76.AO.84 (cat. no. 37).

In THM-Py-GC/MS results for Baltic and Dominican amber standards (figure 2), a total of 69 compounds were identified. Many of these are sesquiterpene and diterpene compounds that are abundant, though not especially characteristic of the type of amber, as well as numerous nonspecific compounds. Succinic acid is the dominant marker compound for Baltic amber, and it appears in the chromatogram as a large peak at 10.3 minutes. In this study, succinic acid was analyzed in the form of the dimethyl ester derivative, and is abbreviated in figures as succinate. In figure 3, which shows THM-Py-GC/MS results for amber object 82.A0.161.285, other Baltic amber marker compounds are present in varying amounts, including fenchol, borneol, camphene, and camphor. Two very small peaks identified as methyl fenchyl succinate and methyl bornyl succinate (Mills et al. 1984) may also appear in THM-Py-GC/MS results for Baltic ambers.

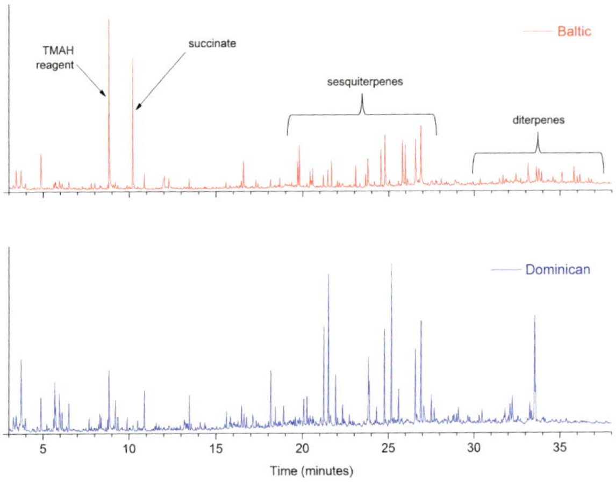

Figure 2 THM-Py-GC/MS results for Baltic and Dominican ambers.

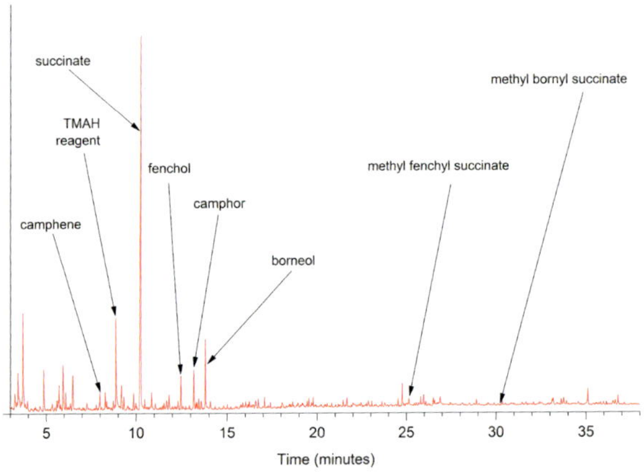

Figure 3 THM-Py-GC/MS results for core sample from amber object 82.AO.161.285.

Tables 1, 2, 3, and 4 list the various classes of compounds identified in the THM-Py-GC/MS analysis results of the amber objects and the reference materials. The identifications were based primarily on the results from mass spectral library searching using the NIST MS Search 2.0 program, and supplemented by published data (Mills et al. 1984). Although the NIST results of the nonspecific compounds listed in table 5 were inconclusive, the unknown compounds did appear on a rather consistent basis in the objects.

The THM-Py-GC/MS results for the reference samples appear in table 6. In this and all subsequent tables, the test results are expressed in terms of peak-area percentages relative to the total peak area for all of the compounds listed in tables 1, 2, 3, 4, and 5 (except for methyl fenchyl succinate, methyl bornyl succinate, and dibornyl succinate, which, due to their extremely small peak sizes, did not contribute significantly to the total peak area). Table 6 shows that the succinate content in the single known sample of Baltic amber was high, whereas almost no succinate was detected in the Dominican ambers, copal resin, sandarac resin, or pine resin. The succinate content in the ambers of unknown origin appeared rather variable, but the presence of the other markers in table 1 placed them firmly in the Baltic category. The "amber varnish" was found to contain a high concentration of a drying oil with no detectable succinate content. Fortunately, the test results for dried amber oil residue showed no significant amount of any of the Baltic marker compounds listed in table 1 except for borneol, indicating that amber oil treatment should not produce a "false positive" identification for Baltic amber.

In the THM-Py-GC/MS results for the core samples from the untreated amber objects (table 7), the most striking feature is the remarkably broad range for the succinate content compared to the composition of the standards. In an overlay of FTIR spectra for some of these samples (figure 4), the main trend is the shift of the carbonyl peak to a lower wavenumber with increasing succinate content, which is characteristic of the conversion of esters to carboxylic acids. These results provide evidence that partial hydrolysis of the succinate esters in the objects has occurred, which is a reaction that would enrich the residual amber in succinic acid.

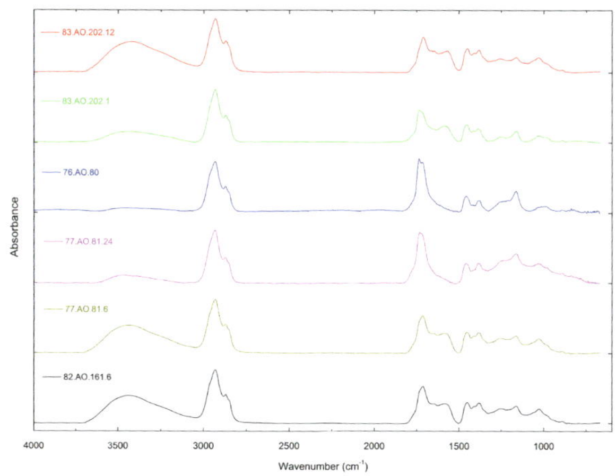

Figure 4 Variation in FTIR spectrum with succinate content for amber objects (core samples).

One concern in this study was that the composition of the surface crusts of the amber objects might be considerably different from that of the inner cores, due to hydrolysis, weathering, handling, and treatment. This is why core samples were removed from the objects by microdrilling. In figure 5, the THM-Py-GC/MS results for the dark surface and inner core of a large piece of reference amber, it is clear that the surface has become partially depleted in succinate, with few other changes apparent. Table 8 shows the results for pairs of surface and core samples from the amber objects, and figure 6 shows a typical chromatographic result (for 83.AO.202.1, cat. no. 12). The surfaces of these objects have also been depleted in succinate, but the sesquiterpenes and diterpenes also have been radically reduced. These compounds are not chemically bound to the polymeric network of the amber, which would make them more susceptible to leaching during burial.

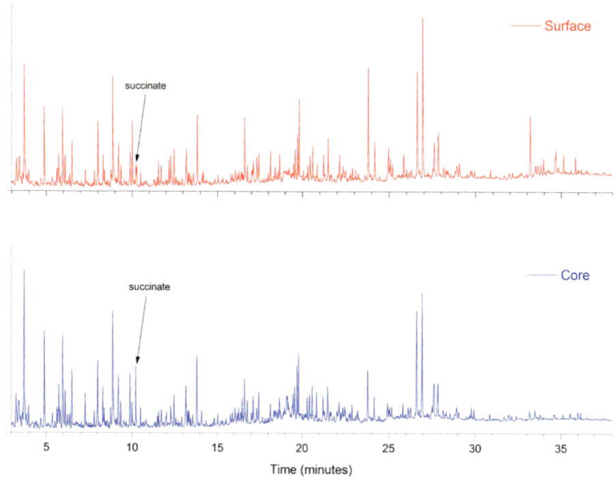

Figure 5 THM-Py-GC/MS results for dark surface and inner core of amber from Verfmolen 'De Kat.'

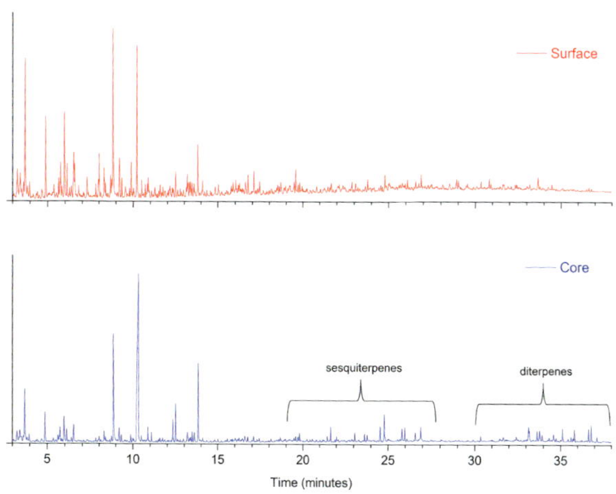

Figure 6 THM-Py-GC/MS results for surface and inner core samples from 83.AO.202.1 (cat. no. 12).

FTIR analysis also reveals important details about the nature of the surface and core compositions. Figure 7 shows FTIR spectra for surface and core samples from 83.AO.202.1. The saturated C-H bands at 2927 and 2869 cm^{-1} in the spectrum of the surface sample are reduced, whereas the C-O stretching modes at 1159 cm^{1} in the fingerprint region are more intense. This indicates that the surface is more highly oxidized than the core. The other important peak appears at 1574 cm^{-1}, which is due to salts of succinic acid. There is a much higher concentration of succinate salts in the surface sample, which is consistent with exposure to alkaline conditions during some period of time.[18] This might have occurred during burial, or resulted from harsh cleaning with alkaline chemicals. FTIR spectra of two surface samples and a core from 82.AO.161.7 (cat. no. 24) (figure 8) show an increased O-H stretching band in the surface sample, with a shift in the C=O band to lower wavenumbers, indicating the prevalence of carboxylic acids. However, the succinate salt peak at 1574 cm^{-1} is only a slight shoulder on the carbonyl peak, indicating that this object was not exposed to the same harsh alkaline conditions as 83.AO.202.1.

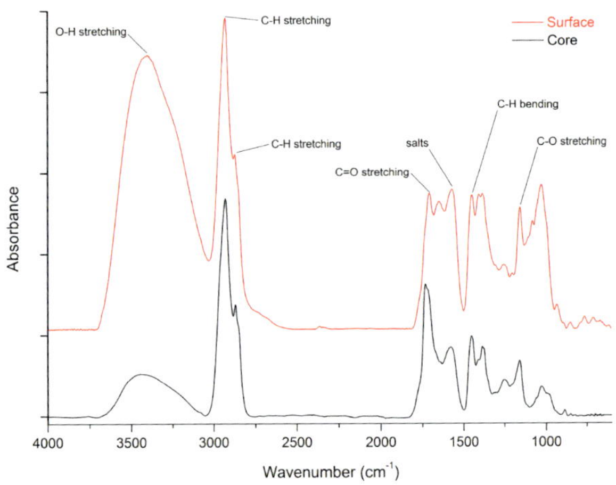

Figure 7 FTIR spectra for surface and core samples from 83.AO.202.1 (cat. no. 12).

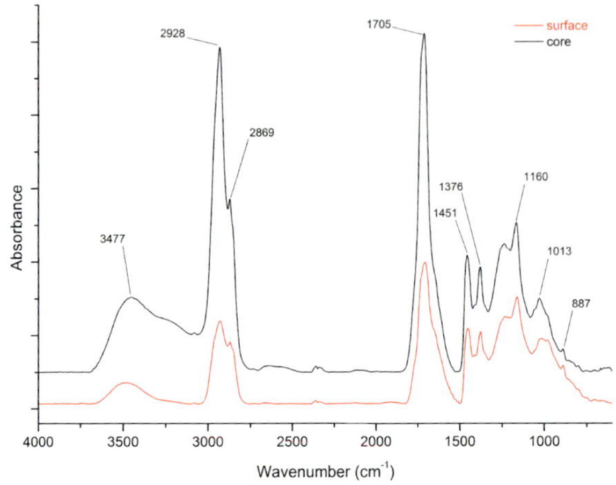

Figure 8 FTIR spectra for surface and core samples from 82.AO.161.7 (cat. no. 24).

Table 9 lists the THM-Py-GC/MS results for the treated amber objects, and representative chromatograms are shown in figure 9. Azelaic acid was detected in three of the objects: 77.AO.81.29 (cat. no. 16), 77.AO.81.5 (cat. no. 23), and 77.AO.81.30 (cat. no. 25). This is a common marker compound for cross-linked drying oils, and its presence along with palmitic acid and stearic acid indicates that drying oils may have been applied to these objects in an alternative type of conservation treatment. In 77.AO.81.4 (cat. no. 14), palmitic and stearic acids were detected along with cholesterol, but azelaic acid was absent. This suggests that an animal fat could have been applied to this object as another type of alternative treatment. Three amber objects tested in this study had been previously treated with amber oil: 77.AO.84 (cat. no. 1), 77.AO.83 (cat. no. 38), and 77.AO.81.7 (cat. no. 41). Their extremely high succinate contents suggest that they were highly degraded prior to treatment. In figure 10, the FTIR spectra for selected treated samples show that treatment with drying oil or amber oil does not interfere with the identification of Baltic amber.

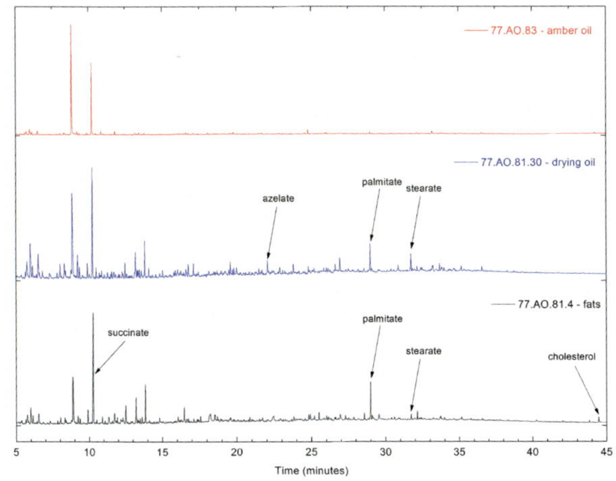

Figure 9 THM-Py-GC/MS results for surface samples from treated amber objects.

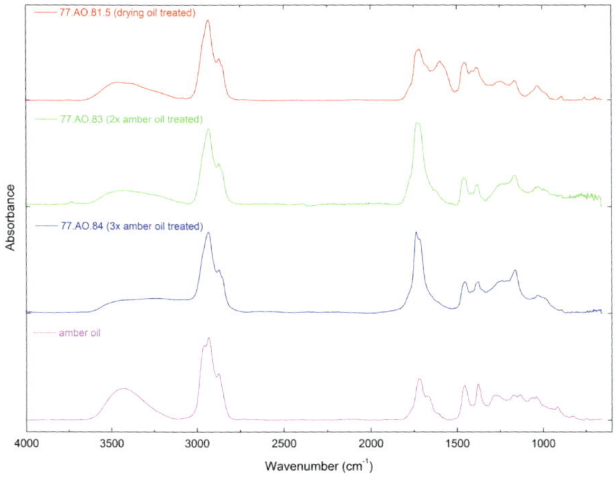

Figure 10 FTIR results for core samples from treated amber objects.

Conclusions

This study has demonstrated that chemical analysis using FTIR and THM-Py-GC/MS can provide rich details concerning the composition of antique amber objects. Fundamentally, the analytical results showed that all of the amber objects in the Getty Museum are classified as Baltic amber. Additional information revealed the nature and extent of deterioration, and provided tantalizing hints about the nature of the burial conditions to which some of these objects may have been exposed. Finally, detection of certain marker compounds has shown that a number of amber objects were treated with drying oils and fats and, furthermore, that amber oil treatment does not interfere with the provenancing process.

Tables

Table 1: Marker Compounds from THM-Py-GC/MS Analysis of Amber

IUPAC Name	Synonym	CAS #	Formula	MW	Retention Time (min)
Bicyclo[2.2.1]heptane, 2,2-dimethyl-3-methylene-	Camphene	79-92-5	$C_{10}H_{16}$	136	8.04
Butanedioic acid, dimethyl ester	Succinic acid, dimethyl ester	106-65-0	$C_6H_{10}O_4$	146	10.27
Bicyclo[2.2.1]heptane, 2-methoxy-1,3,3-trimethyl-	Methyl fenchyl ether	N/A	$C_{11}H_{20}O$	168	12.35
Bicyclo[2.2.1]heptan-2-ol, 1,3,3-trimethyl-	Fenchyl alcohol	1632-73-1	$C_{10}H_{18}O$	154	12.50
Bicyclo[2.2.1]heptan-2-one, 1,7,7-trimethyl-, (1S)-	L-camphor	464-48-2	$C_{10}H_{16}O$	152	13.20
Bicyclo[2.2.1]heptan-2-methoxy, 1,7,7-trimethyl-, (1S-endo)-	Methyl bornyl ether	N/A	$C_{11}H_{20}O$	168	13.48
Bicyclo[2.2.1]heptan-2-ol, 1,7,7-trimethyl-, (1S-endo)-	L-borneol	464-45-9	$C_{10}H_{18}O$	154	13.83
Naphthalene, 1,2,3,4-tetrahydro-1,8-dimethyl-		25419-33-4	$C_{12}H_{16}$	160	17.44
Cedren-13-methoxy, 8-		N/A	$C_{16}H_{26}O$	234	24.80
	Methyl fenchyl succinate	N/A	$C_{15}H_{24}O_4$	268	25.16
	Methyl bornyl succinate	N/A	$C_{15}H_{24}O_4$	268	26.54
Cedren-13-ol, 8-		18319-35-2	$C_{15}H_{24}O$	220	26.92
	Dibornyl succinate	N/A	$C_{24}H_{38}O_4$	390	38.64

Table 2: Diterpenes Identified in Amber Objects Using THM-Py-GC/MS Analysis

IUPAC Name	Synonym	CAS #	Formula	MW	Retention Time (min)
Podocarp-8-en-15-oic acid, 13alpha-methyl-13-vinyl-, methyl ester	Methyl pimara-8,15-dien-18-oate	19907-21-2	$C_{21}H_{32}O_2$	316	33.17
Podocarpa-8,11,13-trien-15-oic acid, 13-isopropyl-, methyl ester	Methyl dehydroabietate	1235-74-1	$C_{21}H_{30}O_2$	314	35.14
Methyl 5-(5,5,8a-trimethyl-2-methylenedecahydro-1-naphthalenyl)-3-methylpentanoate	Labd-8(20)-en-15-oic acid, methyl ester	13008-80-5	$C_{21}H_{36}O_2$	320	33.49
Methyl pimar-7-en-18-oate		72088-13-2	$C_{21}H_{34}O_2$	318	33.80
Labda-8(20),12,14-trien-19-oic acid, methyl ester, (Z)-	Methyl cis-Communate	10178-35-5	$C_{21}H_{32}O_2$	316	33.95
Podocarp-8(14)-en-15-oic acid, 13á-methyl-13-vinyl-, methyl ester	Methyl sandaracopimarate	1686-54-0	$C_{21}H_{32}O_2$	316	33.95
Podocarp-7-en-15-oic acid, 13á-methyl-13-vinyl-, methyl ester	Methyl isopimarate	1686-62-0	$C_{21}H_{32}O_2$	316	34.61
Podocarpa-7,13-dien-15-oic acid, 13-isopropyl-, methyl ester	Methyl abietate	127-25-3	$C_{21}H_{32}O_2$	316	35.84
Podocarpa-6,8,11,13-tetraen-15-oic acid, 13-isopropyl-, methyl ester	Methyl 6-dehydrodehydroabietate	18492-76-7	$C_{21}H_{28}O_2$	312	36.43

Table 3: Fatty Acids in THM-Py-GC/MS Analysis of Lipids

IUPAC Name	Synonym	CAS #	Formula	MW	Retention Time (min)
Hexanedioic acid, dimethyl ester	dimethyl adipate	627-93-0	$C_8H_{28}O_4$	174	15.47
Heptanedioic acid, dimethyl ester	dimethyl pimelate	1732-08-7	$C_9H_{16}O_4$	188	17.78
Octanedioic acid, dimethyl ester	dimethyl suberate	1732-09-8	$C_{10}H_{18}O_4$	202	20.02
Dodecanoic acid, methyl ester	methyl laurate	111-82-0	$C_{13}H_{26}O_2$	214	21.63
Nonanedioic acid, dimethyl ester	dimethyl azelate	1732-10-1	$C_{11}H_{20}O_4$	216	22.10
Decanedioic acid, dimethyl ester	dimethyl sebacate	106-79-6	$C_{12}H_{22}O_4$	230	24.03
Tetradecanoic acid, methyl ester	methyl myristate	124-10-7	$C_{15}H_{30}O_2$	242	25.48
Hexadecanoic acid, methyl ester	methyl palmitate	112-39-0	$C_{17}H_{34}O_2$	270	28.96
9-Octadecenoic acid (Z)-, methyl ester	methyl oleate	112-62-9	$C_{19}H_{36}O_2$	296	31.71
Octadecanoic acid, methyl ester	methyl stearate	112-61-8	$C_{19}H_{38}O_2$	298	32.13
Eicosanoic acid, methyl ester	methyl arachidate	1120-28-1	$C_{21}H_{42}O_2$	326	35.01

Table 4: Nonspecific Compounds Identified in THM-Py-GC/MS Analysis of Amber Objects

IUPAC Name	Synonym	CAS #	Formula	MW	Retention Time (min)
Methyl benzene	Toluene	108-88-3	C_7H_8	92	3.76
1,3-Dimethyl-1-cyclohexene		2808-76-6	C_8H_{14}	110	4.92
Benzene, 1,4-dimethyl-	p-Xylene	106-42-3	C_8H_{10}	106	5.99
Benzene, 1,3-dimethyl-	m-Xylene	108-38-3	C_8H_{10}	106	5.99
2-Propenoic acid, 2-methyl-, methyl ester	Methyl methacrylate	80-62-6	$C_5H_8O_2$	100	2.87
Ethylbenzene		100-41-4	C_8H_{10}	106	5.77
Benzene, 1,2-dimethyl-	o-Xylene	95-47-6	C_8H_{10}	106	6.53
Benzene, 1-ethyl-3-methyl-	Toluene, m-ethyl-	620-14-4	C_9H_{12}	120	8.33
Benzene, 1-ethenyl-2-methyl-	o-Vinyltoluene	611-15-4	C_9H_{10}	118	9.24
Benzenemethanol, 2,5-dimethyl-	2,5-Dimethylbenzyl alcohol	53957-33-8	$C_9H_{12}O$	136	9.36
Benzene, 1,2,4-trimethyl-	Pseudocumene	95-63-6	C_9H_{12}	120	9.90
3,3,5,5-Tetramethylcyclopentene		38667-10-6	C_9H_{16}	124	10.89
Benzoic acid, methyl ester	Methyl benzoate	93-58-3	$C_8H_8O_2$	136	11.85
Naphthalene		91-20-3	$C_{10}H_8$	128	14.12
Naphthalene, 1-methyl-		90-12-0	$C_{11}H_{10}$	142	16.62
Naphthalene, 2-methyl-		91-57-6	$C_{11}H_{10}$	142	17.11
1,2,3-Trimethylindene		4773-83-5	$C_{12}H_{14}$	158	18.68
Naphthalene, 1,6,7-trimethyl-		2245-38-7	$C_{13}H_{14}$	170	22.93

Table 5: Nonspecific Compounds Tentatively Identified in THM-Py-GC/MS Analysis of Amber Objects

IUPAC Name	CAS #	Formula	MW	Retention Time (min)
Methyltricyclo[2.2.1.0(2,6)]heptane	4601-85-8	C_8H_{12}	108	6.15
Cyclopentane, 2-ethylidene-1,1-dimethyl-	56324-66-4	C_9H_{16}	124	7.84
1,3,3-Trimethyl-2-(2-methyl-cyclopropyl)-cyclohexene	285129-06-8	$C_{13}H_{22}$	178	16.62
2-Buten-1-one, 1-(2,6,6-trimethyl-1-cyclohexen-1-yl)-	35044-68-9	$C_{13}H_{20}O$	192	19.73
1,5,9,9-Tetramethyl-2-methylene-spiro[3.5]non-5-ene	N/A	$C_{14}H_{22}$	190	19.82
Bicyclo[4.1.0]heptan-2-ol, 1beta-(3-methyl-1,3-butadienyl)-2alpha, 6beta-dimethyl-3beta-acetoxy-	N/A	$C_{16}H_{24}O_3$	264	21.49
2-Methyl-4-(2,6,6-trimethylcyclohex-1-enyl)but-2-en-1-ol	62924-17-8	$C_{14}H_{24}O$	208	21.70
8-Acetyl-5,5-dimethyl-nona-2,3,8-trienoic acid, methyl ester	68799-74-6	$C_{14}H_{20}O_3$	236	23.09
2-Methyl-4-(2,6,6-trimethylcyclohex-1-enyl)but-2-en-1-ol	62924-17-8	$C_{14}H_{24}O$	208	23.83
7a-Isopropenyl-4,5-dimethyloctahydroindene-4-carboxylic acid	N/A	$C_{15}H_{24}O_2$	236	24.56
2-[5-(2,2-Dimethyl-6-methylene-cyclohexyl)-3-methyl-pent-2-enyl]-[1,4]benzoquinone	N/A	$C_{21}H_{28}O_2$	312	25.17
Acetic acid, (1,2,3,4,5,6,7,8-octahydro-3,8,8-trimethylnaphth-2-yl)methyl ester	314773-27-8	$C_{16}H_{26}O_2$	250	25.83
Acetic acid, 3-(6,6-dimethyl-2-methylenecyclohex-3-enylidene)-1-methylbutyl ester	N/A	$C_{16}H_{24}O_2$	248	26.00

Table 6: THM-Py-GC/MS Results for Reference Samples

Sample	Supplier	GCI Identifier	Peak Area Percentages		
			Succinate	Diterpenes	Fatty Acids
Amber varnish	Zecchi	VARN0084	0	0	77
Copal resin	JPGM		0	36	0
Pine resin	GCI	NRES0244	0	74	0
Sandarac resin	Verfmolen 'De Kat'	NRES0295	0	19	0
Amber oil	JPGM		1	1	3
Dominican amber	JPGM		0	14	1
Amber (Dominican?)	JPGM	NRES0095	2	1	0
Amber	Kremer	NRES0005	3	2	1
Yellow amber	Verfmolen 'De Kat'	NRES0296	3	2	1
Amber	Zecchi		10	2	2
Baltic amber	JPGM		12	12	0.4
Amber	Zecchi	NRES0305	14	4	1
Amber	Kremer	NRES0004	20	6	1
Amber	Kremer	NRES0171	20	4	1

Table 7: THM-Py-GC/MS Results for Untreated Amber Object Core Samples

Accession #	**Peak Area Percentages**		
	Succinate	**Diterpenes**	**Fatty Acids**
(cat. no. 21)	13	1	0
	20	10	0
	24	5	0.4
(cat. no. 18)	25	12	0.4
(cat. no. 11)	27	11	0
	28	12	0.6
(cat. no. 22)	34	7	0.7
	34	9	0.0
82.AC.161.285	35	4	0.7
(cat. no. 33)	36	5	0.6
(cat. no. 56)	38	22	0.5
(cat. no. 28)	38	8	0.3
(cat. no. 53)	39	9	1.1
(cat. no. 29)	39	16	0
(cat. no. 20)	41	8	0.9
(cat. no. 17)	41	9	0.6
(cat. no. 13)	46	7	0.3
(cat. no. 55)	47	16	0.4
(cat. no. 27)	53	4	0.3
(cat. no. 9)	65	6	0.4
average	36	9.1	0.4
standard deviation	12	5.0	0.3

Table 8: THM-Py-GC/MS Results for Amber Object Core & Surface Samples

Accession #	Sample Location	Peak Area Percentages		
		Succinate	Diterpenes	Fatty Acids
	core	20	9.6	0.0
	surface	4.2	9.1	0.8
	core	25	12	0.4
	surface	16	4.6	1.6
(cat. no. 19)	core	27	11	0.0
	surface	11	1.7	1.5
	core	34	9.1	0.0
	surface	11	2.0	1.5
	core	38	8.1	0.3
	surface	29	4.3	0.7
	core	41	8.0	0.9
	surface	12	2.0	1.2

Table 9: THM-Py-GC/MS Results for Treated Amber Objects

Accession #	Sample Location	Treatment	Peak Area Percentages		
			Succinate	Diterpenes	Fatty Acids
	core	drying oil	34	5.2	18
	surface	drying oil	13	4.6	7.7
	core	drying oil	29	6.4	6.6
	core	fatty substance	39	6.2	0.7
	surface		23	2.7	8.1
	core	amber oil, once	72	5.1	0.6
	surface		44	7.6	0.4
	core	amber oil, twice	38	5.6	1.9
	core	amber oil, three times	54	2.9	0.4

NOTES

1. See N. Kalsbeek and K. Botfeldt, "Identification of Amber and Amber Imitations by Infrared Spectroscopy," *Meddelelser om konservering* no. 1 (2007): 3–11. Imitations have included materials such as Bakelite, nitrocellulose, polystyrene, and plant resins.

2. See Rice 2006 for a discussion of amber and its terminology.

3. E. Stout, C. Beck, and B. Kosmowska-Ceranowicz, for example, used infrared spectroscopy (IR) to compare and separate gedano-succinite from succinite: see "Gedanite and Gedano-Succinite," in Anderson and Crelling 1995, pp. 130–48.

4. See J. Waddington and J. Fenn, "Preventive Conservation of Amber: Some Preliminary Investigations," *Collection Forum* 4, no. 2, (Fall 1988): 25–31; and Y. Shashoua, National Museum of Denmark, 2002, http://www.natmus.dk/cons/reports/2002/amber/amber.pdf.

5. G. Pastorelli, "Archaeological Baltic Amber: Degradation Mechanisms and Conservation Measures" (Ph.D. diss., University of Bologna, 2009).

6. F. Preusser, "Zur Restaurierung von stark korrodiertem Bernstein" ("The Restoration of Badly Weathered Amber"), *Arbeitsblätter für Restauratoren* 9, no. 2 (1976): 75–77.

7. See N. Bromelle, C. Beck, and G. Thomson, "Authentication and Conservation of Amber: Conflict of Interests," in *Science and Technology in the Service of Conservation: Preprints of the Contributions to the Washington Congress, 3–9 September 1982*, ed. N. Bromelle and G. Thomson (London, 1982), pp. 104–7.

8. The relationship of Baltic amber and succinic acid was recognized in the early nineteenth century by chemists in Germany, where succinic acid was isolated using strong acids and bases. One study identified the amber constituent camphor (borneol) through smell.

9. For an excellent overview of amber and resin studies see I. Angelini in G. Artioli, *Scientific Methods and Cultural Heritage: An Introduction to the Application of Materials Science to Archaeometry and Conservation Science* (New York, 2010).

10. Brommelle et al. 1982 (in n. 7, above) first warned that most conservation materials will interfere with infrared spectra. D. Thickett discusses the problems of consolidant removal and the effects of solvents on amber in "The Influence of Solvents on the Analysis of Amber," in *Conservation Science in the UK: Preprints of the Meeting Held in Glasgow, May 1993*, ed. N. J. Tennent (London, 1993), pp. 49–56.

11. The earliest IR studies of amber were carried out most notably by J. Langenheim, C. Beck, and R. Rottländer. See, for example, C. Beck and H. Hartnett, "Sicilian Amber," in Beck and Bouzek 1993, pp. 36–47; and C. Beck, "Spectroscopic Investigations of Amber," *Applied Spectroscopy Reviews* 22, no. 1 (1986): 57–110. Amber studies have benefited from further developments such as Fourier-transform infrared spectroscopy (FTIR), diffuse-reflectance infrared Fourier-transform (DRIFT), and Attenuated Total Reflection (ATR) FTIR. For the DRIFT method, see I. Angelini and P. Bellintani, "Archaeological Ambers from Northern Italy: An FTIR-DRIFT Study of Provenance by Comparison with the Geological Amber Database," *Archaeometry* 47, no. 2 (2005): 441–54. For the ATR method, see M. Guiliano, L. Asia, G. Onoratini, and G. Mille, "Applications of Diamond Crystal ATR FTIR Spectroscopy to the Characterization of Ambers," *Spectrochimica Acta Part A: Molecular and Biomolecular Spectroscopy* 67, no. 5 (2007): 1407–11.

12. For a summary of the use of the nuclear magnetic resonance method in amber analysis, see Artioli, *Scientific Methods and Cultural Heritage*, p. 383 (in n. 9, above).

13. See Y. Shashoua et al., "Raman and ATR-FTIR Spectroscopies Applied to the Conservation of Archaeological Baltic Amber," *Journal of Raman Spectroscopy* 37, no. 10 (September 2006).

14. J. Mills, R. White, and L. Gough, "The Chemical Composition of Baltic Amber," *Chemical Geology* 47 (1984): 15–39.

15. See A. Shedrinsky et al., "The Use of Pyrolysis Gas Chromatography (PyGC) in the Identification of Oils and Resins Found in Art and Archaeology," *Conservation of Cultural Property in India* 21 (1988): 35–41; and J. Boon, A. Tom, and J. Purveen, "Microgram Scale Pyrolysis Mass Spectrometric and Pyrolysis Gas Chromatographic Characterization of Geological and Archaeological Amber and Resin Samples," in Beck and Bouzek 1993, pp. 9–27. For comprehensive overviews of the method, see "The Application of Analytical Pyrolysis to the Study of Cultural Materials," chap. 6 in *Applied Pyrolysis Handbook*, 2nd ed., ed. Thomas Wampler (Boca Raton, FL, 2007), pp. 105–31.

16. Gas chromatography effectively separates volatile, solvent-extractable components, which are subsequently detected and analyzed by the mass spectrometer. The limit to analyzing only the extractable components is mostly overcome by pyrolyzing the sample before the analysis. Pyrolysis uses thermal energy to break down polymeric and nonvolatile materials into small volatile molecules that are amenable to gas chromatographic analysis.

17. E. Stout, C. W. Beck, and K. B. Anderson, "Identification of Rumanite (Romanian Amber) as Thermally Succinite (Baltic Amber)," *Physics and Chemistry of Minerals* 27, no. 9 (2000): 665–78.

18. Pastorelli 2009 (in n. 5, above).

Acknowledgments

The long period of research and writing has generated an equally long list of scholarly debts. The first and most important thanks go to the Antiquities Department at the J. Paul Getty Museum, Kenneth Lapatin, Claire Lyons, Mary Louise Hart, Jens Daehner, and David Saunders; and to former department staff Jiří Frel, Kenneth Hamma, Marit Jentoft-Nilsen, Karol Wight, and Marion True. Recently, Alexandra Sofroniew, an Antiquities Department assistant curator, with keen eye and deft hand, has greatly aided the transformation of the manuscript into the book and this online catalogue. In Antiquities Conservation, heartfelt appreciation goes to Jerry Podany, Jeffrey Maish, and former staff member Maya Elston. Michael Shilling and Herant Khanjian of the Getty Conservation Institute provided critical scientific analysis. In the director's office, David Bomford, Stephen Garrett, and John Walsh in turn offered support at important junctures. The beautiful photography by Ellen Rosenbery speaks for itself. Benedicte Gilman and Mark Greenberg, the original supporters at Getty Publications, are owed more than appreciative words. Ann Lucke and Rob Flynn deftly picked up where they left off. The online catalogue is the result of the talents of designer Kurt Hauser, production coordinators Elizabeth Zozom and Elizabeth Kahn, assistant editor Ruth Evans Lane, and software architects Brenda Podemski and Roger Howard. The expertise of the Web Group, particularly Jack Ludden and Susan Edwards, has been invaluable. The anonymous reader offered priceless comments. Cindy Bohn, the manuscript editor, deserves commendation for her skill and patience, as do Laura Harger and Greg Dobie. Last, but certainly not least, without Marina Belozerskaya, guide and lantern, the project would have been benighted long ago.

Two great libraries and their staffs call for special mention, the library at the National Gallery of Art in Washington, DC, my home institution, and that of the Getty Research Institute, where I have been a Reader since it opened. Neal Turtell, Lamia Doumato, Thomas McGill, Jr., and Ted Dalziel (in Washington), and Anne-Mieke Halbrook and Susan Allen (at the Getty) represent the best in art librarianship. I also recognize the help given to me over the years by the library staff of the American Academy in Rome; the University of California, Los Angeles; the University of California, Riverside; the University of California, Santa Barbara; the Sterling and Francine Clark Art Institute; and Edinburgh University.

Throughout my tenure at the National Gallery of Art, I have had superb administrative support, and for it I thank Rusty Powell, Alan Shestack, Franklin Kelly, Elizabeth Pochter, and Lynn Russell. The help offered by colleagues while I was on sabbatical in Edinburgh during an Ailsa Mellon Bruce Curatorial Fellowship at the Center for Advanced Study in the Visual Arts at the National Gallery of Art and during a Summer Fellowship at the Sterling and Francine Clark Art Institute greatly affected the outcome of the publication. In Williamstown, special thanks are offered to Michael Ann Holly, Mark Ledbury, and Richard Rand; and in Washington, to deans past and present Henry A. Millon, Elizabeth Cropper, Therese O'Malley, and Peter Lukehart. The grants and fellowships supporting research abroad began with a Samuel H. Kress Traveling Fellowship. Summer grants from California State University, Long Beach, made possible the first Italian sojourns. Two Robert H. Smith Fellowships awarded by the National Gallery of Art were critical to research in Greek and Italian museums.

Any museum-based endeavor is a complex project, and institutions housing archaeological finds, especially fragile collections, present special challenges. I owe a debt of gratitude to the many people around the world who aided me in the firsthand study of amber objects. At museums in Argentina, Chile, Costa Rica, Croatia, the Czech Republic, Denmark, Egypt, France, Germany (once East and West), Greece, Hungary, Italy, Mexico, Russia, Serbia, Slovenia, Turkey, the United Kingdom, and the United States, curators, conservators, registrars, scientists, photographers, and other professionals were instrumental to the research. Particular appreciation is owed to Joan Mertens and Carlos Picón of The Metropolitan Museum of Art; Denis Haynes and Dyfri Williams of the British Museum; Alain Pasquier of the Louvre; Mary B. Comstock, John Herrmann, and Cornelius C. Vermeule of the Museum of Fine Arts, Boston; Jan Bouzek of the Classical Institute of Archaeology, Charles University, Prague; János György Szilágyi of the Fine Arts Museum, Budapest; Elena Khodza at the State Hermitage Museum, Saint Petersburg; and David Grimaldi of the American Museum of Natural History, New York. Gavin Hamilton, Leon Levy and Shelby White, and Judy and Michael Steinhardt kindly welcomed study of their private collections. Without the generous help of specific scholars and curators in Italy, this project would not have progressed. Among Italian colleagues, I extend particular thanks to Sebastiano Bianco, Nuccia Negroni Catacchio, Giulia Rocco, and Marcello Tagliente.

Substantial recognition goes to my Doktorvater, Mario Del Chiaro, and to my former graduate students and research assistants in California and Washington. First on the list are the Angelenos Maureen Burns, John Tucker, René Ninaber, and Anna Zagorski. Alexis Castor and Bjoern Ewalt were ideal research associates at the Center for Advanced Study in the Visual Arts. In Washington, Patrick Resing and Jack Shepherd performed key editorial work. Lee B. Ewing, photographer, National Gallery of Art, offered new insights. Amber life and work were much ameliorated by my Academic Programs Department colleagues Ana Maria Zavala Kozuch, Ben Masri-Cohen, Ali Peil, Rachel Schulze, and Jennifer Wagelie.

Without the support of family and friends, this work would not have reached daylight. Selma Holo and Fred Croton read and listened to the unfolding story. Alan Conisbee and Marie Clarke offered more than engineering insight. My grandmother and parents, Elizabeth Lees, Willard Causey, and Catherine Causey-Lees, and my siblings C. Andrew Causey, Douglas Causey, Kay Marie Kuder, and Laurel Nelson provided not only encouragement over the years but also keen editorial eyes. Andrew's contributions go far beyond art and anthropology. To my son, Jan, and to the memory of Philip, this book is dedicated.

Bibliography

Abbreviations

AA	*Archäologischer Anzeiger*
ABV	Beazley, J. D. *Attic Black-Figure Vase-Painters.* Oxford, 1956
AJA	*American Journal of Archaeology*
AM	*Mitteilungen des Deutschen Archäologischen Instituts, Athenische Abteilung*
AntK	*Antike Kunst*
ArchCl	*Archeologia classica*
BAR	*British Archaeological Reports*
BCH	*Bulletin de correspondance hellénique*
BdA	*Bollettino d'arte*
CVA	*Corpus vasorum antiquorum*
ES	Gerhard, E., G. Körte, and A. Klugman. *Etruskische Spiegel.* 5 vols. Berlin, 1843–97
Jdi	*Jahrbuch des Deutschen Archäologischen Instituts*
JHS	*Journal of Hellenic Studies*
LIMC	*Lexicon iconographicum mythologiae classicae.* Zurich and Munich, 1974–2013
MonAnt	*Monumenti antichi*
NSc	*Notizie degli scavi di antichità*
RA	*Revue archéologique*
RE	Pauly-Wissowa. *Real-Encyclopädie der klassischen Altertumswissenschaft.* Stuttgart, 1893–2013
RM	*Mitteilungen des Deutschen Archäologischen Instituts, Römische Abteilung*
StEtr	*Studi etruschi*

Frequently Cited Sources

Adam 1984
Adam, A.M. *Bronzes étrusques et Italiques.* Paris, 1984.

Akürgal 1961
Akurgal, E. *Die Kunst Anatoliens von Homer bis Alexander.* Berlin, 1961.

Albizzatti 1919
Albizzatti, C. "Un' ambra scolpita d'arte ionica nella raccolta Morgan." *Rassegna d'arte antica e moderna* 10 (1919): 183–200.

***Ambre* 2007**
Ambre: Trasparenze dall'antico. Exh. cat. Naples, 2007.

***Ancient Italy* 2000**
Ridgway, D., et al., eds. *Ancient Italy in Its Mediterranean Setting: Studies in Honour of Ellen Macnamara.* London, 2000.

Anderson and Crelling 1995
Anderson, K. B., and J. C. Crelling, eds. *Amber, Resinite, and Fossil Resins.* Washington, DC, 1995.

Andrée 1937
Andrée, K. *Der Bernstein und seine Bedeutung in Natur- und Geisteswissenschaften, Kunst und Kunstgewerbe, Technik, Industrie und Handel.* Königsberg, 1937.

Andrée 1951
Andrée, K. *Der Bernstein: Das Bernsteinland und sein Leben.* Stuttgart, 1951.

Andrews 1994
Andrews, C. *Amulets of Ancient Egypt.* Austin, TX, 1994.

***Art of the Classical World* 2007**
Picón, C. A., et al. *Art of the Classical World in the Metropolitan Museum of Art.* New York, 2007.

***Athen* 1986**
Archaische und klassische griechische Plastik: Akten des internationalen Kolloquiums vom 22.–25. April 1985 in Athen. 2 vols. Mainz, 1986.

Avanzini 1997
Avanzini, A., ed. *Profumi d'Arabia: Atti del Convegno.* Rome, 1997.

Baltica 2000
Baltica 2000: International Folklore Festival. Exh. cat. Riga, 2000.

Barfod 1996
Barfod, J. "Bernstein in Volksglauben und Volksmedizin." In _Bernstein: Tränen der Götter_, edited by M. Ganzelewski and R. Slotta, pp. 453–56. Exh. cat. Bochum, 1996.

Barfod 2005
Barfod, J. _Bernstein._ Husum, Germany, 2005.

Barfod et al. 1989
Barfod, J., F. Jacobs, and S. Ritzkowski. _Bernstein: Schätze in Niedersachsen._ Seelze, 1989.

Barringer 2001
Barringer, J. M. _The Hunt in Ancient Greece._ Baltimore and London, 2001.

Bartoloni et al. 2000
Bartoloni, G., et al., eds. _Principi etruschi: Tra Mediterraneo ed Europa._ Exh. cat. Bologna, 2000.

Becatti 1955
Becatti, G. _Oreficerie antiche dalle minoiche alle barbariche_. Rome, 1955.

Beck and Bouzek 1993
Beck, C. W., and J. Bouzek, eds. _Amber in Archaeology: Proceedings of the Second International Conference on Amber in Archaeology, Liblice, 1990._ Prague, 1993.

Beck and Shennan 1991
Beck, C. W., and S. Shennan. _Amber in Prehistoric Britain._ Oxford, 1991.

Beck et al. 2003
Beck, C. W., I. B. Loze, and J. M. Todd, eds. _Amber in Archaeology: Proceedings of the Fourth International Conference on Amber in Archaeology, Talsi, 2001._ Riga, 2003.

Berkin 2003
Berkin, J. M. _The Orientalizing Bucchero from the Lower Building at Poggio Civitate (Murlo)._ Philadelphia, 2003.

Bernstein 1996
Ganzelewski, M., and R. Slotta, eds. _Bernstein: Tränen der Götter_. Exh. cat. Bochum, 1996.

Bevan 1986
Bevan, E. _Representations of Animals in Sanctuaries of Artemis and Other Olympian Deities._ BAR International Series 315. Oxford, 1986.

Beyond Babylon 2008
J. Aruz, K. Benzel, and J. M. Evans, eds. _Beyond Babylon: Art, Trade, and Diplomacy in the Second Millennium B.C._ Exh. cat. New York, 2008.

Bianchi Bandinelli and Giuliano 1973
Bianchi Bandinelli, R., and A. Giuliano. _Les étrusques et l'Italie avant Rome._ Paris, 1973.

Bissing 1931
Bissing, F. W. Freiherr von. "Studien zur ältesten Kultur Italiens II: Etruskische Skarabäen und Skaraböide aus Bernstein." StEtr 5 (1931): 49–69.

Black and Green 1992
Black, J., and A. Green. _Gods, Demons and Symbols of Ancient Mesopotamia: An Illustrated Dictionary._ Austin, TX, 1992.

Boardman 1967
Boardman, J. _Pre-classical: From Crete to Archaic Greece._ Harmondsworth, 1967.

Boardman 1968
Boardman, J. _Archaic Greek Gems._ London, 1968.

Boardman 1980
Boardman, J. _The Greeks Overseas: Their Early Colonies and Trade._ New and enlarged ed. London, 1980.

Boardman 2001
Boardman, J. _Greek Gems and Finger Rings: Early Bronze Age to Late Classical._ New and enlarged ed. New York, 2001.

Bonfante 1985
Bonfante, L. "Amber, Women, and Situla Art." _Journal of Baltic Studies_ 16, no. 3 (1985): 276–91.

Bonfante 1986
Bonfante, L., ed. _Etruscan Life and Afterlife: A Handbook of Etruscan Studies._ Detroit, 1986.

Bonfante 1989
Bonfante, L. "Nudity as a Costume in Classical Art." AJA 93 (1989): 543–90.

Bonfante 1996
Bonfante, L. "Etruscan Sexuality and Funerary Art." In _Sexuality in Ancient Art: Near East, Egypt, Greece, and Italy_, edited by N. B. Kampen, pp. 155–69. Cambridge, 1996.

Bonfante 2003
Bonfante, L. _Etruscan Dress._ Updated ed. Baltimore, 2003.

Bonner 1950
Bonner, C. _Studies in Magical Amulets, Chiefly Graeco-Egyptian._ Ann Arbor, 1950.

Bonner 1954
Bonner, C. "A Miscellany of Engraved Stones." _Hesperia_ 23, no. 2 (1954): 138–57.

Bottini 1987
Bottini, A. "Ambre a protome umana dal Melfese." BdA 41 (1987): 1–16.

Bottini 1990
Bottini, A. "Le ambre intagliate a figura umana del Museo Archeologico Nazionale di Melfi." _Archeologia: Wrocław Zakład Narodowy im. Ossoli'nskich_ 41 (1990): 57–66.

Bottini 1993
Bottini, A. "La Tomba 955 di Lavello-*Forentum*." In *Due donne dell'Italia antica: Corredi da Spina e Forentum*, edited by D. Baldoni, pp. 63–69. Exh. cat. Padua, 1993.

Bottini and Setari 1992
Bottini, A., and E. Setari. "Basileis? I più recenti rinvenimenti a Braida di Serra di Vaglio: Risultati, prospettive e problemi." Prepublication of *Bollettino di Archeologia* 16–18 (1992): 1–3.

Bottini and Setari 1995
Bottini, A., and E. Setari. *Basileis: Antichi re in Basilicata.* Naples, 1995.

Bottini and Setari 1998
Bottini, A., and E. Setari. "L'artigianato arcaico dell'ambra alla luce dei più recenti rinvenimenti in Basilicata." In "Amber in Archeology," edited by N. Negroni Catacchio and C. W. Beck, in *Atti del XIII Congresso: Unione internazionale delle scienze preistoriche e protostoriche, Forlì, 8–14 settembre 1996*, pp. 469–77. Rome, 1998.

Bottini and Setari 2003
Bottini, A., and E. Setari. *La necropoli italica di Braida di Vaglio in Basilicata: Materiali dallo scavo del 1994*. With appendix by M. Torelli. MonAnt 60, serie miscellanea 7. Rome, 2003.

Bouzek 1993
Bouzek, J. "The Shifts of the Amber Route." In *Amber in Archaeology: Proceedings of the Second International Conference on Amber in Archaeology, Liblice, 1990*, edited by C. W. Beck and J. Bouzek, pp. 141–46. Prague, 1993.

Brendel 1995
Brendel, O. J. *Etruscan Art.* New Haven, 1995.

Briquel and Gaultier 1997
Briquel, D., and F. Gaultier, eds. *Les plus religieux des hommes: État de la recherche sur la religion étrusque; Actes du colloque international, Galleries nationales du Grand Palais, 17–19 novembre 1992.* Paris, 1997.

Brown 1960
Brown, W. L. *The Etruscan Lion.* Oxford, 1960.

Buffum 1900
Buffum, W. A. *The Tears of the Heliades: Or, Amber as a Gem.* London, 1897. Repr., London, 1900.

Bulté 1991
Bulté, J. *Talismans égyptiens d'heureuse maternité: "Faïence" bleu vert à pois foncés.* Paris, 1991.

Burkert 1992
Burkert, W. *The Orientalizing Revolution: Near Eastern Influence on Greek Culture in the Early Archaic Age.* Cambridge, MA, 1992.

Cagianelli and Sannibale 1999
Cagianelli, C., and M. Sannibale. *Bronzi a figura umana.* Museo Gregoriano Etrusco Cataloghi 5. Vatican City, 1999.

Carpenter 1986
Carpenter, T. H. *Dionysian Imagery in Archaic Greek Art: Its Development in Black-Figure Vase Painting.* Oxford, 1986.

Carter 1985
Carter, J. B. *Greek Ivory-Carving in the Orientalizing and Archaic Periods.* New York, 1985.

Carter and Morris 1995
Carter, J. B., and S. P. Morris, eds. *The Ages of Homer: A Tribute to Emily Townsend Vermeule.* Austin, TX, 1995.

Caubet 1999
Caubet, A., ed. *Cornaline et pierres précieuses: La Méditerranée, de l'Antiquité à l'Islam; Actes du colloque organisé au Musée du Louvre les 24–25 novembre 1995.* Paris, 1999.

Causey 1990
Causey, F. "Pendant with a Sphinx in Relief." In *Glories of the Past: Ancient Art from the Shelby White and Leon Levy Collection,* edited by D. von Bothmer, pp. 129–30. Exh. cat. New York, 1990.

Causey 1993
Causey, F. "Two Amber Pendants in Malibu: East Greek Craftsmanship?" In *Amber in Archaeology: Proceedings of the Second International Conference on Amber in Archaeology, Liblice, 1990*, edited by C. W. Beck and J. Bouzek, pp. 212–18. Prague, 1993.

Causey 2002
Causey, F. "An Amber Pendant in the Form of a Ship with Sailors." In *Essays in Honor of Dietrich von Bothmer,* edited by J. Clark and J. Gaunt, with B. Gilman, pp. 63–67. Amsterdam, 2002.

Causey 2004
Causey, F. "Amber." With J. Shepherd. In *Colour in the Ancient Mediterranean World: Papers from the International Conference "Colours in Antiquity" Held September 2001 at Edinburgh University,* edited by L. Cleland et al., pp. 70–72. BAR International Series 1267. Oxford, 2004.

Causey 2007
Causey, F. "Anbar, Amber, Bernstein, Jantar, Karabe." In *Bernstein, Sigmar Polke, Amber.* Exh. cat. New York, 2007.

Causey 2010
Causey, F. "A Kore in Amber." In *Archaic Greek Culture: History, Archaeology, Art and Museology; Proceedings of the International Round-Table Conference June 2005, St. Petersburg, Russia,* edited by S. Solovyov, pp. 6–9. BAR International Series 2061. Oxford, 2010.

***Centaur's Smile* 2003**
Padgett, J. M., et al. *The Centaur's Smile: The Human Animal in Early Greek Art.* Exh. cat. Princeton, 2003.

***Civiltà degli Etruschi* 1985**
Cristofani, M, ed. *Civiltà degli Etruschi.* Exh. cat. Florence, 1985.

Clark and Gaunt 2002
Clark, J., and J. Gaunt, eds. *Essays in Honor of Dietrich von Bothmer.* With B. Gilman. Amsterdam, 2002.

Cole 1998
Cole, S. G. "Domesticating Artemis." In *The Sacred and the Feminine in Ancient Greece,* edited by S. Blundell and M. Williamson, pp. 27–44. New York, 1998.

Collon 1997
Collon, D., ed. *7000 Years of Seals.* London, 1997.

Colonna 1985
Colonna, G., ed. *Santuari d'Etruria.* Exh. cat. Milan, 1985.

Comstock and Vermeule 1971
Comstock, M., and C. Vermeule. *Greek, Etruscan, and Roman Bronzes in the Museum of Fine Arts, Boston.* Greenwich, CT, 1971.

Cristofani 1978
Cristofani, M. *L'arte degli Etruschi: Produzione e consumo.* Turin, 1978.

Cristofani 1996
Cristofani, M. *Etruschi e altre genti nell'Italia preromana: Mobilità in età arcaica.* Rome, 1996.

Cristofani and Martelli 1983
Cristofani, M., and M. Martelli, eds. *L'oro degli Etruschi.* Novara, 1983.

Croissant 1983
Croissant, F. *Les protomés féminines archaïques: Recherches sur les représentations du visage dans la plastique grecque de 550 à 480 av. J.-C.* 2 vols. Paris, 1983.

Damgaard Andersen 1996
Damgaard Andersen, H. "The Origin of Potnia Theron in Central Italy." In *Die Akten des Internationalen Kolloquiums "Interactions in the Iron Age: Phoenicians, Greeks and the Indigenous Peoples of the Western Mediterranean" in Amsterdam am 26. und 27. März 1992,* Hamburger Beiträge zur Archäologie 19–20 (Mainz, 1996), pp. 73–113.

Dawkins 1929
Dawkins, R. M., ed. *The Sanctuary of Artemis Orthia.* JHS, supplement 5. London, 1929.

De Grummond 1982
De Grummond, N. T., ed. *A Guide to Etruscan Mirrors.* Tallahassee, FL, 1982.

De Grummond and Simon 2006
De Grummond, N. T., and E. Simon, eds. *The Religion of the Etruscans.* Austin, TX, 2006.

De Juliis 1992
De Juliis, E. M. "Le ambre intagliate." In *Principi imperatori vescovi: Duemila anni di storia a Canosa,* edited by R. Cassano, pp. 128–30. Marsilio, 1992.

De Puma and Small 1994
De Puma, R., and J. P. Small, eds. *Murlo and the Etruscans: Art and Society in Ancient Etruria.* Madison, WI, 1994.

D'Ercole 1995
D'Ercole, M.-C. "Observations sur quelques ambres sculptés archaïques d'Italie méridionale." RA 1995: 265–90.

D'Ercole 2008
D'Ercole, M.-C. *Ambres gravés du département des Monnaies, Médailles et Antiques.* Paris, 2008.

Desroches-Noblecourt 2006
Desroches-Noblecourt, C. *Le secret des découvertes.* Paris, 2006.

Dickie 2001
Dickie, M. *Magic and Magicians in the Greco-Roman World.* New York, 2001.

Dopp 1997
Dopp, S. "Die Tränen von Phaethons Schwestern wurden zu Bernstein: Der Phaethon-Mythos in Ovid's 'Metamorphosen.'" In *Bernstein: Tränen der Götter,* edited by M. Ganzelewski and R. Slotta, pp. 1–10. Exh. cat. Bochum, 1996.

***Due donne* 1993**
Baldoni, D., ed. *Due donne dell'Italia antica: Corredi da Spina e Forentum.* Exh. cat. Padua, 1993.

Eichholz 1962
Pliny. *Natural History: Books 36–37.* Translated by D. E. Eichholz. Loeb Classical Library 419. Cambridge, 1962.

Emiliozzi 1997
Emiliozzi, A., ed. *Carri da guerra e principi etruschi.* Exh. cat. Viterbo, 1997.

***Eroi e regine* 2001**
Eroi e regine: Picene popolo d'Europa. Exh. cat. Rome, 2001.

Evely 2000
Evely, R. D. G. *Minoan Crafts: Tools and Techniques.* 2 vols. Gothenburg, 2000.

Faraone 1991
Faraone, C. A. "The Agonistic Context of Early Greek Binding Spells." In *Magika Hiera: Ancient Greek Magic and Religion,* edited by C. A. Faraone and D. Obbink, pp. 3–32. New York, 1991.

Faraone 1992
Faraone, C. A. *Talismans and Trojan Horses: Guardian Statues in Ancient Greek Myth and Ritual.* New York, 1992.

Faraone and Obbink 1991
Faraone, C. A., and D. Obbink, eds. *Magika Hiera: Ancient Greek Magic and Religion.* New York, 1991.

***First Cities* 2003**
Aruz, J., ed. *Art of the First Cities: The Third Millennium B.C. from the Mediterranean to the Indus.* With R. Wallenfels. Exh. cat. New York, 2003.

Fitton 1992
Fitton, J. L., ed. *Ivory in Greece and the Eastern Mediterranean from the Bronze Age to the Hellenistic Period.* London, 1992.

Franchi Dell'Orto 1999
Franchi Dell'Orto, L. "Le 'anatrelle': Sopravvivenza di una simbologia religiosa dell'età del Bronzo europea." In *Piceni: Popolo d'Europa*, pp. 91–92. Exh. cat. Rome, 1999.

Freyer-Schauenburg 1974
Freyer-Schauenburg, B. *Bildwerke der archaischen Zeit und des strengen Stils: Samos.* Vol. 11. Bonn, 1974.

Friedman 1998
Friedman, F. D., ed. *Gifts of the Nile: Ancient Egyptian Faience.* Exh. cat. Providence, RI, 1998.

Frontisi-Ducroux 1991
Frontisi-Ducroux, F. *Le Dieu-Masque: Une figure du Dionysos d'Athènes.* Paris and Rome, 1991.

Fuchs and Floren 1987
Fuchs, W., and J. Floren. *Die griechische Plastik.* Vol. 1, *Die geometrische und archaische Plastik.* Munich, 1987.

Fuscagni 1982
Fuscagni, S. "Il pianto ambrato delle Eliadi, l'Eridano e la nuova stazione preistorica di Frattesina Polesine." *Quaderni urbanati di cultura classica* 12 (1982): 101–13.

Gage 1993
Gage, J. *Color and Culture: Practice and Meaning from Antiquity to Abstraction.* Boston, 1993.

Gager 1992
Gager, J. *Curse Tablets and Binding Spells from the Ancient World.* New York, 1992.

Gaslain 2005
Gaslain, C. "Dalle rive del mare alle tombe principesche: La circolazione dell'ambra nel bacino del Mediterraneo nell'età del Bronzo e nell'età del Ferro." In *Magie d'ambra: Amuleti e gioielli della Basilicata antica*, pp. 55–70. Exh. cat. Potenza, 2005.

Geerlings 1996
Geerlings, W. "Die Tränen der Schwestern des Phaethon-Bernstein im Altertum." In *Bernstein: Tränen der Götter*, edited by M. Ganzelewski and R. Slotta, pp. 395–400. Exh. cat. Bochum, 1996.

Gempeler 1974
Gempeler, R. D. *Die Etruskischen Kanopen: Herstellung, Typologie, Entwicklungsgeschichte.* Einsiedeln, Switzerland, 1974.

Green 1983
Green, A. "Neo-Assyrian Apotropaic Figures: Figurines, Rituals, and Monumental Art, with Special Reference to the Figurines from the Excavations of the British School of Archaeology in Iraq at Nimrud." *Iraq* 45 (1983): 87–96.

Grimaldi 1996
Grimaldi, D. A. *Amber: Window to the Past.* New York, 1996.

Guzzo 1993
Guzzo, P. G. *Oreficerie dalla Magna Grecia: Ornamenti in oro e argento dall'Italia Meridionale tra l'VIII ed il I secolo.* Taranto, 1993.

Guzzo 1996
Guzzo, P. G. "The Jewelry of the Western Greeks." In *I Greci in Occidente*, edited by G. Pugliese Carratelli, pp. 471–80. Exh. cat. Venice, 1996. Published in English as *The Greek World: Art and Civilization in Magna Graecia and Sicily.* New York, 1996.

Hamiaux 1992
Hamiaux, M. *Les Sculptures grecques: Musée du Louvre.* Vol. 1. Paris, 1992.

Hampe and Simon 1981
Hampe, R., and E. Simon. *The Birth of Greek Art: From the Mycenaean to the Archaic Period.* London, 1981.

Harding and Hughes-Brock 1974
Harding, A., and H. Hughes-Brock. "Amber in the Mycenaean World." *Annual of the British School at Athens* 69 (1974): 145–72.

***Hatshepsut* 2005**
Roehrig, C. H., ed. *Hatshepsut: From Queen to Pharaoh.* With R. Dreyfus and C. A. Keller. New York, 2005.

Haynes 1985
Haynes, S. *Etruscan Bronzes.* London, 1985.

Haynes 1993
Haynes, S. "Thoughts on the Winged Female Figure in the Funerary Sculpture of Chiusi." In *La civiltà di Chiusi e del suo territorio: Atti del XVII convegno di studi etruschi ed italici, Chianciano Terme, May 28–June 1, 1988*, edited by G. Maetzke, pp. 297–309. Florence, 1993.

Haynes 2000
Haynes, S. *Etruscan Civilization: A Cultural History.* Los Angeles, 2000.

Hedreen 1992
Hedreen, G. M. *Silens in Attic Black-Figure Vase-Painting: Myth and Performance.* Ann Arbor, 1992.

Heidenreich 1968
Heidenreich, R. "Über einige Bernsteinarbeiten." In *Festschrift Gottfried von Lücken = Wissenschaftliche Zeitschrift der Universität Rostock* 17 (1968): 655–59.

Helms 1988
Helms, M. W. *Ulysses' Sail: An Ethnographic Odyssey of Power, Knowledge, and Geographical Distance.* Princeton, 1988.

Helms 1993
Helms, M. W. *Craft and the Kingly Ideal: Art, Trade, and Power.* Austin, TX, 1993.

Herfort-Koch 1986
Herfort-Koch, M. *Archaische Bronzeplastik Lakoniens.* Münster, 1986.

Higgins 1980
Higgins, R. A. *Greek and Roman Jewellery.* 2nd ed. London, 1980.

Höckmann 1982
Höckmann, U. *Die Bronzen aus dem Fürstengrab von Castel San Mariano bei Perugia: Staatliche Antikensammlungen München, Katalog der Bronzen.* Vol. 1. Munich, 1982.

Hölbl 1979
Hölbl, G. *Beziehungen der ägyptischen Kultur zu Altitalien*. 2 vols. Leiden, 1979.

Holloway and Nabers 1982
Holloway, R. R., and N. Nabers. "The Princely Burial of Roscigno (Monte Pruno) Salerno." *Revue des archéologues et historiens d'art de Louvain* 15 (1982): 97–163.

Houlihan 1986
Houlihan, P. F. *The Birds of Ancient Egypt.* Warminster, 1986.

Hughes-Brock 1985
Hughes-Brock, H. "Amber and the Mycenaeans." *Journal of Baltic Studies* 16, no. 3 (1985): 257–67.

Hughes-Brock 1993
Hughes-Brock, H. "Amber in the Aegean in the Late Bronze Age: Some Problems and Perspectives." In *Amber in Archaeology: Proceedings of the Second International Conference on Amber in Archaeology, Liblice, 1990*, edited by C. W. Beck and J. Bouzek, pp. 219–29. Prague, 1993.

Hus 1961
Hus, A. *Recherches sur la statuaire en pierre étrusque archaïque.* Paris, 1961.

***I Fenici* 1988**
Moscati, S., ed. *I Fenici.* Exh. cat. Milan, 1988.

Jannot 2005
Jannot, J.-R. *Religion in Ancient Etruria.* Translated by J. Whitehead. Madison, WI, 2005.

Jantzen 1972
Jantzen, U. *Ägyptische und orientalische Bronzen aus dem Heraion von Samos.* Vol. 8. Bonn, 1972.

Johansen 1994
Johansen, F. *Greece in the Archaic Period.* Cat., Ny Carlsberg Glyptotek. Cophenhagen, 1994.

Johnston 1995
Johnston, S. I. "Defining the Dreadful: Remarks on the Greek Child-Killing Demon." In *Ancient Magic and Ritual Power*, edited by M. Meyer and P. Mirecki, pp. 361–87. Leiden, 1995.

Jordan et al. 1999
Jordan, D. R., H. Montgomery, and E. Thomassen, eds. *The World of Ancient Magic: Papers from the First International Samson Eitrem Seminar at the Norwegian Institute at Athens, 4–8 May 1997.* Bergen, 1999.

***J. Paul Getty Museum* 2002**
The J. Paul Getty Museum, Handbook of the Antiquities Collection. Los Angeles, 2002.

***J. Paul Getty Museum* 2010**
The J. Paul Getty Museum, Handbook of the Antiquities Collection. Los Angeles, 2010.

Jürgeit 1999
Jürgeit, F. *Die etruskischen und italischen Bronzen sowie Gegenstände aus Eisen, Blei und Leder im Badischen Landesmuseum Karlsruhe.* 2 vols. Pisa, 1999.

Kampen 1996
Kampen, N. B., ed. *Sexuality in Ancient Art: Near East, Egypt, Greece, and Italy.* Cambridge, 1996.

Karakasi 2003
Karakasi, K. *Archaic Korai.* Los Angeles, 2003.

Keesling 2003
Keesling, C. M. *The Votive Statues of the Athenian Acropolis.* New York, 2003.

Kotansky 1991
Kotansky, R. "Incantations and Prayers for Salvation on Inscribed Greek Amulets." In *Magika Hiera: Ancient Greek Magic and Religion*, edited by C. A. Faraone and D. Obbink, pp. 107–37. New York, 1991.

Kotansky 1994
Kotansky, R. *Greek Magical Amulets: The Inscribed Gold, Silver, Copper, and Bronze "Lamellae."* Part I, *Published Texts of Known Provenance.* Opladen, 1994.

Kozloff 1981
Kozloff, A. P., ed. *Animals in Ancient Art from the Leo Mildenberg Collection.* Cleveland, 1981.

Kozloff and Mitten 1988
Kozloff, A. P., and D. G. Mitten, eds. *The Gods Delight: The Human Figure in Classical Bronze*. Exh. cat. Cleveland, 1988.

Kraay 1976
Kraay, C. M. *Archaic and Classical Greek Coins.* London, 1976.

Krauskopf 1995
Krauskopf, I. *Heroen, Götter und Dämonen auf etruskischen Skarabäen.* Mannheim, 1995.

Krauskopf 1997
Krauskopf, I. "Influences grecques et orientales sur les représentations de dieux étrusques." In *Les Étrusques: Les plus religieux des hommes; État de la recherche sur la religion étrusque; Actes du colloque international, Galeries nationales du Grand Palais, 17-18-19 novembre 1992*, edited by Dominique Briquel and Françoise Gaultier, pp. 25–36. Paris, 1997.

Krauskopf 1998
Krauskopf, I. "Artemis." *Annali della Fondazione per Museo "Claudio Faina"* 5 (1998): 171–206.

Kredel 1923–24
Kredel, F. "Ein archaisches Schmuckstück aus Bernstein." Jdi 38–39 (1923–24): 169–80.

Kyrieleis 1995
Kyrieleis, H. "Eine neue Kore des Cheramyes." *Antike Plastik* 24 (1995): 7–36.

Laffineur 1978
Laffineur, R. *L'orfèvrerie rhodienne orientalisante.* Athens, 1978.

Laffineur 1980
Laffineur, R. "L'orfèvrerie rhodienne orientalisante." In *Études sur l'orfèvrerie antique,* edited by T. Hackens, pp. 13–29. Louvain-la-Neuve, 1980.

La Genière 1961
La Genière, J. de. "Ambre intagliate del Museo di Salerno." *Apollo* 1 (1961): 73–88.

La Genière 1967
La Genière, J. de. "À propos du catalogue des ambres sculptés du British Museum." RA (1967): 297–304.

La Genière 1968
La Genière, J. de. *Recherches sur l'âge du fer en Italie meridionale: Sala Consilina.* Naples, 1968.

Langdon 1993
Langdon, S., ed. *From Pasture to Polis: Art in the Age of Homer.* Exh. cat. Columbia, MO, 1993.

Langenheim 2003
Langenheim, J. H. *Plant Resins: Chemistry, Evolution, Ecology, and Ethnobotany.* Portland, OR, 2003.

Lapatin 2001
Lapatin, K. D. S. *Chryselephantine Statuary in the Ancient Mediterranean World.* New York, 2001.

Lippolis 1996
Lippolis, E., ed. *Arte e artigianato in Magna Grecia.* Naples, 1996.

Llewellyn-Jones 2002
Llewellyn-Jones, L., ed. *Women's Dress in the Ancient Greek World.* Swansea, 2002.

Losi et al. 1993
Losi, M., B. Raposso, and G. Ruggiero. "The Production of Amber Female Heads in Pre-Roman Italy." In *Amber in Archaeology: Proceedings of the Second International Conference on Amber in Archaeology, Liblice, 1990*, edited by C. W. Beck and J. Bouzek, pp. 203–11. Prague, 1993.

***Magie d'ambra* 2005**
Magie d'ambra: Amuleti e gioielli della Basilicata antica. Exh. cat. Potenza, 2005.

***Magna Graecia* 2002**
Bennett, M. J., and A. J. Paul. *Magna Graecia: Greek Art from South Italy and Sicily.* Exh. cat. Cleveland, 2002.

Mano-Zisi and Popović 1969
Mano-Zisi, D. von, and L. Popović. "Der Fund von Novi Pazar." *Bericht der Römisch-Germanischen Kommission* 50 (1969): 191–208.

Marangou 1969
Marangou, L. *Lakonische Elfenbein- und Beinschnitzereien.* Tübingen, 1969.

Marconi 1933
Marconi, P. "La cultura orientalizzante nel Piceno." MonAnt 35 (1933): cols. 265–454.

Marshall 1911
Marshall, F. H. *Catalogue of the Jewelry, Greek, Etruscan and Roman, in the Departments of Antiquities, British Museum.* London, 1911.

Martelli Cristofani 1985
Martelli Cristofani, M. "Gli avori tardo-arcaici: Botteghe e aree di diffusione." In *Il commercio etrusco arcaico: Atti dell'Incontro di Studi, 5–7 dicembre 1983*, pp. 207–48. Rome, 1985.

Massaro 1943
Massaro, D. "Le ambre di Vetulonia." StEtr 18 (1943): 31–46.

Mastrocinque 1991
Mastrocinque, A. *L'ambra e l'Eridano: Studi sulla letteratura e sul commercio dell'ambra in età preromana.* Este, 1991.

Menichetti 1994
Menichetti, M. *Archeologia del potere: Re, immagini e miti a Roma e in Etruria in età arcaica.* Milan, 1994.

Metzger 1991
Metzger, C. "Noeud d'Héraclès et torsades: Trois bijoux hellénistiques." *Revue du Louvre* 41 (1991): 18–23.

Meyer and Mirecki 1995
Meyer, M., and P. Mirecki, eds. *Ancient Magic and Ritual Power.* Leiden, 1995.

Mitten and Doeringer 1967
Mitten, D. G., and S. F. Doeringer, eds. *Master Bronzes from the Classical World.* Mainz am Rhein, 1967.

Moorey 2003
Moorey, R. S. *Idols of the People: Miniature Images of Clay in the Ancient Near East.* Schweich Lectures. Oxford and New York, 2003.

Moorey 2004
Moorey, R. S. *Catalogue of the Ancient Near Eastern Terracottas in the Ashmolean.* Oxford, 2004.

Mottahedeh 1979
Erhart [Mottahedeh], K. P. *The Development of the Facing Head Motif on Greek Coins and Its Relation to Classical Art.* New York, 1979.

Musti et al. 1992
Musti, D., et al. *L'oro dei Greci.* Novara, 1992.

Nagy 1997
Nagy, G. *The Best of the Achaeans: Concepts of the Hero in Archaic Greek Poetry*. Rev. ed., Baltimore, 1997. http://www.press.jhu.edu/books/nagy/BofA.html.

Nava 1988
Nava, M. L. *Le stele della Daunia.* Milan, 1988.

Negroni Catacchio 1972
Negroni Catacchio, N. "La problematica dell'ambra nella protostoria italiana." *Padusa* 8 (1972): 1–8.

Negroni Catacchio 1975
Negroni Catacchio, N. "Le ambre garganiche nel quadro della problematica dell'ambra nella protostoria italiana." In *Civiltà preistoriche e protostoriche della Daunia: Atti del colloquio internazionale di preistoria e protostoria della Daunia, Foggia, Italia, 1973,* pp. 310–19. Florence, 1975.

Negroni Catacchio 1989
Negroni Catacchio, N. "L'ambra: Produzione e commerci nell'Italia preromana." In *Italia, omnium terrarum parens: La civiltà degli Enotri, Chone, Ausoni, Sanniti, Lucani, Brettii, Sicani, Siculi, Elimi,* edited by C. Ampolo et al., pp. 659–96. Milan, 1989.

Negroni Catacchio 1993
Negroni Catacchio, N. "The Production of Amber Figurines in Italy from the 8th to 4th Centuries B.C." In *Amber in Archaeology: Proceedings of the Second International Conference on Amber in Archaeology, Liblice, 1990*, edited by C. W. Beck and J. Bouzek, pp. 191–202. Prague, 1993.

Negroni Catacchio 1999
Negroni Catacchio, N. "Alcune ambre figurate preromane di provenienza italiana in collezioni private di New York." In *Koina: Miscellanea di studi archeologici in onore di Piero Orlandini,* edited by M. Castoldi, pp. 279–90. Milan, 1999.

Negroni Catacchio and Beck 1998
Negroni Catacchio, N., and C. W. Beck. "Amber in Archeology." In *Atti del XIII Congresso: Unione internazionale delle scienze preistoriche e protostoriche, Forlì, 8–14 settembre 1996.* Rome, 1998.

Nicholson and Shaw 2000
Nicholson, P. T., and I. Shaw, eds. *Ancient Egyptian Materials and Technology.* New York, 2000.

***Norbert Schimmel Collection* 1974**
Muscarella, O. W. *Ancient Art: The Norbert Schimmel Collection.* Mainz, 1974.

Olbrich 1979
Olbrich, G. *Archaisches Statuetten eines metapontiner Heiligtums.* Rome, 1979.

***Ornamenti e lusso* 2000**
Nava, M. L., and M. N. Santi, eds. *Ornamenti e lusso: La donna nella Basilicata antica.* Exh. cat. Rome, 2000.

Özgen and Öztürk 1996
Özgen, İ., and J. Öztürk. *Heritage Recovered: The Lydian Treasure.* Istanbul, 1996.

Palavestra 2003
Palavestra, A. "A Composite Amber Jewelry Set from Novi Pazar." In *Amber in Archaeology: Proceedings of the Fourth International Conference on Amber in Archaeology, Talsi, 2001*, edited by C. W. Beck, I. B. Loze, and J. M. Todd, pp. 213–23. Riga, 2003.

Palavestra and Krstić 2006
Palavestra, A., and V. Krstić. *The Magic of Amber.* Belgrade, 2006.

Paoletti and Settis 1981
Paoletti, M., and S. Settis. *Medma e il suo territorio: Materiali per una carta archeologica.* Bari, 1981.

Pelka 1920
Pelka, O. *Bernstein.* Berlin, 1920.

Phillips 1993
Phillips, K. M., Jr. *In the Hills of Tuscany: Recent Excavations at the Etruscan Site of Poggio Civitate (Murlo, Siena).* Philadelphia, 1993.

***Piceni* 1999**
Piceni: Popolo d'Europa. Exh. cat. Rome, 1999.

Pinch 1994
Pinch, G. *Magic in Ancient Egypt.* Austin, TX, 1994.

Plantzos 1997
Plantzos, D. "Crystals and Lenses in the Graeco-Roman World." AJA 101 (1997): 451–64.

Plantzos 1999
Plantzos, D. *Hellenistic Engraved Gems.* New York, 1999.

Poinar 1992
Poinar, G. O., Jr. *Life in Amber.* Stanford, CA, 1992.

Poinar and Poinar 1999
Poinar, G. O., Jr., and R. Poinar. *The Amber Forest: A Reconstruction of a Vanished World.* Princeton, 1999.

Pollak and Muñoz 1912
Pollak, L., and A. Muñoz. *Pièces de choix de la collection du comte Grégoire Stroganoff à Rome.* 2 vols. Rome, 1912.

Pollard 1977
Pollard, J. *Birds in Greek Life and Myth.* London, 1977.

Pontin and Celi 2000
Pontin, C., and M. Celi, eds. *Ambra: Scrigno del tempo.* Exh. cat. Montebelluna (Treviso), 2000.

Pontrandolfo Greco 1977
Pontrandolfo Greco, A. "Su alcune tombe pestane: Proposte di una lettura." *Mélanges de l'École française de Rome* 89 (1977): 31–98.

***Popoli anellenici* 1971**
Popoli anellenici in Basilicata. Exh. cat. Potenza, 1971.

Popović 1994
Popović, L. B. *Collection of Greek Antiquities.* Belgrade, 1994.

Poursat 1977
Poursat, J. C. *Catalogue des ivoires mycéniens du Musée National d'Athènes: Les ivoires mycéniens.* Athens, 1977.

Prayon and Röllig 2000
Prayon, F., and W. Röllig, eds. *Akten des Kolloquiums zum Thema der Orient und Etrurien: Zum Phänomen des "Orientalisierens" im westlichen Mittelmeerraum (10.–6. Jh. v. Chr.), Tübingen, 12.–13. Juni 1997.* Pisa, 2000.

Prier 1989
Prier, R. A. *Thauma Idesthai: The Phenomenology of Sight and Appearance in Archaic Greek.* Tallahassee, FL, 1989.

Pugliese Carratelli 1996
Pugliese Carratelli, G., ed. *I Greci in Occidente*. Exh. cat. Venice, 1996. Published in English as *The Greek World: Art and Civilization in Magna Graecia and Sicily.* New York, 1996.

Rice 2006
Rice, P. C. *Amber, the Golden Gem of the Ages.* Bloomington, IN, 2006.

Richardson 1983
Richardson, E. H. *Etruscan Votive Bronzes: Geometric, Orientalizing, Archaic.* 2 vols. Mainz am Rhein, 1983.

Richter 1940
Richter, G. M. A. *Handbook of the Etruscan Collection*. Metropolitan Museum of Art. New York, 1940.

Richter 1968
Richter, G. M. A. *Korai.* London, 1968.

Riddle 1965
Riddle, J. M. "Amber and Ambergris in Materia Medica during Antiquity and the Middle Ages." Ph.D. diss., University of North Carolina, Chapel Hill, 1965.

Ridgway 1977
Ridgway, B. S. *The Archaic Style in Greek Sculpture.* Princeton, 1977.

Ridgway 1990
Ridgway, B. S. "Birds, 'Meniskoi,' and Head Attributes in Archaic Greece." AJA 44 (1990): 583–612. Reprinted in B. S. Ridgway, *Second Chance: Greek Sculptural Studies Revisited*. London, 1994.

Ridgway and Ridgway 1979
Ridgway, D., and F. R. Ridgway, eds. *Italy before the Romans: The Iron Age, Orientalising and Etruscan Periods.* New York, 1979.

Riis 1938
Riis, P. J. "Some Campanian Types of Heads." In *From the Collections of the Ny Carlsberg Glyptotek,*, vol. 2, pp. 140–68. Copenhagen, 1938.

Riis 1981
Riis, P. J. *Types of Heads: A Revised Chronology of the Archaic and Classical Terracottas of Etruscan Campania and Central Italy.* Copenhagen, 1981.

Riis 1998
Riis, P. J. *Vulcientia Vetustiora: A Study of Archaic Vulcian Bronzes.* Copenhagen, 1998.

Ritner 1993
Ritner, R. K. *The Mechanics of Ancient Egyptian Magical Practice.* Chicago, 1993.

Rizzo 1988
Rizzo, M. A. *Un artista etrusco e il suo mondo: Il pittore di Micali.* Exh. cat. Rome, 1988.

Rocco 1999
Rocco, G. *Avori e ossi dal Piceno.* Rome, 1999.

Rolley 1996
Rolley, C. "Sculpture in Magna Grecia." In *I Greci in Occidente*, edited by G. Pugliese Carratelli, pp. 369–98. Exh. cat. Venice, 1996. Published in English as *The Greek World: Art and Civilization in Magna Grecia and Sicily.* New York, 1996.

Roncalli 1998
Roncalli, F. "Una imagine femminile di culto dalla 'tomba d'Iside' di Vulci." *Annali della Fondazione per il Museo "Claudio Faina"* 5 (1998): 15–39.

Ross 1998
Ross, A. *Amber: The Natural Time Capsule.* London, 1998.

Ruuskanen 1992
Ruuskanen, J.-P. *Birds on Aegean Bronze Age Seals: A Study of Representation.* Rovaniemi, Finland, 1992.

Säflund 1993
Säflund, G. *Etruscan Imagery: Symbol and Meaning.* Jonsered, 1993.

Salvini 1996
Salvini, M. "Verucchio: Centro etrusco 'di frontiera.'" *Ocnus* 4 (1996): 249–71.

Schwarzenberg 2002
Schwarzenberg, E. "L'ambre: Du mythe à l'épigramme." In *L'épigramme de l'Antiquité au XVIIe siècle ou Du ciseau à la pointe*, edited Jeanne Dion, pp. 33–67. Collection Études anciennes, 25. Nancy and Paris, 2002.

Sebesta and Bonfante 1994
Sebesta, J. L., and L. Bonfante, eds. *The World of Roman Costume.* Madison, WI, 1994.

Serpico 2000
Serpico, M. "Resins, Amber and Vitumen." With contribution by R. White. In *Ancient Egyptian Materials and Technology*, edited by P. T. Nicholson and I. Shaw, pp. 430–74. New York, 2000.

Settis 1987
Settis, S. *Archeologia in Calabria: Figure e temi.* Rome, 1987.

Settis 2001
Settis, S., ed. *I Greci: Storia Cultura Arte Società.* Vol. 3, *Greci oltre la Grecia.* Turin, 2001.

Shennan 1993
Shennan, S. "Amber and Its Value in the British Bronze Age." In *Amber in Archaeology: Proceedings of the Second International Conference on Amber in Archaeology, Liblice, 1990*, edited by C. W. Beck and J. Bouzek, pp. 59–66. Prague, 1993.

Simon 1998
Simon, E. "Apollo in Etruria." *Annali della Fondazione per il Museo "Claudio Faina"* 5 (1998): 119–43.

Siviero 1959
Siviero, R. *Jewelry and Amber of Italy: A Collection in the National Museum of Naples.* New York, 1959.

Smithers 1988
Smithers, S. G. "The Typology and Iconography of Etruscan Terracotta Curotrophic Votives: The Heads and Bambini." Ph.D. diss., University of Iowa, 1988.

Spivey 1987
Spivey, N. *The Micali Painter and His Followers.* New York, 1987.

Sprenger and Bartoloni 1981
Sprenger, M., and G. Bartoloni. *The Etruscans: Their History, Art and Architecture.* Translated by R. E. Wolf. Milan, 1981.

Staccioli et al. 1991
Staccioli, R., et al., eds. *Miscellanea etrusca e italica in onore di Massimo Pallottino.* 2 vols. Rome, 1991.

Steiner 2001
Steiner, D. T. *Images in Mind: Statues in Archaic and Classical Greek Literature and Thought.* Princeton, 2001.

Steingräber 1986
Steingräber, S., ed. *Etruscan Painting: Catalogue Raisonné of Etruscan Wall Paintings.* English ed., edited by D. and F. Ridgway. New York, 1986.

Steingräber 2006
Steingräber, S. *Abundance of Life: Etruscan Wall Painting from the Geometric Period to the Hellenistic Period.* Translated by R. Stockman. Los Angeles, 2006.

Stewart 1997
Stewart, A. F. *Art, Desire and the Body in Ancient Greece.* New York, 1997.

Stibbe 2000
Stibbe, C. *The Sons of Hephaistos.* Rome, 2000.

Stibbe 2006
Stibbe, C. M. *Agalmata: Studien zur griechische-archaischen Bronzekunst. Bulletin antieke beschaving* supplement. Leuven, 2006.

Strøm 1971
Strøm, I. *Problems Concerning the Origin and Early Development of the Etruscan Orientalizing Style.* 2 vols. Odense, 1971.

Strong 1966
Strong, D. E. *Catalogue of the Carved Amber in the Department of the Greek and Roman Antiquities.* London, 1966.

Szilágyi 1992
Szilágyi, J. G. *Ceramica etrusco-corinzia figurata: Parte I, 630–580 a.C.* Florence, 1992.

Todd and Eichel 1974
Todd, J. M., and M. F. Eichel. "A Reappraisal of the Prehistoric and Classical Amber Trade in the Light of New Evidence." *Journal of Baltic Studies* 5, no. 4 (1974): 295–314.

Torelli 2000
Torelli, M., ed. *Gli Etruschi.* Exh. cat. Venice, 2000. Published in English as *The Etruscans.* Milan, 2000.

***Treasures* 1998**
Treasures from the South of Italy: Greeks and Indigenous People in Basilicata. Exh. cat. Milan, 1998.

True and Hamma 1994
True, M., and K. Hamma, eds. *A Passion for Antiquities: Ancient Art from the Collection of Barbara and Lawrence Fleischman.* Malibu, 1994.

Tsetskhladze et al. 2000
Tsetskhladze, G., A. J. N. W. Prag, and A. M. Snodgrass, eds. *Periplous: Papers on Classical Art and Archaeology Presented to Sir John Boardman.* London, 2000.

Tuchelt 1970
Tuchelt, K. *Die archaischen Skulpturen von Didyma: Beiträge zur frühgriechischen Plastik in Kleinasien (Istanbuler Forschungen 27).* Berlin, 1970.

Turfa 2005
Turfa, J. Macintosh. *Catalogue of the Etruscan Gallery of the University of Pennsylvania Museum of Archaeology and Anthropology.* Philadelphia, 2005.

Uberti 1988
Uberti, M. L. "Gli avori e gli ossi." In *I Fenici*, edited by S. Moscati, pp. 394–403. Exh. cat. Milan, 1988.

Valentini 1969
Valentini, G. "Il motivo della Potnia theron sui vasi di bucchero." StEtr 37 (1969): 413–40.

Vermeule 1979
Vermeule, E. *Aspects of Death in Early Greek Art and Poetry.* Berkeley, 1979.

***Verucchio* 1994**
Il dono delle Eliadi: Ambre e oreficerie dei principi etruschi di Verucchio. Exh. cat. Rimini, 1994.

Waarsenburg 1995
Waarsenburg, D. J. *The Northwest Necropolis of Satricum: An Iron Age Cemetery in Latium Vetus.* Amsterdam, 1995.

Waarsenburg 1996
Waarsenburg, D. J. "Astarte and Monkey Representations in the Italian Orientalizing Period: The Amber Sculptures from Satricum." In *Die Akten des Internationalen Kolloquiums "Interactions in the Iron Age: Phoenicians, Greeks and the Indigenous Peoples of the Western Mediterranean" in Amsterdam am 26. und 27. März 1992*, Hamburger Beiträge zur Archäologie 19–20 (Mainz, 1996), pp. 33–71.

Walter-Karydi 1973
Walter-Karydi, E. *Samische Gefässe des 6. Jahrhunderts v. Chr. Landschaftsstile ostgriechischer Gefässe*. Vol. 6.1, *Samos*. Bonn, 1973.

Warden 1994
Warden, P. G. "Amber, Ivory, and the Diffusion of the Orientalizing Style along the Adriatic Coast: Italic Amber in the University Museum (Philadelphia)." In *Murlo and the Etruscans: Art and Society in Ancient Etruria*, edited by R. De Puma and J. P. Small, pp. 134–43. Madison, WI, 1994.

Weitshaft and Wichard 2002
Weitshaft, W., and W. Wichard. *Atlas of Plants and Animals in Baltic Amber.* Munich, 2002.

White 1992
White, R. "Beyond Art: Towards an Understanding of the Origins of Material Representations in Europe." *Annual Review of Anthropology* 21 (1992): 537–64.

Wilkinson 1992
Wilkinson, R. H. *Reading Egyptian Art: A Hieroglyphic Guide to Ancient Egyptian Painting and Sculpture.* London, 1992.

Wilkinson 1994
Wilkinson, R. H. *Symbol and Magic in Egyptian Art.* London, 1994.

Williamson 1932
Williamson, G. C. *A Book of Amber.* London, 1932.

Winter 1994
Winter, I. J. "Radiance as an Aesthetic Value in the Art of Mesopotamia (with Some Indian Parallels)." In *Art: The Integral Vision; A Volume of Essays in Felicitation of Kapila Vatsyayan,* edited by B. N. Saraswati et al., pp. 123–31. New Delhi, 1994.

Winter 1999
Winter, I. J. "The Aesthetic Value of Lapis Lazuli in Mesopotamia." In *Cornaline et pierres précieuses: La Méditerranée, de l'Antiquité à l'Islam; Actes du colloque organisé au Musée du Louvre les 24–25 novembre 1995*, edited by A. Caubet, pp. 45–68. Paris, 1999.

Winter 2000
Winter, I. J. "The Eyes Have It: Votive Statuary, Gilgamesh's Axe, and Cathected Viewing in the Ancient Near East." In *Visuality before and beyond the Renaissance: Seeing as Others Saw,* edited by R. S. Nelson, pp. 22–44. Cambridge, 2000.

***World of the Etruscans* 2001**
The World of the Etruscans. Exh. cat. Florence, 2001.

Yon 1977
Yon, M. "Les bijoux d'ambre de la nécropole d'Aléria." *Mélanges de l'École française de Rome* 89 (1997): 589–619.

Zamarchi Grassi 1992
Zamarchi Grassi, P., ed. *La Cortona dei principes.* Exh. cat. Cortona, 1992.

Zazoff 1966
Zazoff, P. "Zur ältesten Glyptik Etruriens." Jdi 81 (1966): 63–78.

Zazoff 1968
Zazoff, P. *Etruschische Skarabäen.* Mainz, 1968.

Zazoff 1983
Zazoff, P. *Die antiken Gemmen.* Munich, 1983.

About the Authors

Faya Causey

Faya Causey is the head of the academic programs department at the National Gallery of Art, Washington, DC. She was educated at the University of California, receiving her BA at UC Riverside and her MA and PhD degrees at UC Santa Barbara. A lifelong fascination with the ancient resin began on a student trip to the British Museum, where she first encountered Etruscan carved ambers. Causey began her professional life as an academic, teaching at the Art Center College of Design and California State University, Long Beach. She has lectured and published internationally on antiquity, contemporary art, and museums. She is the author of *Amber and the Ancient World* (J. Paul Getty Museum, 2012).

Herant Khanjian

Herant Khanjian is an assistant scientist at the Getty Conservation Institute, where has worked since 1988. He specializes in the study of organic material found in traditional and modern works of art, using Fourier-transform infrared spectroscopy.
Technical Essay: Analysis of Selected Ambers from the Collections of the J. Paul Getty Museum

Jeff Maish

Jeff Maish is a conservator of antiquities at the J. Paul Getty Museum at the Getty Villa, where he has worked since 1988. He has conducted research and worked on the treatment of Attic ceramics and bronzes in the Getty's collection and in cooperation with other institutions, in addition to his work with ambers.
Technical Essay: Analysis of Selected Ambers from the Collections of the J. Paul Getty Museum

Michael R. Schilling

Michael R. Schilling is a senior scientist in charge of the organic materials laboratory of the Getty Conservation Institute, where he has worked since 1983.
Technical Essay: Analysis of Selected Ambers from the Collections of the J. Paul Getty Museum